Genetics and Mutagenesis of Fish

Edited by J. H. Schröder

With Contributions by

F. W. Allendorf · A. Anders · F. Anders · N. Egami · W. Engel
H. O. Hodgins · S. Holzberg · Y. Hyodo-Taguchi · A. G. Johnson
K. D. Kallman · V. S. Kirpichnikov · K. Klinke · C. Kosswig
K. Lepper · R. F. Lincoln · W. Lueken · Ch. Meske · J. L. Mighell
E. T. Miller · S. Ohno · J. Parzevall · G. Peters · N. Peters · A. Post
C. E. Purdom · D. L. Pursglove · H.-H. Reichenbach-Klinke
H.-H. Ropers · N. Satoh · J. J. Scheel · E. R. Schmidt · A. Scholl
J. H. Schröder · R. J. Schultz · M. Schwab · J. B. Shaklee · F. M. Utter
J. Vielkind · U. Vielkind · G. S. Whitt · H. Wilkens · U. Wolf
D. S. Woodhead

With 132 Figures

Springer-Verlag · Berlin · Heidelberg · New York 1973

Dr. Johannes Horst Schröder
Institut für Biologie
Gesellschaft für Strahlen- und Umweltforschung
D 8042 Neuherberg/Munich

ISBN-13: 978-3-642-65702-3 e-ISBN-13: 978-3-642-65700-9
DOI: 10.1007/978-3-642-65700-9

Softcover reprint of the hardcover 1st edition 1973

Genetics and Mutagenesis of Fish

Edited by J. H. Schröder

With Contributions by

F. W. Allendorf · A. Anders · F. Anders · N. Egami · W. Engel
H. O. Hodgins · S. Holzberg · Y. Hyodo-Taguchi · A. G. Johnson
K. D. Kallman · V. S. Kirpichnikov · K. Klinke · C. Kosswig
K. Lepper · R. F. Lincoln · W. Lueken · Ch. Meske · J. L. Mighell
E. T. Miller · S. Ohno · J. Parzevall · G. Peters · N. Peters · A. Post
C. E. Purdom · D. L. Pursglove · H.-H. Reichenbach-Klinke
H.-H. Ropers · N. Satoh · J. J. Scheel · E. R. Schmidt · A. Scholl
J. H. Schröder · R. J. Schultz · M. Schwab · J. B. Shaklee · F. M. Utter
J. Vielkind · U. Vielkind · G. S. Whitt · H. Wilkens · U. Wolf
D. S. Woodhead

With 132 Figures

Springer-Verlag · New York · Heidelberg · Berlin 1973

Dr. Johannes Horst Schröder
Institut für Biologie
Gesellschaft für Strahlen- und Umweltforschung
D 8042 Neuherberg/Munich

ISBN-13: 978-3-642-65702-3 e-ISBN-13: 978-3-642-65700-9
DOI: 10.1007/978-3-642-65700-9

Softcover reprint of the hardcover 1st edition 1973

Preface

The present volume contains papers presented on the occasion of the Ichthyological Symposium on Genetics and Mutagenesis held on October 13 through 15, 1972 at the Biology Institute of the Association for Radiation and Environmental Research in Neuherberg near Munich, Germany. These symposia have been held annually since 1955 by former students of Prof. C. Kosswig and Prof. F. Anders in Hamburg or Giessen. In the last two years attendance has increased beyond the national German basis, but the Neuherberg meeting of fish geneticists was the first that could be called "international", the participants coming from twelve different countries. The organization of this meeting was made possible by the support of the Association for Radiation and Environmental Research, which provided the financial backing; special thanks are due to Min.-Rat. H. Costa and Dr. R. Wittenzellner. The scientific advice and encouragement of the Director of the Biology Institute, Prof. O. Hug, was also much appreciated, as was the helpful assistance of Dr. B. Betz and Dr. K. Göttel. In addition, I have to thank my colleagues and coworkers Mdmes. E. Neubner, I.S. Otten, and K. Peters, Messrs. M. Murr and M. Wiestner, and Dr. S. Holzberg for their help in organizing the meeting.

Apart from the papers presented at the above symposium, we have included some papers by colleagues who were unable to attend the meeting. The contents of this volume may thus be considered representative of current research on the genetics and mutagenesis of fish.

First there is a historical review indicating the importance of fish in the study of genetics, by Prof. Dr. C. Kosswig, himself one of the founders of ichthyogenetics in a broad sense. Subsequent contributions are arranged according to the main features of ichthyogenetics: sex determination and melanoma formation; mutagenesis; chromosomes and cytology; ethology; evolution; and biochemistry. Such an arrangement is, of course artificial, but helpful for locating certain topics. For example, the paper by Holzberg on inherited changes in male aggressiveness in postirradiation generations of convict cichlids could have been placed in either the mutagenesis or the ethology chapter of this book, since aggressiveness is a behavioural trait.

Some of the papers give monograph-type summaries of certain areas in fish genetics, while others go into detail, for instance, Kallman's paper on the geographical distribution of homo- and heterogametic females and males within the species, *Xiphophorus maculatus*. Japanese fish geneticists present a preliminary report on sex determination in the germ cells of transplanted gonads as compared to normal gonads of the medaka, *Oryzias latipes*. Anders and coworkers summarize their recent studies concerning the inheritance and mutagenesis of the regulation system of melanomas in xiphophorin fish.

The next chapter deals with mutagenesis in fish, a topic already mentioned in the preceding chapter on sex determination and melanoma formation. Purdom and Woodhead survey genetic and somatic radiation damage in fish, ionizing radiation being the only mutagenetic agent so far used systematically in fish mutagenesis. Egami and Hyodo-Taguchi describe dose-mutation relationships in dominant lethal mutations after irradiation of either female or male germ cells of the medaka. Purdom and Lincoln discuss radiation-induced gynogenesis in marine flatfish, which can even lead to triploid forms. These results are not merely of theoretical interest, they also have practical significance, because it is now possible to obtain homozygous strains without several generations of inbreeding. Genetically well-defined strains are a prerequisite for fish breeding. Intensive farming of both sea and freshwater fish is the only way to ensure increased supplies of this important source of animal protein. Exploitation of marine fish alone would soon lead to a complete exhaustion of natural populations and endanger the nutritional basis of the constantly growing human population.

My own paper is a short survey of fish mutagenesis with the aim of encouraging geneticists to use teleostean fish which have proved an excellent tool for studying mutagenesis.

In the chapter on chromosomes and cytology, Post describes unusual karyotypes of two species and Scheel deals with the chromosomes of some anuran species. This author's earlier work on chromosomes has been very important in elucidating the evolutionary processes of African cyprinodontiform fish. He is now continuing his work with frogs, so providing more information on the evolution of karyotypes in lower vertebrates. Vielkind et al. describe the incorporation of tritiated thymidine into the DNA of melanotic and amelanotic melanoma explants and the fate of bacterial DNA in embryos of poeciliid fish. Lueken et al. discuss the arrangement of pigment cells in pure-breeding strains and interspecific hybrids of xiphophorin fish. Meske describes the spermatozoa of eels bred to sexual maturity under experimental conditions.

The chapter on ethology contains only two contributions. As already mentioned, Holzberg reports his preliminary results on the aggressiveness of F_2 cichlid males after ancestral irradiation in connection with the male aggressiveness of the precedent generation, whereas Parzefall deals with the sexual behaviour of *Poecilia (Mollienesia) sphenops*.
The chapter on evolution exposes some problems of evolutionary processes in vertebrates. Poeciliid fish are very useful for such studies since they produce fertile inter-specific hybrids. The regressive evolution of cavernicolous forms was investigated by crossing the normal-eyed river fish with its subterristic blind and depigmented derivatives in the *Astyanax* complex. Stimulated by Kosswig, both Peters and Wilkens successfully used these characinid fish as model for the genetics of regressive evolution. Schultz discovered all-female populations which propagate by "hybridogenesis" and was successful in synthesizing an all-female "species" under laboratory conditions, thus justifying his hypothesis on the natural origin of hybridogenetic populations in the *Poeciliopsis* group of poeciliid fish. Ohno reviews recent results on evolution at the molecular level and explains his now widely accepted view on the conservative nature of selection and the importance of gene duplication and "junk" DNA for evolutionary processes in vertebrates.

The final chapter covers biochemical studies of fish, a topic which has grown in importance during the last decade. Kirpichnikov describes biochemical polymorphism from the evolutionary point of view, whereas Whitt et al., discussing dehydrogenase isozymes of fish, adopt the standpoint of developmental and biochemical genetics. Scholl confines himself to the biochemical evolution and tissue-specific expression of 6-phosphogluconate dehydrogenase within the genus *Xiphophorus*. Reichenbach-Klinke deals with the possible application of serum polymorphism in fish for the purpose of breeding strains resistant to various bacteria. Wolf and coworkers discuss the inheritance of NADP-dependent isocitrate dehydrogenase in the rainbow trout. Finally Utter et al. examine the inheritance and the application of biochemical variants to population studies in sea-water fish.

Prof. Kosswig's 70th birthday falls on 30 October 1973, and it is a pleasure to dedicate this volume on Genetics and Mutagenesis of Fish to him. I have to thank all the colleagues who unanimously agreed with this dedication. The appreciation of Curt Kosswig's scientific life by F. Anders is gratefully acknowledged.

Neuherberg, April 1973

Johannes Horst Schröder

Contents

4. Chromosomes and Cytology

5. Ethology

6. Evolution

7. Biochemistry

Curt Kosswig

Curt Kosswig on his 70th Birthday

Curt Kosswig will complete his 70th year on the 30th of October 1973. Friends and students wish him the heartiest birthday and dedicate to him this volume as a token of their respect and gratitude. At the same time they congratulate him on his successful work as a researcher and teacher and thank him for the untiring attention to detail which he paid to the basis of fish genetics in yearly workshops. An American colleague formulated appropriately: "Kosswig is the nucleus of the European Ichthyology".

Curt Kosswig was born in Berlin, studied there, and, in Dahlem early in his scientific career, worked as a student in a unique circle of friends with Erwin Baur on the problems of the genetics of animal husbandry in rabbits and swine. He has completely retained the sense for the practical aspect of genetics as have all other former students of Baur. He can be as much at home in applied science as he is in basic research. It is a credit to his outstanding gift as a teacher that he was appointed as director of the Zoological Institute and the Museum of Natural History in Braunschweig when he was just 29, two years after the "Habilitation" by Leopold von Ubisch in Münster. In the period following this came his first and most important research in Germany, first disputed, then recognized world-wide, with genetic sex determination without sex chromosomes and with the evolution of sex chromosomes for which he found an outstanding object in platyfish and swordtails. At the same time came also his first work on melanomas in platyfish/swordtail hybrids, discovered by him, which he accomplished partly alone, partly with his friend and colleague Hans Breider. The interpretation at that time, - that certain genes which normally determine the color pattern of the platyfish would fall to an abnormally strong action in the hybrid genome and from this originates the formation of melanomas, - is valid still today. With that, tumor formation in the platyfish/swordtail hybrids became a problem of gene regulation, and all cancer researchers today who perceive the problem of cancerogenesis as one of gene regulation rely on the work of Curt Kosswig, whether they realize it or not.

The fruitful years in Braunschweig were suddenly interrupted in 1937 when he, like many other important men, was forced at that time to flee from Germany. Turkey accepted him and gave him and his family a second home. In Istanbul he founded new laboratories and institutes; he gathered new students around him and advanced their talents; he wrote three biology text books; and he accepted guest professorships in Bagdad, Damascus, Cairo, Alexandria and Jerusalem. At the same time he shifted the emphasis of his work and of that suggested by him to historical zoogeography and hydrobiology in the Middle and Near East, in the Mediterranian Ocean and Black Sea, especially though, in the far Anatolias from Izmir to Erzurum in the headwaters of the Euphrates where he was involved in the foundation of Atatürk University and was engaged as a guest professor from 1968 until a short time ago. The study of zoogeography has undoubtedly stimulated him to further work with speciation and degenerative evolution which had already interested him for a long time in his experiments with woodlice, fish, and birds. It is characteristic for Curt Kosswig that he undertook these projects with all available resources in large-scale plans by expedition to the Near East, Peru, Bolivia, the Philipines and Mexico.

In the mean time he was appointed in 1955 as director of the Institute of Zoology and the Zoological Museum in Hamburg, and here continued, with new colleagues and circle of friends the work that he had started in Braunschweig and Istanbul. It is a sign of his allegiance to his second home, Anatolia, that he did not give up his house on the Bosporus but uses it as an intermediate station for excursions and expeditions with his students and colleagues from Hamburg. After his self-proposed retirement at 65 he lives partly at the Bosporus and partly at the Elba.

In a rich life like Curt Kosswig lived and lives, despite several humiliations, honors are a natural occurrence. To these belong honorary memberships in respected scientific societies, honorary doctorates and other honorary distinctions. For Curt Kosswig there are many other small honors from his friends when they worked up the material of his expeditions and found he had discovered with reliable instinct animal species unknown up to now: 5 genera and more than 40 species and subspecies are named after Curt Kosswig.

We wish that Curt Kosswig whose birthday we are celebrating will be able to continue his work in the service of our science, with his robust nature and creative power, for many many years.

Fritz Anders

1. Historical Review

The Role of Fish in Research on Genetics and Evolution

C. Kosswig

For over 50 years fish have been used as experimental animals in the field of classical genetics. To quote only a few examples: the investigations of J. Schmidt on *Zoarces* (Schmidt, 1919) and *Lebistes* (Schmidt, 1919), Winge's now famous analysis of sex determination in the latter (Winge, 1930), and similar work on *Oryzias* by Aida (1930). In 1914, far ahead of his time, Gerschler (1914) published his sensational results of crosses between *Xiphophorus* and *Platypoecilus*. It is incomprehensible that studies on fish genetics are so seldom found in the summarizing literature despite the fact that the smaller fish species provide by far the best material, at least among the vertebrates, for studying inheritance. No other group of vertebrates is so easy to keep and breed in large numbers as the different orders of small fish. Moreover, bony fish of a number of species can be successfully crossed to produce fertile offspring, and in this respect fish are even superior to the otherwise ideal *Drosophila*. The goldfish, in the course of more than 1 000 years of domestication, has produced a hitherto unexploited wealth of mutants. Genetic laboratories in the USA, Europe and Russia have, over the past decades, bred many hundreds of thousands, and probably even millions, of controlled progeny, numbers which have never been attained in other vertebrates (excluding the mammals bred for medical research purposes, such as mice, guinea pigs and rats) and only rarely among insects.

The following survey of results of genetic and ichthyologic research is by no means exhaustive: its primary aim is to emphasize results obtained exclusively or mainly on fish. The literature cited is limited to review articles with the help of which further publications can readily be found.

Geologically, teleosts are a very old group. In the Eocene and Oligocene highly specialized genera of bony fish already occurred, from both salt- and freshwater, that are still in existence today. Nevertheless, they are still in a state of active evolution and are generally held to be the most species-rich class of vertebrates. The figure of 20 000, often cited in the literature for the number of morphospecies, is far too low. Each year new genera and often even families are discovered in regions difficult of access. Furthermore, if it is taken into consideration that, in many cases, the term morphospecies covers several biospecies (Scheel, 1968) whose members, on account of extensive phenotypic similarities, can only experimentally be shown to be sexually isolated, it becomes obvious how promising is the study of the genetics of bony fish.

In the past two decades, which have brought close contact between classical genetics and evolutionary research, population genetics has occupied the centre of interest. Panmictic populations are more or less heterozygous, two or more alleles of numerous genes occurring in their gene pools. Thanks to the possibilities offered by recombination, it is possible to produce individuals that vary in their fitness with respect to different external conditions. Territorial expansion and subsequent isolation in hitherto unoccupied ecological niches is, according to the concepts of population genetics, achieved by isolated individuals of a certain genotype. Emigrants that successfully founded new populations possess, in all probability, only a fraction of the alleles of the mother population, and those in a modified proportion. So far, this has not been demonstrated unequivocally in fish, and obviously a different principle is involved here, which has probably been overlooked most readily in other groups of organisms. It seems that the conquest of new areas and of new ecological niches is achieved not by specially predestined genotype carriers, but can be carried out by any individual of a population or species thanks to a very wide reaction norm or range of adaptability which is part of the general genetic make-up of the species (Kosswig, 1972). A wide reaction norm guarantees the plasticity necessary to enable fish to adapt to permanent or temporary changes in environmental conditions. Since 1900, as a result of the opening of the Suez Canal in 1869, an ever-increasing number of fish and other animals have migrated from the Red Sea to the Mediterranean, have dispersed and reproduced there. Some of these erythraean immigrants have already reached the Aegean Sea and represent a considerable part of the fish caught there: in some cases the intruders even appear to be crowding out the original Mediterranean fauna. A very wide range of ecological situations is presented by the 100 km length of the Suez Canal. Temperature and salinity are not the same as in the Red Sea, the salt content in particular changing considerably: in the south it is higher than in the Red Sea, in the north lower, but in the Mediterranean it is again higher than in the northern part of the canal. If, in the course of its life span, an erythraean fish swims the length of the canal, this indicates that it has inherited a range of adaptability large enough for it to achieve the necessary physiological adjustments. According to population genetics migration along the canal in the course of more than one generation would have to be accompanied by selection for survival first in water of raised salinity, then in lowered salinity. This assumption seems to be less plausible than that of a simple wide reaction norm for all individuals of the immigrating species. This renders possible not only tolerance of a wide range of environmental conditions, but also the development of long-term modifications of a physiological and/or morphological nature. Many examples of extremely different phenotypes are known, especially among the salmonids (e.g. nationes fario, lucustris and marina of *Salmo trutta*), which have been shown to be merely modifications based on the wide reaction norm of the genotype.

Preadaptations, anchored in the genotype, even permit the conquest of extreme biotopes such as caves. From *Astyanax mexicanus*, a large-eyed fish that lives in swarms and exhibits a slightly negative phototaxis, originate typical cavernicolous populations of varying genotype, that have quite incorrectly been given a special genus name and whose various populations, again incorrectly, have been accorded the status of species. In the meantime it has transpired that although the epigeous *Astyanax* can swim actively into cave waters, passive transport is of primary importance in the genesis of its cavernicolous relatives. In karst regions of the Sierra de El Abra river beds may abruptly sink down into old underground river systems. Apparently every surface form of *Astyanax* is preadapted to find its way (Schemmel, 1967) and to reproduce (Wilkens, 1968) in perpetual darkness. Although optical orientation appears to be of decisive importance for a fish living in surface water, an experimentally blinded fish or one transferred to and kept in permanent darkness can thrive and reproduce even under the new conditions. All that is lost, is the optically governed swarming behaviour: the blinded fish behaves in this respect like its blind cave derivative and wanders about ceaselessly in the aquarium. In this case the morphological basis of the preadaptation to perpetual darkness is known to be a system of neuromasts, sense organs of current perception, distributed over the entire body. This system is so well developed in epigeous fish that no further increase is necessary in the blind, poorly pigmented subterranean populations. Preadaptation has been demonstrated experimentally in a younger cave-dwelling form of *Poecilia sphenops* (Zeiske, 1968).

Nothing whatever is known about the genetic basis of this wide reaction norm or of preadaptability. Their large degree of stability, on the one hand, and their occurrence in all individuals of a species lead to the assumption that a highly polygenic system is involved in which numerous genes and + and - modifiers collaborate, and in which, besides additive and non-additive polymery, hierarchies of genes are involved, the entire system being held together by pleiotropy of its elements. The genotype of a higher organism, which a bony fish certainly is, is thus revealed to be more than the sum of its constituent genes. It is tempting to put forward the hypothesis that it is just that part of the genotype of every individual that cannot be analysed in a crossing experiment on account of its polygeny, which contains the characteristics determining those of the taxon, i.e. of the "type", in the morphological sense.

It was mentioned above that in teleosts, in contrast to almost all other groups in the animal kingdom, species crosses can be carried out. This means that genomes that have undergone divergent evolution over thousands of years can be combined with one another in the hybrid. If such hybrids prove to be fertile, inferences can be drawn as to the number of genes that have been modified since the separation of the two species from a common ancestor. The information which, thanks to the fertility of the F_1 offspring, is provided by crosses between two different species, is discussed in more detail below. Even complete or partial sterility of the resultant hybrids can, under certain circumstances, reveal some information as to the genetical basis of the species-separating mechanism. Various degrees of sterility of species hybrids can be distinguished:

1. The gonads of the F_1 hybrids contain scarcely any or no gonogonia. Whether the latter are not formed in hybrids or whether they simply do not enter the gonads at the right time is unknown. This type of sterility can only be attributed to the incompatibility of foreign genomes in general (Öztan).

2. In some crosses, such as, for example, the backcross of an F_1 hybrid of *Xiphophorus helleri* x *X. maculatus* with *X. helleri* the individuals that possess the gonosome of their *maculatus* grandparent remain sterile for years, irrespective of whether they are ♂♂ or ♀♀. Treatment of the otherwise sterile ♀♀ with gonadotropic hormone and of the ♂♂ with androsterone can render them fertile (Öztan, 1963). Thus it appears that certain combinations of genes can elicit hormonal imbalance, in this case not until F_2R, and that this can be overcome by appropriate treatment.

3. In some crosses, the gonads and sometimes even the phenotype (Villwock, 1958) of the ♀♀ can be shown to be intersexual. Animals of this type are able to produce fertilizable eggs whilst young. Later, spermatogenic tissue develops in the ovaries and disturbances in spermatogenesis typical for certain kinds of ♂♂ sterility then occur.

4. Sterility of ♂♂ following normal synapsis is due to irregular distribution of chromosomes on the two secondary spermatocytes and subsequent premature termination of spermatogenesis (Karbe, 1961). In many such cases of ♂♂ sterility the F_1 hybrid ♀♀ are normally fertile and can be backcrossed with the ♂♂ of the two original species. With an increasing number of such backcrosses the number of fertile ♂♂ progeny rises, from which it can be concluded that the normal course of spermatogenesis in both parent species is under different polygenic control.

5. In this type of cross ♂♂ occur in which spermatogenesis is normal but which are nevertheless sterile. It has been shown that, in such cases, hybridization results in a genetically conditioned lack of mobility of the spermatozoa (Karbe, 1961).

6. Within the *Xiphophorini* sympatric or allopatric species can be crossed with one another if no ♂♂ of the same species are present. It has been demonstrated in competitive experiments, and this is important in sympatric species, that the spermatozoa of foreign species are much inferior to those of the same species. Given simultaneous insemination with both types of spermatozoa the former have a smaller chance of achieving fertilization and a weaker chance of surviving in the oviduct of the ♀ of another species as compared to the spermatozoa of the same species (Zander, 1962). Since the 'foreign' spermatozoa in allopatric populations of the same species already exhibit inhibitory phenomena of a comparable nature, it can be assumed that an existing system of incompatible genes has been reinforced by new genes in the course of phylogenetic divergence.

The analysis of sex determination in viviparous cyprinodontids has brought new aspects to light and has shown that (Kosswig, 1964a) a far greater diversity of genetic principles is involved than had been assumed by M. Hartmann (1956). In many species sex determination is of the so-called monogenic type, for which Correns' classical interpretation holds true on principle. It was surprising to learn that among close relatives, in fact within one population of the same species, male and female heterogamety occur side by side. In *X. maculatus* XY and YY ♂♂, and XX, XW and YW ♀♀ are to be found (Kallman, 1965). Winge's (1930) results that in *Lebistes reticulatus* the decisive male-determining factor is in the Y holds for many other species. A potent female-determining factor (gene or complex of genes) is located in the W of the *maculatus* ♀♀. This principle of localization of the decisive gene for the heterogametic sex in the Y or W chromosome seems to be the rule in vertebrates. Winge (1934) knew already, however, that genes influencing sex differentiation occur not only on the gonosomes but on the autosomes as well. An increase in number of such genes is responsible, for example, for the occurrence of ♂♂ of XX constitution in *Lebistes*. A similar situation was later brought to light in *X. maculatus* (Öktay) and was termed polygenic sex determination. Years ago, sex determination was shown in *X. helleri* to be exclusively polygenic and was described in detail (Kosswig, 1964b). The same principle of polygenic sex determination has been revealed in a large number of invertebrates.

For reasons unknown, a differentiation of the opposite sex to that of the animal's gonosomic constitution occasionally occurs, and further the normal monogenic sex-determining mechanism can be put out of action by species crosses (Kosswig, 1936; Rust, 1941). With the aid of such exceptions even WW-♀♀, for example, can be produced by appropriate crossing. In the same way it is possible to obtain YY homogametic ♂♂ by crossing exceptional XY ♀♀ with normal XY ♂♂. For a long time it had seemed that in an animal group that, despite clear-cut differentiation into ♂♂ and ♀♀, is approaching hermaphroditism, morphologically differentiated gonosomes could not be expected. However, increasing numbers of reports are available of well differentiated gonosomes in various not particularly closely related Cyprinodonts. It is interesting that such cases appear in just those groups that are notable for a large degree of constancy in their typical chromosome number (n=24) (Miller and Walters, 1972).

Treatment with sex hormones during a critical period of early developing stages results in a complete inversion of the sex. Treatment of *Lebistes* (Zander and Dzwillo, 1969) and of *Oryzias* (Yamamoto, 1967) with androsterone transforms ♀♀ (XX) into functional ♂♂ and vice versa ♂♂ (XY) can become functional ♀♀ under the influence of female sex hormone. In suitable crosses (e.g. transformed XY-♀ crossed with normal XY-♂) homogametic YY-♂♂ can be obtained which become functional ♀♀ by treatment with female hormone, etc.

These experiments have shown that the genetic constitution XY or XX of the gonogonia does not, as such, determine sex, but rather that the primary gonogonia in the young gonads are exposed to the influence of sex-determining substances, normally produced by the soma of the gonads. Experimentally, the effect of these substances is masked by that of the sex hormones added. This interpretation agrees with that proposed by Witschi (1929) for amphibia. Occasional, spontaneous cases of faulty sex differentiation, i.e. one contrary to the gonosome constitution, are interesting in regard to the phylogeny of various modes of sex determination among closely related animals (Kosswig, 1936).

Finally, it should also be mentioned that, among the teleosts, hermaphroiditism occurs in many species from various families: it can be either protandrous, protogynous or synchronous. The latter leads to cases of self fertilization, an example of which is provided by the small toothed carp *Rivulus marmoratus*.

A unique sensation at the time was afforded by Hubbs' finding that *Mollienesia* (now *Poecilia*) *formosa* is a constant gynogenetic hybrid between *P. sphenops* and *P. latipinna*. In order to produce fertile offspring the *P. formosa* ♀♀ has to be fertilized by a ♂ of one of the two original species. Gynogenetic development with the diploid maternal chromosome complement normally takes place following fertilization of the egg by a *sphenops* or a *latipinna* sperm: sperms of other species have the same effect. Normally, however, the genome of the sperm is not involved in the development of the gynogenetic progeny, which all possess the same genotype as the individuals of an asexually reproducing clone (Kallman, 1962). Under what particular conditions the gynogenetic *formosa* arose in nature is unknown. Crossing both parent species in the aquarium produces in F_1 *formosa*-like offspring of both sexes which, after further crossing, result in progeny of mixed polygenic recombination. In recent years similar cases of gynogenetic reproduction have been reported in several species of the genus *Poeciliopsis*, coupled, however, with triploidy (Schultz, 1969). It is noteworthy that in this genus the tendency is apparently widespread for species crosses to result in triploid, gynogenetic ♀♀, the eggs of which require fertilization although the genome of the sperm plays no part in development.

In passing, it should be mentioned that the presence of heterogametic males and females within one and the same species, as described in *X. maculatus*, has been proved to be of practical importance as well as of theoretical interest. The genus *Tilapia* is nowadays bred in fish ponds in all tropical and subtropical countries. In populations consisting of both sexes the growth of the ♂♂ slows down rapidly as soon as they begin to mate and breed in the presence of ♀♀. If, however, in the breeding pond a homogametic ♀ (XX) is paired with a homogametic ♂ (YY) the entire progeny is ♂♂ (XY), which in the absence of female sexual partners exhibit normal growth.

Many populations of the most varied species of small fish exhibit a more or less pronounced polymorphism for certain characteristics connected with form and particularly for characteristics of colour (Kosswig, 1964). In some cases the polymorphism, being humorally controlled, is restricted to one sex, as in *Lebistes*, whereas in other cases both sexes are to a greater or lesser extent obviously polymorph. Upon detailed investigation the polymorphism is seen to be due to genes or supergenes that may occur in several alleles or pseudoalleles, as is the case with the colour genes in the Y of *Lebistes* or with those carried in the X or Y of *X. maculatus* and other species of *Xiphophorus*. Since the frequency with which some of these colour genes occur in natural populations does not change over a long period of time (Gordon, 1947) the possibility has been discussed of this being perhaps a case of balanced polymorphism as understood in population genetics.

Two or more quite distinct phenotypes, or the genes responsible for them, often occur in natural populations: examples are provided by albinism, xanthorism and other colour variants. In *Gasterosteus aculeatus* the extent of the scutaceous body covering is monogenically controlled (Münzing, 1963). The phenotype that is completely covered with scutes, trachurus, is homozygous TT for one pair of genes, the largely naked form, leiurus, is tt, and the heterozygous form with an intermediate degree of covering Tt is termed semiarmatus. In anadromic populations between the English Channel and the North Sea the frequency of tt and Tt individuals decreases from W to E, and from W to NE. Whether such a gradient is of adaptive value or merely the result of postglacial contact of a north european scutate 'subspecies' and a mediterranean naked 'subspecies' is unknown. Experimental crosses have revealed that the heterozygous semiarmatus does not always have the same type of body covering, which suggests the involvement of modifier genes. It is quite certain that the european *Gasterosteus* is a single species, whereas the situation on the Pacific coast of North America is probably different (Hagen and McPhail, 1970; Miller and Hubbs, 1969) and more complicated.

Polygenic differences are sometimes encountered in crosses of populations from separate river systems, and even occur regularly when species that have been crossed with one another produce a fertile F_1 so that F_2- and backcross generations can be obtained. In such hybrids genomes that have developed along different lines over longer or shorter periods of isolation from one another are united and later recombined. Some insight is thus provided into the extent of diversification of the genotypes since their separation from a common ancestor. Segregation in the F_2 and backcross generations only rarely permits of an approximate estimate of the number of allele pairs involved, and as a rule the verdict of 'higher polygeny' has to suffice. The inheritance of meristic characters and of body proportions was the subject of a classical investigation almost forty years ago (Breider, 1936) involving F_2 and F_2R analyses of hybrids between species of the genus *Limia* (now *Poecilia*): *nigrofasciata*, *vittata* and *caudofasciata*. Quite independently, in New York and in Istanbul, the mode of inheritance of differences in the structure of the gonopodia of *X. helleri* and *X. maculatus* was investigated. Fine differences in the individual anal fin rays that form a part of the gonopodium are probably polygenic (Sengün, 1950; Gordon and Rosen, 1951) and show free recombination. It has been demonstrated that one species of the genus *Macropodus*, *concolor*, possesses latent genes for a vertically striped pattern that is regularly seen in another species, *opercularis*.

In polymorphic populations of the killifish *Aphanius anatoliae* varying degrees of degeneration of the scaly covering can be observed. This is not the result of a gradual loss of scales but is due rather to a pronounced multiplication of the scale primordia in two main gradients, the one dorsad to the lateral line, and the other ventrad. Repeated division of the morphogenetic field of the individual scales results in delayed and imperfect formation of miniature scales or, in the end, in their total absence. Just how complicated is the genetic background of the reduction process can be demonstrated by crossing two individuals from certain populations with reduced scales which can result in progeny with more or less normal scales, whilst the mating of two individuals with normal scales can lead to offspring with reduced scales. The number of genes involved and their mode of action is unknown (Aksiray, 1952; Villwock, 1958; Franz and Villwock, 1972). More is known about the genetic background of the phyletic reduction process leading to the differences between the surface form of *Astyanax mexicanus* and its cavernicolous relatives. (For systematicists this would be a genus cross *Astyanax* x *"Anoptichthys"*). The distribution in F_2 shows (Peters and Peters, 1966; Wilkens, 1968) that at least four and perhaps even seven genes or supergenes are responsible for eye reduction. The reduction of the number of melanophores depends in one population (Pachon) upon two alleles and in the other (Sabinos) upon one allele to those of the river fish (Wilkens, 1970a). In another allele pair one allele governs the presence of many melanin granula in the individual melanophores of surface fish, whilst the other is responsible for there being fewer in cavernicolous fish. It is probable that mutations at two loci led to the loss of the ability of cavernicolous fish to react to alarm substances. The number of genes responsible for the loss of morphological colour change in cave-dwelling fish is still unknown. Some populations of the latter possess a gene for albinism, which is usually regarded as a characteristic of domestication (Wilkens, 1970b). Partially, at least, the regression process of the eye of cavernicolous fish takes place by means of different, non allelic genes: a cross between populations of Sabinos and Pachon produced hybrids whose eyes, although blind, were better developed than those of their parents (Wilkens, 1970). Apart from this the genes for eyes and those for loss of pigment are to a large extent able to combine freely. Thus in the F_2 both well pigmented, blind individuals as well as poorly pigmented specimens with eyes are encountered among others. It can be assumed that following the transition of preadapted epigeous fish to caves, all of those genes that had no biological significance in the new environment could mutate. In other words, the typical formation of a normal eye capable of sight in surface fish is subject to the control of preservatory or stabilizing selection processes (Wilkens, 1968). Removal of the pressure of selection gives mutation pressure a free hand, with the result that all structures, and only those, that have become biologically meaningless can be lost. This holds true also for the aggressivity of ♂♂, which becomes superfluous in permanent darkness (Parzefall, 1969). Hearing (Popper, 1970) and sense of smell remain as well developed in the cavernicolous form as in its surface-dwelling relative. There is no evidence for the assumption that loss of the eye and of body pigment is of any selective value. Thus the phylogenetic age of cave populations can be regarded exclusively as the result of mutation pressure. If *Astyanax* was not able to migrate into Mexico until the transition from pliocene to pleistocene (following establishment of a land bridge between South and Central America) it follows that the differentiation of the genotypes of the cavernicolous fish required less than one million years, and perhaps even much less than this (Parzefall and Wilkens, 1972). A secondary sex characteristic of the males of certain species and subspecies of *Xiphophorini* is a black-edged sword, of up to body length, with an iridescent centre. The possession of such a sword can be regarded as a derived character. Those species of *Xiphophorus* possessing no such adornment can be termed primitive. Between the two extremes are species with the mere beginnings of a colourless sword, whilst in *X. pygmaeus nigrensis* the sword length is variable. There are ♂♂, for example, with a slight suggestion of a sword whereas in others it has attained at least half the length of the body and has a black ventral border. That the sword length is a polygenic character has been demonstrated in numerous crosses between eight different species of *Xiphophorini* (Zander and Dzwillo). Similarly, there are special genes responsible for the coloration of different portions of the sword. In the F_1 of swordless species with *X. helleri* the swords vary in length: the F_1 ♂♂ from *X. couchianus* and *X. helleri* have very short swords whereas those of the hybrids from *X. maculatus* and *X. helleri* are of intermediate length. A clue as to the reason for this is provided by treating *couchianus* and *maculatus* with androsterone, which in the latter leads to a pronounced increase in the length of those caudal fin rays forming the sword in *helleri* (Zander and Dzwillo, 1969), whereas no such effect is seen in *couchianus*. From these results it can be concluded that the increased hormone level activates normally latent genes for sword formation in *maculatus*. The apparently orthogenetic development from a small protuberance up to the formation of a colourful structure body length has its origin in the possession of latent genes. It is noteworthy that in the F_2 of a cross between two species possessing well developed swords such as *X. helleri* and *X. montezumae cortezi*, ♂♂ occur with a sword shorter than that of the *cortezi*, and in some completely fertile F_2 ♂♂ the sword may be entirely absent. In two closely related species, *helleri* and *m. cortezi*, sword formation is therefore, at least in part, ensured by the presence of different, non-allelic genes (see above, eye genes in two cavernicolous strains of *Astyanax*). The entire problem of parallel evolution and its genetic basis thus gains fresh interest. If, in *Xiphophorini*, the sword is a new acquisition, that is to say, it reflects a constructive evolutionary process, the question immediately arises as to the biological significance of this secondary sexual character. There is no evidence that it has arisen by a process of sexual selection (Kosswig, 1963) and its possession cannot be said to be of any significance either in mating or securing a position within the hierarchy of a group of ♂♂. A premature anthropomorphic interpretation would only serve to obscure the facts. A well developed structure is only of secondary, if any, biological significance. Apart from *X. maculatus* the display of all other *Xiphophorini* is characteristic for the species (Frank, 1964). Gerschler (1914) already pointed out that in the F_2 from *helleri* x *maculatus* ♂♂ occur whose display, despite the absence of a sword, is typical for *helleri*. The genes for sword formation and for display, whereby the role of the sword is only an apparent one, are independently inherited. Certain individual features of display, too, in which, for example, *helleri* and *montezumae cortezi* differ from one another (Frank, 1970) can be recombined in the F_2 generation just like the parts of a morphological structure.

The surface form of *Astyanax* has taste buds only in the mouth zone of the upper and lower jaws. A blinded surface fish, dependent in its search for food upon its sense of taste, stands perpendicularly above the food under investigation. In blind cave-dwelling fish, however, the taste-bud area has, as a result of selection, spread far over the ventral side of the fish so that in its search for and inspection of food it stands at an angle of 45° and not 90°. In the F_2 generations of crosses between hypo- and epigeous fish, individuals occur, in which the genes for increase in number of taste buds have been separated from those responsible for their 'proper' use. This is apparently a polygenic effect involving recombination. As a result, some individuals, although possessing plentiful taste buds on the anterior ventral half of the body 'stand on their head' like a blinded surface fish, whilst others, despite a reduced number of taste buds, search for food in exactly the same manner as typical cave fish (Schemmel, 1967). Another example of constructive evolution in cave fish is provided by the ♀♀ of a cavernicolous population of *Poecilia sphenops*: with the help of a 'genital cushion' larger than that of the surface-dwelling form a ♂-attracting substance is probably produced in increased quantities (Parzefall, 1970).

Perhaps the most exciting chapter in ichthyological genetics is that of changes in the manifestation of colour genes of one species of *Xiphophorini* when combined with the hereditary material of another. So far, as in the genetic analysis of regressive and constructive evolution, comparable investigations in other groups of organisms are lacking. At least two loci for macromelanophore distribution and at least two more for the formation of pterinophores in various regions of the body are situated in close proximity in the gonosomes (X, Y, in W only in domesticated strains of *X. maculatus*) (Kallman, 1970). Whether a 'gene' is present in several allelic forms on each locus or whether it is a supergene with several pseudoalleles cannot be proved with certainty at the moment. For the sake of simplicity, the term 'gene' should be used without any implication as to its structure. It was very soon recognized that each one of these colour genes has its own polygenic system of modifiers (Kosswig, 1929) by means of which the characteristic manifestation of just this colour gene is guaranteed within the species (or even population). In the other species (Anders et al., 1972) and, in some cases, in other populations (Gordon and Gordon, 1957) allelic or non-allelic genes have to be postulated: they influence the degree to which a foreign colour gene is manifested, sometime inhibiting, but usually enhancing its effect. In 6 of the species of a total of 9 *Xiphophorini* investigated so far (*X. clementiae* has not yet been included) gonosomally localized colour genes of this type occur. Even the species that possess neither genes for macromelanophores nor for pterinophores possess specific modifier systems for the different colour genes of the other species (Zander, 1969). In decades of experimental work about 20 of these colour genes, many of them in several allele forms, have been carried over into other species. This could only be achieved due to the availability of these ideal experimental objects. Somebody who is not acquainted with the many details of the results obtained and who is not interested in studying the literature had better neglect the facts and pass over them silently. Investigations into the manifestation of colour genes of one species in a genotypic milieu of another are of particular interest because in certain invariably reproduceable crosses melanomas and occasionally erythromas are formed. The melanomas occurring subsequent to species crossing of *Xiphophorini* do not differ essentially in structure from human melanomas. They are definitely not the result of a virus infection, but rather of the combination of various completely healthy genomes. They can be produced at will and reversed by appropriate crossing. Such investigations have been going on for many years in Giessen under the direction of Anders. Apart from elucidation of their genetics, histological and cytological investigations have revealed the polyploidy (Vielkind et al., 1971) of the macromelanophores and, using the electron microscope, the fine structure of melanomas has been demonstrated (Vielkind et al., 1971). Of special interest are the investigations on amelanotic tumours, made possible by combining the genetic constitution producing the melanoma with that of the albino gene of *X. helleri*. This represents a great advantage as compared with similar investigations on *Homo sapiens*. We owe to Anders (1972) also the pursuance of a useful hypothesis concerning the cooperation of the colour gene with regulator genes of the same species and booster genes of other species. In the past few years this field of research has been extended by two further sets of facts. By means of a mutation on one of the macromelanophore loci the normal manifestation of a particular colour gene can, within the species, be so much enhanced that, in spite of the presence of modifier systems typical of the species, melanosis and even melanoma formation can occur (Anders et al., 1971). The mutation may be either a simple gene mutation or it may be the result of translocation of a macromelanophore gene of the same or another species. In both of the latter cases it still remains to be established whether this is a positional effect in the classical sense of the term or whether, in the course of translocation, the colour gene has been separated from a regulating gene normally adjacent to it. In some ♂♂ of *X. pygmaeus nigrensis* a gene occurs in the Y chromosome that is responsible for yellow coloration (Zander, 1968).

Yellow coloration results from the so-called G-(colourless, granula-bearing) xanthophores: the carotenoid pigment lies in the middle of the cell. Crossing with *X. maculatus* brings about the transformation of these G-xanthophores (Öktay, 1964) into typical erythroxanthophores, the granules of which become occupied by a red drosopterin instead of a colourless pterin. *X. maculatus*, which possesses limiting modifiers for the distribution of its special erythroxanthophore pattern, has, on the other hand, a modifier system that permits the transformation of yellow G-xanthophores (Öktay, 1964) over the entire body into drosopterin-containing erythroxanthophores. Gordon (1950) has described a case where an erythroma occurring in one and the same hybrid together with a melanoma was able to supplant the latter. Even if this is only a case of the devil chasing Beelzebub, the fact is of interest in that it reveals a close connection between melanin and pterin pigments based upon closely coupled genes in the gonosome of *X. maculatus*. So far a detailed analysis of the modifier genes, some of which have an enhancing some an inhibitory effect on the colour genes, has been impossible. Their large numbers and complicated interrelationships will probably also prevent their elucidation in the future. Hundreds of modifier genes for the colour genes are probably contained in each genome of each species. As regards genetic interpretation we are in the same position as Dobzhansky, who was forced to capitulate faced with an analysis of the 'modifier systems' responsible for the heterotic effect of certain inversion heterozygotes within a population of *Drosophila pseudobscura* (Dobzhansky, 1951). We are still far removed from a molecular interpretation of relatively simple gene effects.

A complete contrast to the recognition of the highly polygenic origin of even simple morphological characters such as the colour patterns of the *Xiphophorini* is offered by the results of molecular biological and biochemical investigations, which have come to play an ever-growing role in genetics over recent years. Whereas the systematicist, the morphologist and the classical geneticist aim, albeit with different methods, at elucidating the basis of organic diversity, the biochemist concentrates upon the unifying characteristics in nature. Functionally identical proteins, especially enzymes, are widespread in the organic kingdom. Their more or less largely identical functions are ensured by a remarkably similar sequence of amino-acids in their molecular make-up, at least at the points known as 'hot spots'. These similarities result from a similar coding and sequence of triplets in a gene (cistron), and it appears unlikely that they arose independently in each case. The presence of identical or merely constitutionally similar enzymes in organisms that cannot be crossed with one another (e.g. in bacteria and man) requires the assumption of 'homologous' genes in otherwise very different organisms. On account of the meaning attached to the word 'homology' by morphologists I suggested (Kosswig, 1961) the use of the term 'ubiquitous' genes in such cases some time ago. Present-day knowledge concerning so-called point mutations within a cistron justifies the use of the term allele genes instead of ubiquitous genes even in species between which insuperable sexual barriers were established long ago. The failure of one vital enzyme as a result of mutative changes in a structural gene involved must have a lethal effect. A feasible way out of this dangerous strait might lie in the possibility that by repeated tandem duplications a cistron can be replicated so that in the event of functional failure one of the duplicates can take over its role. What then becomes of the mutant gene is not clear. It might form part of the redundant 'junk' (as it is termed by Ohno, 1972) which is dragged on because of unknown reasons or it might achieve a new functional state by further processes of mutation, so contributing to the division of labour within an enzyme system, or it might even assume a completely new role. It remains to be seen whether or not fish are suitable objects for the investigation of such problems. Although exhibiting a large degree of constancy at all the 'hot spots' of the molecule of certain structural genes they show a remarkable variability in the genes determining the so-called isoenzymes. In *Xiphophorini* (only 3 species so far investigated) 24 loci for the formation of different enzymes could be identified, of which 16 possess up to 4 alleles, each allele of the very same locus controlling a different isoenzyme. The analysis of the frequency of such alleles in natural populations has become a favoured method for tracking morphologically indistinguishable populations. Nevertheless, little is known of the biological significance of such heterozygosity. It could be demonstrated in *Astyanax* that the large surface populations (Avise and Selander, 1972) possess considerably more variability than the cave populations, which in all probability originate in a few individuals. The heterozygosity under discussion here was proved using the sequence of the bands produced by the various isoenzymes in electrophoresis experiments. In determining the species (and population-) specific content of free amino acids the Giessen group (Anders and Klinke, 1966) again encountered a highly polygenic system, of which it can be assumed that the genes play a boosting role in tumor formation following species crosses. This assumption is supported by the observation that elevation of the salt content of the water in the aquarium is paralleled by an increase in tumor formation and a rise in free amino acids. Yet another polygenic system of genes was discovered by Kallman (1964) in the course of transplantation experiments, which are only successful between individuals of isogenic stocks.

Isogenic stocks resulting from years of inbreeding offer ideal material for mutation experiments, which have so far made use of X-irradiation. The Giessen group succeeded in hitting with X-rays one of the polygenic systems influencing the manifestation of melanophore genes in *X. maculatus* ♀♀. Schröder (1969a, 1969b) has carried out extensive series of investigations on *Lebistes*, in which he tested, amongst other points, the effect of irradiation on vitality, differences in reactivity of inbred and hybrid stocks, occurrence of inheritable changes in number of vertebrae, structure of vertebral column and body proportions, frequency of small mutations correlated with the material started with, rise in cross-over value between X- and Y-chromosomes, as well as sensitivity of different stages of gametogenesis to intensity of radiation.

Cytological investigations on a large number of bony fish have brought sensational results, following the reactivation of this topic by Post. Improved methods enabled Scheel (1972) to carry out chromosome studies on somatic tissue in which, although the chromosome number is doubled, the individual chromosomes are considerably larger and more easily recognizable than in spermatogenesis. In contrast to the large degree of uniformity in chromosome numbers in the *Aphaniini* (n=24) or on the whole in the very varied forms of *Poeciliidae* (n=24, rarely 23) other groups of cyprinodontids exhibit amazing variability even among close relatives, such as, for example, the African and South American *Rivulini*, in which even the haploid numbers of different biospecies concealed within one morphospecies are very different. Even if two of these biospecies have the same number of chromosomes this does not necessarily indicate common ancestry. The reduction in chromosome numbers observed in such groups is due to the transformation of originally metacentric chromosomes into acrocentric elements by pericentral inversion and their refusion to form giant metacentric -, or perhaps following a still further transformation by pericentric inversion, giant telocentric-elements. Since such changes do not take place simultaneously, populations (biospecies) of the same morphospecies can form widely differing karyotypes when in isolation. If crosses are possible, meiosis presents the further complication that in the various karyotypes different arms of the old complement have fused with one another. The *Characidae*, just like the *Rivulini*, present an abundance of different karyotypes within a close circle of relatives. No other class of vertebrates exhibit such variability. Polyploidy also arose polyphyletically. In all probability this is always a case of autotetraploidy (Uyeno and Smith, 1972), which is obscured by mutation in two of the four original homologues in a process termed diploidization by Ohno (1970). Clear-cut sexual differentiation of tetraploid individuals is ensured by the strength of the realisator gene on the Y or W chromosome of the heterogametic sex.

For decades innumberable species of ornamental fish have been kept by amateurs in aquaria. Such animals are often the offspring of only a few original imported individuals. It is small wonder that numerous mutations have arisen as a result of unintentional inbreeding under domestication. Some of these mutants have been further cultivated on account of their aberrations. In no case, however, has crossing and planned selection achieved the diversity seen in the goldfish after more than 1 000 years of domestication (Grzimek and Ladiges, 1970). Nevertheless, the number of variants and hybrid progeny of some of the more common ornamental fish is already large. Mutations such as albinism, xanthorism, melanism and elongated fins are especially frequent. In complete contrast to the polygenic nature of many characters of natural forms, these mutations of easily traceable monogenic characters are so nearly identical with their phenotypes that there is scarcely any doubt that they involve "ubiquitous" genes that, just like certain genes for biochemical characters, are part of the general make-up of the genotype of every bony fish or even of every vertebrate. A more detailed phenogenetic investigation would certainly yield results of scientific interest as well as providing pets for the aquarium. An immense diversity of aquarium strains of popular ornamental fish owe their origin to the principle of introgressive hybridization in *Xiphophorini* and *Poeciliini*. Unfortunately, many of these colourful variations are the result of species crosses, multiple backcrossing and perhaps crossing with yet a third species, so that their origins are no longer obvious and the breeders are unwilling to reveal their secrets. Some popular variations, however, have been produced in genetic laboratories and the methods published (Gordon, 1946).

At this point a phenomenon which is unparalleled in other vertebrates should be mentioned: this is selective fertilization, by means of which certain genes can be prevented from occurring in the homozygous state. This holds for the gene Fu in the Y-chromosome, responsible for producing melanosis in the 'pure' domesticated *X. maculatus* (Kosswig, 1938), as well as for autosomally inherited genes for elongated unpaired fins in domesticated strains of "*X. helleri*" (Schröder, 1966). In the ovoviviparous xiphophorine fish fertilization of the egg takes place in the ovary so that the spermatozoa have to penetrate the follicle epithelium. This presents unsolved problems of developmental physiology which can probably better be studied in oviparous forms for which phenotypically similar mutants are available.

Many expeditions in recent years have had the aim of collecting material from well-defined sites and incorporating this material into crossing experiments. Mexico and Central America have been visited repeatedly by investigators from the USA and twice by members of the Hamburg group. Parts of West Africa have often been visited by Scheel, and almost all the waters of Anatolia have been searched for *Aphaniini*. In recent years the latter subfamily has been sought by Villwock (Franz and Villwock, 1972) in the entire Mediterranean region, especially in hitherto unknown regions of N. Africa. In Hamburg extensive fixed material of *Orestiatini* (from Peru) is under investigation, besides endemic *Cyprinidae* from Lake Lanao on the island of Mindanao. These are probably the last collections from both regions, where, thanks to human folly, the endemic species are doomed to extinction, if this has not already taken place. The latter investigations concentrated on sympatric species formation since much of what has appeared on this subject in the literature ignores the genetics involved. For the endemic 'species flocks' the problem of intralacustric speciation is still open to question only in the *Cichlidae* of the large East African lakes. It seems to be supported by some observations on aquarium fish and is only apparently disproved by others. The term 'preferential mating' and perhaps even 'Praegung' offer starting points for experimental solutions. As should already be clear, fish behave differently from other vertebrates in so many respects that conclusions as to evolutionary mechanisms drawn from other groups should only be applied with great caution.

References[1]

Aida, T. (1930): J. Genetics 15, 1-16.
Aksiray, F. (1952): Hidrobiologi (Istanbul) 1, 33-81.
Anders, A., Anders, F. et al. (1969): Zool. Anz. (Suppl.) 33, 333-339.
Anders, A. et al. (1971): Experentia 27, 931-932.
Anders, F. (1967): Experentia 23, 1-10.
Anders, F., Klinke, K. (1966): Verh. dtsch. zool. Ges. 1966, 391-401.
Anders, F. et al. (1972): Biologie in unserer Zeit. 2 Jg., 35-45.
Avise, J.C., Selander, R.K. (1972): Evolution 26, 1-19.
Breider, H. (1936): Z. ind. Abst. Vererbgsl. 71, 441-499.
Dobzhansky, Th. (1951): Genetics and the origin of species. 3. ed. New York: Columbia Univ. Press.
Franck, D. (1964): Zool. Jb. (Physiol.) 71, 117-170.
Franck, D. (1970): Z. Tierpsychol. 27, 1-34.
Franz, R., Villwock W. (1972): Mitt. ham. zool. Mus. Inst. 68, 135-176.
Gerschler, M.W. (1914): Z. ind. Abst. Vererbgsl. 12, 73-96.
Gordon, H., Gordon, M. (1957): J. Genetics 55, 1-44.
Gordon, M. (1946): Zoologica 31, 77-88.
Gordon, M. (1947): Adv. in Genetics 1, 95-132.
Gordon, M. (1950): Endavour 9, 26-34.
Gordon, M., Rosen, D.E. (1951): Bull. Am. Mus. Nat. Hist. 95, 413-464.
Grzimek, X., Ladiges, W. (1970): Grzimek Tierleben 4, 360 ff.
Hagen, D.W., McPhail, J.D. (1970): J. Fish. Res. Bd. Canada 27, 147-155.
Hartmann, M. (1956): Die Sexualität. Stuttgart: G. Fischer.
Holzberg, S., Schröder, J.H. (1972): Mutation Res. 16, 289-296.
Hubbs, C. L., Hubbs, L.C. (1946): The Aquarium Journal 17, 4-6.
Kallman, K.D. (1962): J. Genetics 58, 7-21.
Kallman, K.D. (1964): J. Genetics 50, 583-595.
Kallman, K.D. (1965): Zoologica 50, 151-190.
Kallman, K.D. (1970): Zoologica 55, 1-16.
Karbe, L. (1961): Mitt. hamb. zool. Mus. Inst. 59, 73-104.
Kosswig, C. (1929): Z. ind. Abst. Vererbgsl. 50, 63-73.
Kosswig, C. (1936): Biol. Zbl. 56, 409-414.
Kosswig, C. (1937): Roux' Arch. Entw.-Mech. 136, 491-528.
Kosswig, C. (1938): Rev. Fac. Sci. Instanbul 3, 1-7.
Kosswig, C. (1947): Nature 159, 605-606.
Kosswig, C. (1961): Zool. Anz. 166, 333-356.
Kosswig, C. (1963): Veröff. Inst. f. Meeresforsch. Bremerhaven (Sonderband), 178-196.
Kosswig, C. (1964a): Experentia 20, 1-10.
Kosswig, C. (1964b): Copeia 1964, 65-75.

[1] The number of reference works is so large (over 1 000) that only a sample can be listed here. This sampling, however, will lead the way to further publications.

Kosswig, C. (1972): XVII. Congr. internat. Zoologie Monaco (im Druck).
Kosswig, C., Villwock, W. (1964): Verh. dtsch. zool. Ges. Kiel 1964, 95-102.
Miller, R.R., Hubbs, C.L. (1969): Copeia 1969, 52-69.
Miller, R.R., Walters, V. (1972): Contr. in Science no. 233 (Nat. Hist. Mus., Los Angeles County).
Münzing, J. (1963): Evolution 17, 320-332.
Öktay, M. (1959): Rev. Fac. Sci. Istanbul, B 24, 225-233.
Öktay, M. (1964): Mitt. hamb. zool. Mus. Inst. (Erg.bd.) 61, 133-157.
Öztan, N. (1954): Rev. Fac. Sci. Istanbul, B 19, 245-280.
Öztan, N. (1963): Rev. Fac. Sci. Istanbul, B 25, 27-47.
Ohno, S. (1970): Evolution by gene duplication. Berlin-Heidelberg-New York: Springer.
Ohno, S. (1972): this symposium.
Parzefall, J. (1969): Behaviour 33, 1-37.
Parzefall, J. (1970): Z. Morph. Tiere 68, 323-342.
Parzefall, J., Wilkens, H. (1972): Z. Morph. Tiere 73, 63-79.
Peters, N., Peters, G. (1966): Roux' Arch. Entw.-Mech. 157, 393-414.
Pfeifer, W. (1960): Z. vergl. Phys. 43, 578-614.
Popper, A.N. (1970): Behaviour 18, 552-562.
Rust, W. (1941): Z. ind. Abst. Vererbgsl. 71, 336-395.
Sadoglu, P. (1955): Experentia 13, 394-395.
Scheel, J.J. (1968): T.F.H. publications, 473 p. New York
Scheel, J.J. (1972): Z. zool. Syst.-Evolutionsforsch. 10, 180-203.
Schemmel, Ch. (1967): Z. Morph. Tiere 61, 255-316.
Schmidt, J. (1919): J. Genetics 8, 147-153.
Schröder, J.H. (1966): Zool. Beitr. (NF) 12, 27-42.
Schröder, J.H. (1969a): Mutation Res. 7, 75-90.
Schröder, J.H. (1969b): Zool. Beitr. 15, 237-265.
Schultz, R.J. (1969): Am. Nat. 103, 605-619.
Sengün, A. (1950): Rev. Fac. Sci. Istanbul 15, 110-133.
Siciliano, M.C. et al (1972): XVII. Congr. internat. Zoologie Monaco
Uyeno, T., Smith, G.R. (1972): Science 175, 644-646.
Vielkind, J. et al. (1971): Cancer Reseach 31, 868-875.
Villwock, W. (1958): Mitt. hamb. zool. Mus. Inst. 56, 81-152.
Wilkens, H. (1968): Zool. Anz. 180, 454-464.
Wilkens, H. (1970a): Z. zool. Syst.-Evolutionsforsch. 8, 173-199.
Wilkens, H. (1970b): Roux' Arch. Entw.-Mech. 166, 54-75.
Wilkens, H. (1972): Zool. Anz. 188, 1-11.
Winge, Ö. (1930): J. Genetics 23, 69-76.
Winge, Ö. (1934): C.R. Lab. Carlsberg, Ser. phys. 21, 1
Witschi, E. (1929): Handb. Vererbgswiss., 11/10. Berlin: Bornträger
Yamamoto, T. (1967): Genetics 55, 329-336.
Zander, C.D. (1962): Mitt. hamb. zool. Mus. Inst. 60, 205-264.
Zander, C.D. (1968): Molec. Gen. Genetics 101, 29-42.
Zander, C.D. (1969): Mitt. hamb. zool. Mus. Inst. 66, 241-271.
Zander, C.D., Dzwillo, M. (1969): Z. wiss. Zool. 178, 276-315.
Zeiske, E. (1968): Z. vergl. Phys. 58, 190-222.

2. Sex – Determination and Melanoma Formation

The Sex-Determining Mechanism of the Platyfish, *Xiphophorus maculatus*[1]

K. D. Kallman

Among certain higher categories mechanisms of sex determination are of taxonomic significance. The XX-XY mechanism is present in most mammals. Some species have a multiple sex chromosome system which, however, is clearly derived from the basic type. In birds females are the heterogametic sex. However, in teleosts and amphibians male and female digamety may be found among closely related forms. Of particular interest is the sex chromosome mechanism of the southern platyfish *Xiphophorus maculatus* (Guenther) which is polymorphic for sex chromosomes, W, X and Y.

Bellamy (1924, 1928), Gordon (1927) and Kosswig (1934) found independently that in the early domesticated stocks of platyfish of unknown origin, females were heterogametic and males homogametic. Subsequently, Bellamy (1936) found that in *X. variatus* which replaces *X. maculatus* geographically to the North, the sex-determining system was of the XX ♀♀ - XY ♂♂ kind. This interesting situation in which apparently two different modes of sex determination were present in closely related forms, became further complicated with Gordon's discovery (1947, 1951) that in samples from three Mexican rivers females of *X. maculatus* were XX and males XY while in the Belize River of British Honduras females were WY and males YY. Kallman (1965, 1970) determined later that W and X chromosomes occur together in 90 per cent of the species' range. Within the same natural population three genotypes of females (WY, WX, XX) and two genotypes of males (YY, XY) may occur together. The different kinds of males and females look indistinguishable from each other and, as far as can be determined, breed randomly in natural populations.

X. maculatus ranges in the coastal plains of Middle America from the Rio Jamapa, Veracruz, Mexico, eastwards to the small streams of British Honduras that drain the Maya Mountains. The easternmost (southernmost) station at which platyfish have been looked for (and found) is Mango Creek in British Honduras, 125 kilometers (km) south of the Belize River. Most likely platyfish extend another 100 km to the South as far as the Rio Sarstoon where the mountains meet the coast. In some parts of British Honduras the coastal plain inhabited by platyfish is less than 10 km wide, while in Guatemala this species is found 375 km inland from the Gulf of Mexico.

The sex chromosomes of platyfish have been analyzed from various stations between the Rio Jamapa and August Creek (about 115 km south of the Belize River). Platyfish are found in a variety of habitats. They occur under the quiet banks of streams and in stagnant pools, preferably in tangled vegetation. They live both under a dense canopy of trees and also in shallow, sun-drenched flooded areas of cow pastures and along roadsides. The substrate of all stations is generally mud. In temporary pools, hundreds or thousands of fish may sometimes be found.

[1]This investigation was supported in part by a grant, CA 06665, from The National Cancer Institute, U.S. Public Health Service.

Although the sex chromosomes of *X. maculatus* have not yet been identified cytologically, no doubt exists as to the reality of the W and X. Many codominant pigment factors belonging to several allelic series provide convenient gene markers for the sex chromosomes. Sex ratios vary according to the genotypes of the parents (Gordon, 1952; Kallman, 1965).

Four types of matings lead to an even sex ratio:

XX ♀♀ x XY ♂♂
WX ♀♀ x YY ♂♂
WY ♀♀ x YY ♂♂
WY ♀♀ x XY ♂♂

One kind of cross results in a ratio of 3 ♀♀ : 1♂:

WX ♀♀ x XY ♂

From one mating only male progeny is produced:

XX ♀♀ x YY ♂♂

The identification of the gonosomes of wild-caught fish (P_1) is based upon the analysis of sex ratios and the inheritance of pigment patterns (Kallman, 1965, 1970). It is also possible to identify the sex chromosomes of unknown males that had fertilized the P_1 females before collection, although this method has not been stressed in previous publications. If, for example, a P_1 female with *Dr* (red dorsal fin) gives rise to progeny consisting of + and *Ir* (red iris) females and *Dr* and *IrDr* males, the wild-caught female most likely possessed a W chromosome and must have been fertilized by a male heterozygous for *Ir*. A further cross, involving a F_1 *Dr* male with a XX female of a laboratory stock, may give rise to + female and *Dr* male progeny. From these crosses one can conclude that the genotype of the P_1 female was W-+ Y-*Dr* and that of the unknown male X-+ Y-*Ir*. An equal number of paternal X and Y chromosomes will be identified from females of populations lacking the W chromosome; the frequency of Y will be greater than 50 per cent in populations with a W chromosome.

The sex genotypes of all platyfish collected during the last forty years have been listed in Table 1 (See page 22). The X chromosome is present throughout the range of this species and has been demonstrated in collections both from the eastern- (August Creek) and westernmost (Rio Jamapa) populations. The X has not yet been found in two places in British Honduras (New River, Sibun River), but this can be attributed to sampling error. The inclusion of the sex chromosomes of "unknown males" in Table 1 has lead to the demonstration of the X in four collections in which it was not present among the genotypes of the fish returned to the laboratory (Rio Coatzacoalcos, '71; Bermudian Landing - a, '66; c & f, '69). Apparently the only populations that lack the W chromosome are the ones that inhabit the Rio Jamapa and Rio Papaloapan systems in the West. The sex genotypes of all wild-caught fish (♀♀ - XX, ♂♂ - XY) from the five stations are consistent with this interpretation. Populations with both W and X chromosomes are found near the coast (Belize River stations, Rio Coatzacoalcos '71) and far inland at the foot of the Sierra (Rio de la Pasion '63, the farthest point inland at which platyfish occur). Important differences can exist between local populations in time or space. This was most marked for the collections from the Rio Coatzacoalcos where only XX and XY fish were found in 1948 at a site 75 km inland whereas all females were WY and all but one male YY in 1971 at a location 2 km from the coast. Whether these differences are merely sampling errors or perhaps a reflection of real differences between coastal and inland populations in this river has to be determined.

Most extensive is the sample from the Belize River in British Honduras. The twelve collections are in agreement with each other in the low incidence of XY males and WX females. The frequency of the X chromosome in this population is approximately 0.045. Only 0.4 per cent of the females are expected to possess the XX genotype and this explains readily their absence from all samples. Since three out of five fish from August Creek had X chromosomes, its frequency appears to be quite high in this population south of the Belize River. The wide geographical distribution and the relative abundance of the W chromosome account readily for female heterogamety in the early domesticated stocks (Bellamy, 1924, 1928; Gordon, 1927, 1937; Kosswig, 1934; Breider, 1942).

The sex-linked pigment factors are located on both X and Y chromosomes (Gordon, 1947; Kallman, 1965), but the W chromosome of heterogametic females of natural populations always carries the + alleles at the various loci (Kallman, 1970). Similar comments have been made by Bellamy (1924), Gordon (1927, 1937) and Breider (1942) concerning the linkage of pigment genes in domestic stocks. I have listed in Table 2 the linkages of all 105 WY and WX females obtained in 22 collections and analyzed in the laboratory (See page 24). All 133 pigment factors were either X- or Y-linked. The pigment loci must be present on the W chromosome, since crossing over between W and Y occurs at about the same frequency as between X and Y (Gordon, 1937; Kallman, 1965, 1970). W-linked pigment genes obtained in the laboratory are inherited strictly maternally and the breeding performance of such strains is the same as that for wild stocks (Kallman, unpublished). No crossing over between W and X chromosomes has yet been reported, but this is readily accounted for by the small number of WX females bred. Presumably, crossing over occurs also under natural conditions and in the absence of selection the frequency of marked W, X and Y chromosomes should be the same. Kallman (1970) has suggested that in nature selection may be for brightly colored males and cryptic pigmentation in females. A marked W being inherited by females only would be exposed to negative selection in every generation and eventually lost. Conversely, there would be a high premium on W-+ chromosomes, because females possessing them would on the average be less colorful than those with a marked W or than XX females which could carry pigment genes on both chromosomes. Perhaps this is the explanation for the apparent selective advantage of the W which is now widespread and often more numerous than the X.

There is no evidence that *X. maculatus* was ever seperated into two major populations, one with a XX-XY system in the West (Mexican Rivers) and the other with a WY-YY (WZ-ZZ) system in the East (British Honduras) as proposed by Gordon (1952) and later expanded by Dzwillo and Zander (1967). Gordon who had based his theory upon collections from the Rio Jamapa (1939), Rio Papaloapan (1932), Rio Coatzacoalcos (1948) and Belize River (1949), had thought that originally platyfish had no gonosomes and that sex was determined by many genes scattered over most chromosomes. Subsequently, different sex determining systems evolved independently in the two populations, but the selective forces leading to different mechanisms were not apparent (Gordon, 1952).

It is unlikely that the same chromosome pair (*X. maculatus* has 24 pairs) has differentiated twice into sex chromosomes (Kallman, 1965). Moreover, the sex chromosomes of *X. variatus*, *X. milleri* (Kallman and Atz, 1966) and *X. pygmaeus* (Zander, 1968) are homologous to those of *X. maculatus*. Most likely, the ancestor of all four species already possessed the XX-XY system and the existence of the W in *X. maculatus* is an evolutionary novelty. The original XX-XY condition may still

Table 1. The Sex-determining mechanism of the platyfish, *Xiphophorus maculatus*

River system and year of collection[a]	Sex chromosome constitution of wild-caught platyfish: No. of females XX	X?[b]	WX	WY	WW	W?[b]	No. of males XY	YY	?	Sex chromosomes of unknown males that had fertilized females before collection: Number of X	Y	Minimum number of males involved
Rio Jamapa												
a 1939[c]	6						4					
b 1971[d]	17						26		1[g]	13	13	13
Rio Papaloapan												
a 1932[c]	5						1					
b, c 1971[d]	3						7			2	2	2
Rio Coatzacoalcos												
a 1948[e]	4						6			3	3	3
b 1971[d]				5				1		1	3	3
Rio Tonala												
a, b 1971[d]				3			1	3			1	1
Rio Grijalva												
a 1952[e]	1		1				3	2			1	1
Rio Usumacinta												
Rio de la Pasion 1963[e]	1		1	4			7			4	7	7
Rio San Pedro 1963[e]				7			1	4			5	5
Lake Peten												
a 1954[e]			1				3	1		1	2	2
b 1963[e]	1		1	4		1	4	3		3	5	5
Rio Hondo												
Tikal 1963[e]				8				6			11	8
San Antonio 1954[e]	2			3			1					
Douglas 1963[e]				8	1		1	3			5	5

New River									
a 1954[e]			1			2		1	1
Belize River									
Gabourel Creek 1949[e]			4			2			
Gabourel Creek 1966[f]			5		1	1		1	1
Bermudian- a 1966[f]			14			4	1	4	4
Landing b 1969[d]			12			2		11	11
" c 1969[d]			8				1	7	7
" d 1969[d]			7			1		5	5
" e 1969[d]			7	1		3		6	6
" f 1969[d]			5				1	5	5
" g 1969[d]		1	16		1			15	15
" h 1969[d]			1			2		1	1
" i 1969[d]		1	4		1	5		2	2
" j 1969[d]			2		1	4			
Sibun River									
a 1966[f]			6	2		2		3	2
August Creek									
a 1969[d]	1	1	1		1	1		3	3

[a] a,b,c etc. refer to different collecting sites within the same river system.

[b] An insufficient number of offspring precluded identification of the other sex chromosome.

[c] Data from Gordon, 1947. The number of fish listed for the 1939 collection is less than the number reported by Gordon. All of Gordon's data are consistent with the XX-XY mechanism, but in some cases alternate explanation are possible. Such cases have been omitted from Table 1.

[d] New data.

[e] From Kallman, 1965.

[f] From Kallman, 1970.

[g] This male was mated to a female known to be XX. From this cross 17 ♀♀ and 2 ♂♂ were obtained. No suitable pigment genes were present.

Table 2. Linkage of pigment genes in female *X. maculatus* of genotypes WX and WY

Population and year of collection	Number of WX or WY females collected from natural populations with: one pterinophore or carotenoid gene	one macromelanophore gene	two pterinophore or carotenoid genes	one pterinophore or carotenoid and one macromelanophore gene	Number of pigment genes linked on: W	X or Y
Rio Coatzacoalcos						
1971	2	1			0	3
Rio Usumacinta system						
Rio de la Pasion 1963		1			0	1
Rio San Pedro 1963	1				0	1
Lake Peten						
1963	1	2			0	3
Rio Hondo system						
Tikal 1963		4			0	4
San Antonio 1954		1			0	1
Douglas 1963		4			0	4
New River						
1954		1			0	1
Belize River						
Gabourel Creek 1949		3			0	3
Gabourel Creek 1966	2			3	0	8
Bermudian Landing						
a 1966	4	3	1	6	0	21
b 1969	7		2	2	0	15
c 1969	5		1		0	7
d 1969	3	1		2	0	8
e 1969	5			3	0	11
f 1969	3				0	3
g 1969	6	5	2	1	0	17
h 1969	1				0	1

i 1969					0	5
j 1969			1		0	2
	5					
<u>Sibun River</u>						
1966	3	1	2	2	0	12
<u>August Creek</u>						
1969	2	—	—	—	0	2
TOTAL	50	27	9	19	0	133

exist today in the Rio Jamapa and Rio Papaloapan. It is difficult to decide whether the W chromosome arose as the species expanded eastwards and W and X chromosomes were present among the founders of each river system or whether *X. maculatus* with a XX-XY mechanism once occupied the entire range and the W arose independently in several areas. Populations without a W chromosome should be more randomly distributed, if the W had an independent origin in the different rivers. There is also the possibility that the W arose somewhere in the center of the species' range (Rio Usumacinta, Rio Grijalva) and then spread to adjacent drainages along the coast where altitudinal differences are nonexistent. If this is the case, the X chromosome should be most common in the populations at either end of platyfish distribution and at those stations farthest away from the coast. The data in Table 1 are not inconsistent with this interpretation. The X is most prevalent in eastern (August Creek, few data) and western populations (Rio Jamapa, Rio Papaloapan) and inland (Rio de la Pasion, Rio Coatzacoalcos, 1948). Most instructive, therefore, will be future collections from the coastal area of the Rio Papaloapan drainage (all Papaloapan stations of Table 1 are 90 km inland) and from the rivers south of August Creek in British Honduras. At the present time platyfish occupying major drainages seem to be effectively isolated from each other as indicated by significant differences in the frequencies of certain pigment patterns (Gordon and Gordon, 1957).

While selection for cryptic females (W-+) may rationalize the spread of the W through natural population, it does not explain the origin of the W. Without further evidence it cannot be assumed that the W arose from a X chromosome, just because both are female determining. It is even known that in special crosses the W may behave as if it were a Y. Kallman (1968) has found some autosomal transformer genes that act epistatically to the W leading to males among the WY and WX, rather than XX offspring. Nothing is known about the differences between sex chromosomes at the level of the gene. If sex in poeciliid fishes is determined by a single gene and does not involve major cytological differences between X and Y chromosomes, a single mutation on either X or Y could result in a W. Several instances of atypical sex determination in *X. maculatus* have been studied in which phenotypic sex did not agree with "genotypic" sex. Of the six possible combinations of sex chromosomes (Table 3) all but the YY class can differentiate into normal males or females (Kallman, 1968).

Table 3. Normal and atypical sex determination in *Xiphophorus maculatus*

	Sex Determination	
Genotype	Normal	Atypical
WW[a]	female	male
WY	female	male
WX	female	male
XX	female	male
XY	male	female
YY	male	none

[a]Of the six possible sex chromosome combinations only WW is not usually found in natural populations. WW fish can be obtained by crossing WY females with exceptional WY males (Kallman, 1968). "Normally" WW fish are females except in special genetic crosses (Kallman, 1968).

Evidently, genes for the determination of either sex are present on the W and X chromosomes, but the Y carries predominantly genes for maleness.

It is incorrect to state that *X. maculatus* has two different sex chromosome mechanisms (WY-YY versus XX-XY). The species is polymorphic for those sex chromosomes (X,W) that determine femaleness. There exists only one system consisting of three gonosomes: one, Y (male determining), is present in all populations, while the X or W may be absent from some of them.

Other genera of poeciliid fishes may also have species with both male or female heterogamety. In *Poecilia*, the guppy, *P. reticulata*, is XX-XY, but in a short-finned molly, *P. sphenops*, Schröder (1964) documented the existence of both W, X and Y chromosomes. His work was based exclusively upon fish of unknown geographic origin and cannot, therefore, be directly related to natural populations. This is unfortunate, since several cryptic species may hide under the name of *P. sphenops* (Hubbs, 1961; Schultz and Miller, 1971). The same criticism applies to Chen and Ebeling's work (1968) who demonstrated cytologically female heterogamety in *Gambusia* but used an introduced population of unknown origin. The existence of both WY-YY and XX-XY systems within several genera of poeciliid fishes suggests that a W chromosome arose several times independently and that the step from X or Y to a W is a relatively simple one.

References

Bellamy, A.W. (1924): Bionomic studies on certain teleosts (Poeciliinae). I. Statement of problems, description of material, and general notes on life histories and breeding behavior under laboratory conditions. Genetics 9, 513-529.

Bellamy, A.W. (1928): Bionomic studies on certain teleosts (Poeciliinae). II. Color pattern inheritance and sex in *Platypoecilus maculatus* (Gunth.). Genetics 13, 226-232.

Bellamy, A.W. (1936): Inter-specific hybrids in *Platypoecilus*: one species ZZ-WZ; the other XY-XX. Proc. Nat. Acad. Sci. 22, 531-535.

Breider, H. (1942): ZW-Männchen und WW-Weibchen bei *Platypoecilus maculatus*. Biol. Zbl. 62, 187-195.

Chen, T.R., Ebeling, A.W. (1968): Karyological evidence for female heterogamety in the mosquitofish, *Gambusia affinis*. Copeia (1), 70-75.

Dzwillo, M., Zander, C.D. (1967): Geschlechtsbestimmung und Geschlechtsumstimmung bei Zahnkarpfen (Pisces). Mitt. Hamburg. Zool. Mus. Inst. 64, 147-162

Gordon, H., Gordon, M. (1957): Maintenance of polymorphism by potentially injurious genes in eight natural populations of the platyfish, *Xiphophorus maculatus*. J. Genet. 5, 1-44.

Gordon, M. (1927): The genetics of a viviparous top-minnow *Platypoecilus*: the inheritance of two kinds of melanophores. Genetics 12, 253-283.

Gordon, M. (1937): Genetics of *Platypoecilus*. III. Inheritance of sex and crossing over of the sex chromosomes in the platyfish. Genetics 22, 376-392.

Gordon, M. (1947): Genetics of *Platypoecilus maculatus*. IV. The sex determining mechanism in two wild populations of the Mexican platyfish. Genetics 32, 8-17.

Gordon, M. (1951): Genetics of *Platypoecilus maculatus*. V. Heterogametic sex-determining mechanism in females of a domesticated stocks originally from British Honduras. Zoologica 36, 127-153.

Gordon, M. (1952): Sex determination in *Xiphophorus (Platypoecilus) maculatus*. III. Differentiation of gonads in platyfish from broods having a sex ratio of three females to one male. Zoologica 37, 91-100.

Hubbs, C.L. (1961): Isolating mechanisms in the speciation of fishes. In: Vertebrate Speciation, W. F. Blair, editor, University of Texas Press, Austin, pp. 5-23.

Kallman, K.D. (1965): Genetics and geography of sex determination in the poeciliid fish, *Xiphophorus maculatus*. Zoologica 50, 151-190.
Kallman, K.D. (1968): Evidence for the existence of transformer genes for sex in the teleost *Xiphophorus maculatus*. Genetics 60, 811-828.
Kallman, K.D. (1970): Sex determination and the restriction of sex-linked pigment patterns to the X and Y chromosomes in populations of a poeciliid fish, *Xiphophorus maculatus*, from the Belize and Sibun Rivers of British Honduras. Zoologica 55, 1-16
Kallman, K.D., Atz, J.W. (1966): Gene and chromosome homology in fishes of the genus *Xiphophorus*. Zoologica 51, 107-135.
Kosswig, C. (1934): Farbfaktoren und Geschlechtsbestimmung (nach Untersuchungen an Zahnkarpfen). Der Züchter 6, 40-47.
Schröder, J.H. (1964): Genetische Untersuchungen an domestizierten Stämmen der Gattung *Mollienesia* (Poeciliidae). Zoologische Beiträge 10, 369-463.
Schultz, R.J., Miller, R.R. (1971): Species of the *Poecilia sphenops* complex (Pisces: Poeciliidae) in Mexico. Copeia (2), 282-290.
Zander, C.D. (1968): Über die Vererbung von Y-gebundenen Farbgenen des *Xiphophorus pygmaeus nigrensis* Rosen (Pisces). Molec. Gen. Genetics 101, 29-42.

Preliminary Report on Sex Differentiation in Germ Cells of Normal and Transplanted Gonads in the Fish, *Oryzias latipes*

N. Satoh and N. Egami

In the medaka, *Oryzias latipes*, much data on the mechanism involved in sex differentiation have been accumulated by Yamamoto and his co-investigators. On the basis of the following facts obtained from the oral administration of exogenous steroids, 1. the specific activity of sex steroids as exogenous sex inducers, 2. the very low effective dosage of these sex steroids as the inducers, 3. the selective incorporation of sex steroids into the differentiating gonads, they reached the conclusion that the sex steroids are natural sex inducers (cf. Yamamoto, 1969). In this report some results of our recent experiments are summarized.

1. Microscopic Observations of Sex Differentiation in Germ Cells

The mitotic and meiotic activities of germ cells during normal embryonic development were cytologically and histologically examined (Tsuzuki et al., 1966; Satoh and Egami, 1972). Primordial germ cells (PGCs) were obviously distinguishable from somatic cells 3 days after fertilization and began to proliferate about 8 days after fertilization. The mean number of PGCs increased during a period of 8-10 days after fertilization, reaching about 90 immediately before hatching. Newly-hatched fry could be classified into two types according to the number and the nucleic activity of germ cells in the gonad. One type consisted of fry containing about 100 germ cells and none of the cells in the meiotic prophase. In the other type of fry the number of germ cells increased by mitotic divisions and some of the cells began to enter into the meiotic prophase. During the course of further development the fry of the former type differentiated into males, forming the typical testis about 12 mm in total body length, and the latter into females, containing many auxocytes in the ovary.

Therefore, it can be concluded that the morphological sex differentiation of germ cells occurs at the time of hatching. As a result, the sexuality of each individual is already morphologically differentiated at the time of hatching. However, no sexual differences in the histological structure of somatic elements in the gonad were observable at that time, and the differentiation is still labile.

2. Electron Microscopic Observations of the Larval Gonads

Electron microscope studies of the sex differentiation of germ cells as well as that of the gonad have been carried out (Satoh, unpublished). Both PGCs in embryos and germ cells in newly-hatched fry characteristically contained the so-called germinal dense bodies interspersed within large aggregation of mitochondria in their cytoplasm (Fig. 1). This structural character was the same as that of amphibian oocytes. Every germ cell was surrounded by a few somatic cells. In those somatic cells the nucleus occupied large parts of the cell and long foot of cytoplasm enclosed the germ cell. From the ultrastructural

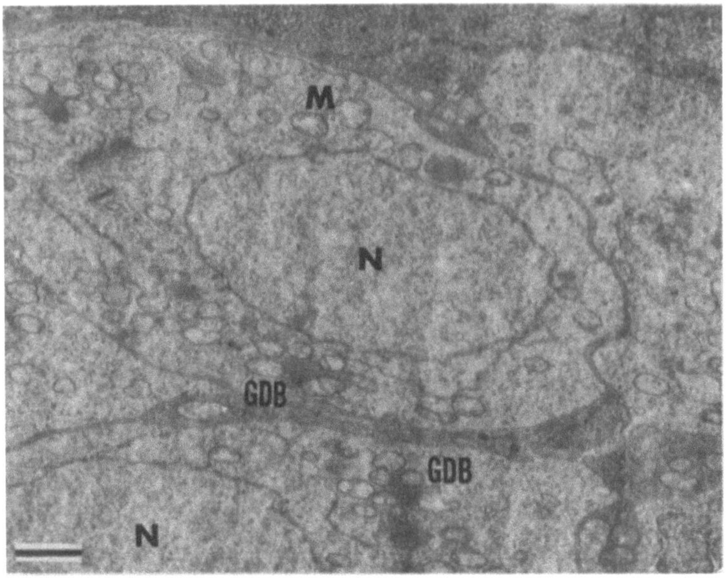

Fig. 1. Electron micrograph of primordial germ cells at 10 days after fertilization. The germinal dense bodies interspersed within large aggregation of mitochondria are conspicuous in the cytoplasm. GDB, germinal dense body; N: nucleus; M: mitochondria. Bottom bar: 1 μm

features of the germ cells it was difficult to identify the sex of the embryo before hatching.

In the ovary of newly-hatched female fry oocytes containing synaptonemal complexes in the nucleus were observed, and a few of desmosomes between the adjacent follicle cells enclosing the germ cell were also conspicuous. On the other hand, in male fry the similar structures of germ cells and the somatic cells, surrounding the germ cell, as those in embryos were merely detectable. Ten days after hatching the difference in the scantiness of organells in the cytoplasm was clear between spermatogonia and oogonia. In the testis of adult fish the steroid-secreting cells were clearly observable by electron microscope. In the newly-hatched fry, however, no such cells were demonstrated.

These results are in good agreement with the histological observations in important points and the similarity in ultrastructural features of male and female PGCs suggests the sexual bipotentiality of the germ cells.

3. Transplantation Experiments

In order to examine whether or not the sex differentiation in the medaka is modified by the physiological level of sex hormones in adult fish, the trunk region, containing the gonad of newly-hatched fry, was transplanted into the anterior chamber of the eyes of adult male and female fish. The grafts could be classified into two types according to the vascularization in the graft. One type of graft developed well; some of them protruded from the eye of host fish. In these grafts the connection of the blood circulation between the graft and the host was detectable without exception. The other type of graft consisted of those samples getting no vascularization. The grafts of this type did not grow in size; this condition seemed to be similar to the so-called in vivo culture in the anterior chamber of the eye in rodents. Most of these grafts, however, degenerated.

Judging from the histology of the gonad in the graft, a genetic male graft transplanted into the eye of a fish developed into a testis, regardless of the sexuality of the host (Fig. 2). This fact was

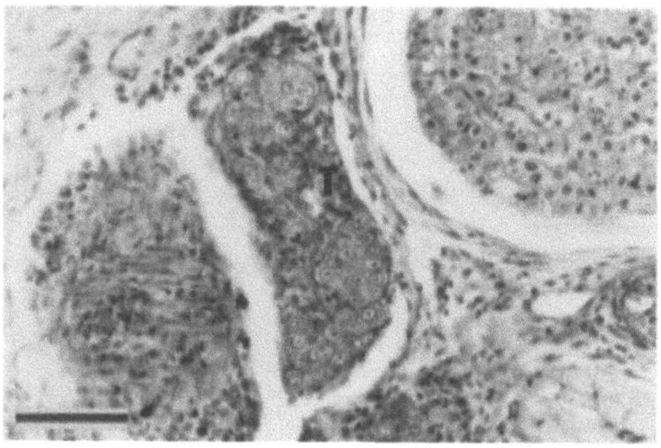

Fig. 2. Differentiated testis of a graft transplanted into the eye of a female host fish 30 days after transplantation. T: testis. Bottom: 50 μm

confirmed by two series of the experiments. The gonad of a genetic female fry developed into an ovary if the graft was transplanted into a female fish. On the other hand, the gonad of a genetic female graft transplanted into a male fish failed to develop into an ovary and formed spermatogenetic cells in the gonad of abnormal structure (Fig. 3).

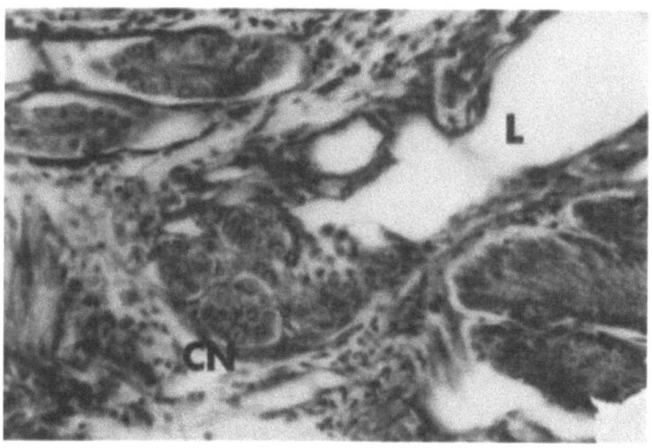

Fig. 3. Abnormal gonad developed in a graft in the eye of a male host fish 25 days after transplantation. Some cell-nests are seen on the periphery of the gonad and a lumen in the center. CN: cell-nest; L: lumen

Therefore, it is highly probable that the reversal of sex differentiation in genetic females into males is accomplished by the physiological level of sex hormones in male fish. On the contrary, the sex reversal of genetic males to females is not induced by the physiological level of female sex hormones (Satoh, unpublished).

From the facts of sex differentiation in germ cells of normal and transplanted gonads in this fish, besides the possibility that the sex hormones act as sex inducers in normal development, the second possibility is suggested that some intracellular mechanisms are involved in sex differentiation of the germ cells under natural condition.

References

Satoh, N., Egami, N. (1972): Sex differentiation of germ cells in the teleost, *Oryzias latipes*, during normal embryonic development. J. Embryol.exp.Morph. 28, 385-395.
Tsuzuki, E., Egami, N., Hyodo, Y. (1966): Multiplication and sex-differentiation of germ cells during development in the medaka, *Oryzias latipes*. Japan.J.Ichthyol. 13, 176-182.
Yamamoto, T. (1969): Sex differentiation. In: *Fish Physiology*, vol. III (ed. W.S. Hoar and D.J. Randalls), pp. 117-175. New York-London: Academic Press.

Regulation of Gene Expression in the Gordon-Kosswig Melanoma System

I. The Distribution of the Controlling Genes in the Genome of the Xiphophorin Fish, *Platypoecilus maculatus* and *Platypoecilus variatus*

A. Anders, F. Anders, and K. Klinke

1. Introduction

The Xiphophorin fish, including platyfish and swordtails, live in brooks, rivers, ponds and pools in Central America, where they have developed many species, subspecies, and populations. In some populations nearly all individuals have certain spot patterns composed of macromelanophores, in others only a certain percentage of the individuals are spotted, then again others exhibit no spots whatsoever (Gordon, 1942, 1947; Gordon and Gordon, 1950, 1957; Kallman, personal communication). Such spot patterns are, without a doubt, phylogenetic late acquirements which succeed in overcoming ecological and geographical barriers from population to population, and species to species.

Kosswig (1927), Gordon (1927, 1931) and Häussler (1928) found that the spots are determined by Mendelian genes (macromelanophore genes). Furthermore, they found that these same genes are responsible for melanoma formation in certain hybrids and they judged that this melanoma formation is based on the enhancement in gene expression, which is caused by an inbalance of modifier genes in the hybrid genome (Kosswig, 1929; Gordon, 1931). Breider (1952) proposed a classification of these modifiers in two types, which were later called "stimulation genes", which stimulate the expression of the macromelanophore gene, and "repression genes", which repress it (Anders, 1967, 1968).

The present paper deals with the regulation of the expression of five macromelanophore genes. In doing so the repression genes for these loci are brought into view. Because these have proved themselves to be much more heterogeneous than was originally thought, the terms "controlling genes" and "controlling elements" will, by preference, be used here. A part of the controlling elements could correspond to the regulator genes and operators of molecular genetics (see Part II, Anders and Klinke, 1973, this issue).

There is no reason to suppose that viruses play a part in the production of melanomas; nevertheless, it is easy to formally equate the macromelanophore loci with the oncogenes of the tumor viruses according to Bentvelzen (1968), Paine and Chubb (1968), Huebner and Gilden (1972), Todaro and Huebner (1972) and others (see Emmelot and Bentvelzen, 1972). Both are more or less only a loose constituent of the host genome which can also be transferred to other hosts and for whose expression genetic control systems must be built. Therefore, this paper could also describe a model which is useful in researching the regulation of the viral oncogenes, which are thought to be inherited and regulated as part of the natural gene pools.

2. Material and Methods

As experimental animals that posses loci for the determination of the macromelanophores were used:

Platypoecilus maculatus, from Rio Jamapa, Mexico, and *Platypoecilus variatus*, from Rio Panuco (probably), Mexico.

Both species are known to the aquarist as platyfish. In zoological systematics they are referred to as *Xiphophorus maculatus* (Guenther, 1866) and *Xiphophorus variatus* (Meek, 1904).

From these platyfish the following codominant sex chromosome-linked macromelanophore loci were chosen for the examination of the regulation of their expression.

Macromelanophore locus	Phenotype	Sex chromosome	Species
Sp (spotted)	Spots on the side of the body	*X*	*P. maculatus* (Fig. 1a)
Sd (spotted dorsal)	Spots on the dorsal fin	*X*	*P. maculatus* (Fig. 2a)
Li (lineatus)	Stripes on the side of the body	*X*	*P. variatus* (Fig. 4a)
Sr (stripe sided)	Stripes on the side of the body	*Y*	*P. maculatus* (Fig. 5a)
Pu (punctatus)	Small spots on the back	*Y*	*P. variatus* (Fig. 6a)

These loci will be dealt with in the given order which forestalls any partial results.

Both, *P. maculatus* and *P. variatus*, were crossbred with *Xiphophorus helleri guentheri* (Jordan and Evermann, 1896) from Rio Lancetilla. This is a subspecies of Xiphophorin fish, which are known to aquarist as swordtail. Despite being relatively closely related to the platyfish in their evolution, the swordtails have developed no sex chromosomes, and also no macromelanophore loci which could remind one of similarities with the sex chromosome-linked macromelanophore genes of platyfish.

In order to be able to make statements about the function and distribution of the controlling genes which are non-linked with the macromelanophore loci, the chromosomes of both platyfish were replaced step by step with those of the swordtail. This was done by suitable crossbreeding as Gordon, Kosswig, Breider, Kallman, Zander and many others in principle had previously accomplished (see *Xiphophorus* bibliography in Wolf and Anders, 1973). When confronted with difficulties in breeding artificial insemination was used in the method of Zander (1961), whereby all interspecific crossbreeds in the Xiphophorin fish could be extorted.

Next, the gene dosage effect or the gene dosage compensation of the respective loci was tested for the proof of the macromelanophore locus-linked controlling genes. Later, X-ray-induced mutations were used for subdividing the sex chromosome-linked controlling elements.

The fish were irradiated in a metal basin filled to 2 cm with water. This basin was placed 80 cm from the focus of a Röntgen Miller apparatus MG 150. X-rays were emitted at a dose rate of 22 R/min, 150 kV, 12 mA and filtered through 0.2 mm Cu and 0.5 mm Al.

The breeding of the fish proceeded according to Gordon (1963). For details of systematics, origin, ecology, propagation, and spot patterns of the platyfish and swordtails, see Gordon (1951), Rosen (1960), Kallman and Atz (1966), Zander (1964, 1967) and Wolf and Anders (1973).

3. Results

Controlling Genes for the *X* Chromosome-Linked Macromelanophore Locus *Sp* (Spotted) from *Platypoecilus maculatus*

Non-linked controlling genes. *Sp*, which only causes some spots on the side of the body of *P. maculatus*, shows a jump in gene expression in the F_1-generation after hybridization with *X. helleri*, which leads to premelanoma formation (Fig. 1a and b). On the other hand, the *Sp* group of the F_2-generation shows a continuous spectrum of the *Sp* expression from spots to malignant melanomas, of which the ones growing particularly fast certainly must be associated with the *Sp/Sp* genotype, and, therefore are inapplicable to the judgement of the controlling genes. Crossing of the F_1 and F_2 hybrids with *P. maculatus* likewise leads to interfluent variability of expression in the offspring. As was to be expected, the degrees of expression lie under those of the parent hybrids employed for these special crosses. Through further backcrossing of the backcross hybrids with *P. maculatus*, the *maculatus*-specific *Sp* expression is reproduced in all descendants.

These results can be explained by the facts that

a) the step from the spots of platyfish to the premelanomas of the F_1-generation is caused by the substitution of a chromosome set ($n = 24$) from *P. maculatus* by chromosomes from *X. helleri*; that

b) the continuous spectrum of macromelanophore gene expression in F_2 and in the backcross generations, in which *P. maculatus* was the recurrent parent, occurs through independent assortment of the chromosomes of both species; and that

c) the normalization of the *Sp* expression in the further backcross generations is based on a reintroduction of all chromosomes of *P. maculatus*.

We conclude from these results, that the *Sp* locus is normally regulated by a large quantity of controlling genes which are distributed over many chromosomes of *P. maculatus*. As it will later be shown, this controlling gene-system is *Sp* specific. To be sure, it is experimentally not excluded that the chromosomes of *X. helleri* also carry controling genes for *maculatus*-specific loci; however, this appears to be highly unlikely, because such controlling genes in the genome of *X. helleri* were superfluous.

These results are specified by backcrossing the F_1 with *X. helleri*: in the first backcross generation (Bc_1) once more a jump in gene expression occurs which exceeds the premelanomas of the F_1-generation. Thereby appear two classes of *Sp* phenotypes that are usually confluent which, however, seperate from each other more sharply under conditions of stress (50 to 100 fish for 10 l water; food shortages) and then show a clear 1 : 1 Mendelian segregation. These two phenotypes are represented by premelanoma carriers, which develop during the last third of their lives (life expectancy: 2 years) benign melanomas (Fig. 1c), and melanoma carriers that are destroyed by the melanoma in the juvenile stage (Fig. 1d).

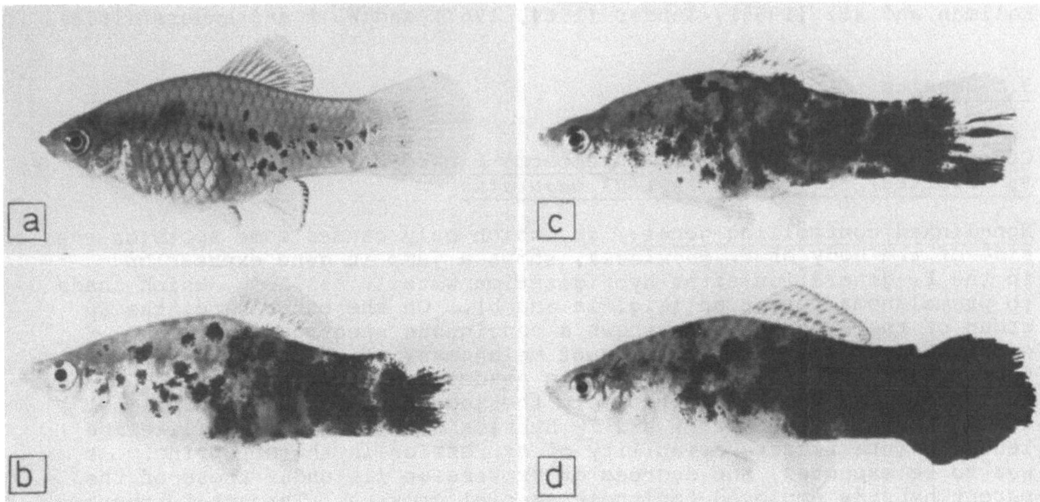

Fig. 1a-d. Expression of *Sp* (spotted), a) in the purebred *P. maculatus*, b) in *P. maculatus/X. helleri*-F_1-hybrids, c) and d) in backcross hybrids (recurrent parent = *X. helleri*), exhibiting the 1:1 segregation of these two phenotypes (premelanoma and melanoma bearing animals)

Backcrossings of the premelanoma carriers with *X. helleri* allows systematic cultivation of the tumor fish which in all Bc-generations segregate into these two phenotypes (Table 1, first line). This 1 : 1 ratio is certainly based on the segregation of a certain chromosome pair consisting of one autosome from *P. maculatus* and one from *X. helleri* (A_1^{mac}/A_1^{hell}). Under the assumption that *X. helleri* contains no *maculatus*-specific controlling genes, this platyfish autosome has more or stronger controlling genes for the expression of *Sp* than the others. The premelanoma phenotype has accordingly, in regard to these autosomal controls, the constitution A_1^{mac}/A_1^{hell}, while the melanoma phenotype has the constitution A_1^{hell}/A_1^{hell}.

A further specificity of the regulation of *Sp* results from the fact that an X-ray mutation, which is localized on the *Y* chromosome of *P. maculatus*, causes an extra spot deep under the skin in front of the dorsal fin. It has not yet been decided whether *X* chromosomes, which have no *Sp*, influence the expression of an *Sp* or not.

<u>Linked controlling genes.</u> In the genome of *P. maculatus*, *Sp* shows <u>gene dosage compensation</u>; sex chromatin was not observed. This dosage compensation can be interpreted that each X^{Sp} chromosome on the one hand carries at least one compensator gene for the *Sp* locus of its partner chromosome, and that it is, on the other hand, also equipped with a controlling element for the reception of the compensator gene signals of the partner chromosome. Due to this, the compensator gene

Table 1. Segregation of gene expression of platyfish macromelanophore loci in the genome of *Xiphophorus helleri*

Macromelanophore locus	Backcross generation	No. of offspring	No. of macromelanophore-bearing offspring	Segregation of gene expression: premelanoma (melanoma in *Sd';Sd"*)	Segregation of gene expression: melanoma (strong melanoma in *Sd';Sd"*)
Sp	Bc_1-Bc_4	ca. 800	380	173	207
Sd	Bc_1-Bc_4	ca. 6000	2702	1358	1344
Sd' *Sd"*	Bc_1	1395	646	313	333

signals can be sent and received reciprocally (Part II, Anders and Klinke, 1973, this issue).

Kallman and Schreibman (1971) report of an *Sp* linked controlling gene. After an unequal crossing over in which *Sp* is translocated from the *X* to the *Y* chromosome, this macromelanophore locus shows a rise in expression. The authors attribute this position effect to a separation of *Sp* from a special controlling element which is originally linked with it.

Subresult. The macromelanophore locus *Sp* is regulated by a high polygenic controlling system, which is distributed over many chromosomes (including the *X* chromosome, on which it is located). Nevertheless, more or stronger controlling genes are located on a certain autosome (A_1) than on the rest of the chromosomes.

Controlling Genes for the *X* Chromosome-Linked Macromelanophore Locus *Sd* (Spotted Dorsal) from *Platypoecilus maculatus*

Non-linked controlling genes. In these studies the same experiments were carried out as they were described for *Sp* and, in principle, the same results were obtained (Fig. 2); however, a few particulars arise.

After backcrossing F_1-hybrids with *X. helleri* and after further backcrossing the Bc-hybrids with the same species, there is seen a 1 : 1 segregation of premelanoma and melanoma bearing animals (Table 1, second line and Fig. 2c and d) by which both the phenotypes are more clearly separated from each other as it was the case in the corresponding *Sp* hybrids. From this it can be concluded that the autosomal controlling elements for *Sd* are more strongly concentrated on a certain *maculatus* autosome than are those for *Sp*.

This concentration is shown also in the offspring of premelanoma-carrying Bc-hybrids which are crossed between themselves. In the *Sd* offspring beside premelanoma and melanoma carriers including the *Sd/Sd*-genotype (Fig. 2f), individuals also appear, which show only an insignificant risen *Sd* expression (Fig. 2e), despite the fact that most of its chromosomes originate from *X. helleri*. In crossing with *X. helleri* , they produce in the *Sd* group premelanoma carriers only and prove themselves thereby, to be homozygous for the above mentioned autosome.

In order to test whether it is the same chromosome which plays a special role in the controlling of *Sp* and *Sd*, premelanoma carrying *Sp* and *Sd* Bc-hybrids were crossed with each other. The *Sp/Sd* offspring show a corresponding degree of expression in both *Sp* and *Sd* and thereby indicate that it is the same autosome which is more strongly involved in the controlling of both these macromelanophore loci.

The above mentioned X-ray mutation in the *Y* chromosome of *P. maculatus* influences *Sd* as well as *Sp*.

Linked controlling genes. Like *Sp*, *Sd* shows gene dosage compensation. Accordingly, an X^{Sd} chromosome carries at least one compensator gene, whose signals are received and processed by the *Sd* of the other *X* chromosome.

Kallman and Schreibman (1971) postulated a controlling gene linked to *Sd* when they found that a *Sd*, translocated to a *Y* chromosome, causes premelanomas. They assume that *Sd* was separated through this process of its regulator gene *R*, with which it is normally tightly linked.

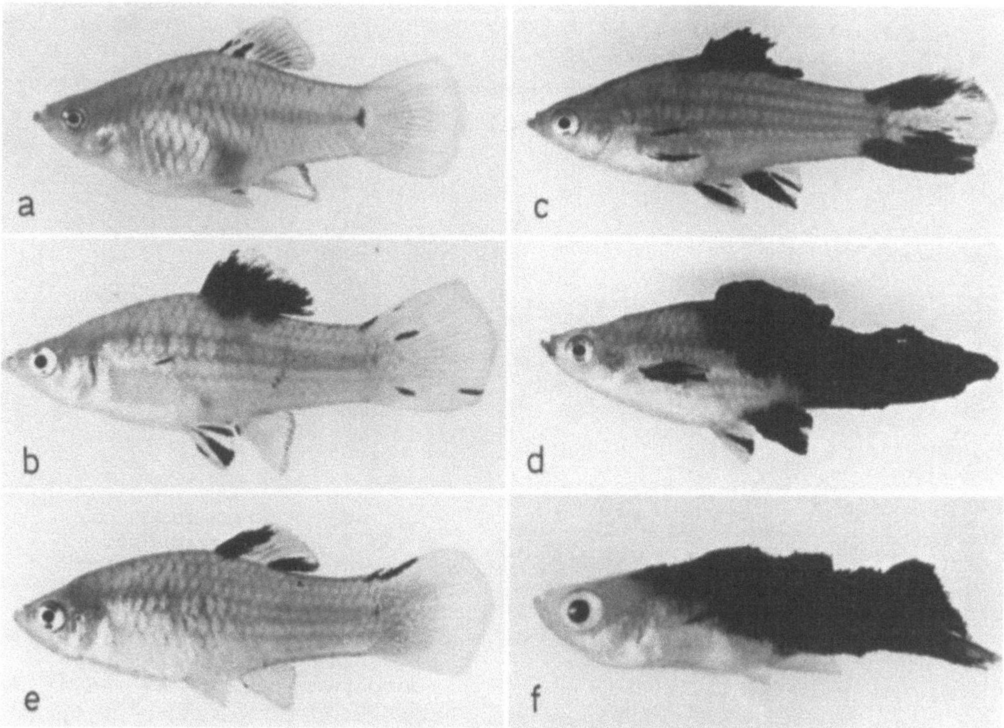

Fig. 2a-f. Expression of *Sd* (spotted dorsal), a) in the purebred *P. maculatus*, b) in *P. maculatus/X. helleri*-F_1-hybrids, c) and d) in backcross hybrids (recurrent parent = *X. helleri*), exhibiting the 1:1 segregation of these two phenotypes (premelanoma and melanoma bearing animals), e) and f) in the offspring of backcross hybrids of type c which were crossed with each other. Note that f represents a young fish (3.5 cm in length), the shape of which was enlarged to the same size as that of the adults (normal length 5-8 cm). For the genetic conditions which are responsible for these two extremely different phenotypes see text

From several X-irradiated *Sd* breeds of *P. maculatus*, three different *Sd* mutants (*Sd'*, *Sd"*, and *Sd'''*) were isolated and bred as substrains. These animals show in the homozygote state, besides the normal spot patterns on the dorsal fin, additional spots on the side of the body. In *Sd'''* individuals these additional spots have the character of premelanomas (Fig. 3a). In combination with a standard X^{Sd} chromosome, the additional expression on the side of the body is reduced or completely repressed. The standard X^{Sd} chromosome, therefore, transmits controlling signals, which are received and processed by the mutated X^{Sd}.

The following description of experiments will be limited to *Sd'* and *Sd"*, because only for these substrains could a sufficient amount of animals be bred.

In *P. maculatus/X. helleri*-F_1-hybrids, both *Sd* mutations cause the rise in expression in the dorsal fin typical for the standard *Sd*. Additionally, yet, *Sd'* causes premelanomas on the side of the body, and *Sd"* melanomas. The first Bc-generation, the backcross parent of which is *X. helleri*, shows a 1 : 1 Mendelian segregation of macro-

Fig. 3a-c. Expression of a) *Sd'''* in the purebred *P. maculatus* (compare with Fig. 2a), b) *Sd"* in *P. maculatus/X. helleri*-F_1-hybrids (compare with Fig. 2b), c) *Sd"* in backcross hybrids (compare with Fig. 2c and d)

melanophore gene expression (Table 1, third line), by which the expression is considerably risen in comparison to the standard *Sd*. The phenotype with the stronger *Sd'* or *Sd"* expression is killed by the melanoma in the earlier juvenile stage, and that with the weaker in the later one. From the relatively unimportant mutational rise in expression in the *maculatus* genome, and from the clear controlling function of the autosome in the Bc-generation we conclude that the mutated *Sd*, as well as the standard *Sd*, is controlled by autosomal factors. That component of the *Sd* locus which receives and processes the autosomal controlling signals, remains obviously unchanged. The excessive melanoma formation of the mutated *Sd* in the genome of *X. helleri* shows clearly, however, that the unaltered *Sd*-linked controling elements are strong ones.

Subresults. Like *Sp*, *Sd* is regulated by a high polygenic controlling system, whose constituents are distributed over many chromosomes. Varying from *Sp*, however, a stronger concentration of controlling elements is recognizable on a certain autosome and on the X^{Sd} chromosome.

Controlling Genes for the *Y* Chromosome-Linked Macromelanophore Locus *Li* (Lineatus) from *Platypoecilus variatus*

Non-linked controlling genes. *Li*, which is responsible for longitudinal lines on the side of the body in *P. variatus* (Fig. 4a), shows a rise in expression in *P. variatus/X. helleri*-F_1-hybrids, by which the longitudinal lines flow together to an united area (Fig. 4b). Differently from *Sp* and *Sd*, the expression of *Li* after backcrossings with *X. helleri* exceeds the expression of the F_1 only slightly. In F_2 there is an interfluent variability which ranges from the longitudinal lines to the enhanced expression present in the Bc-hybrids;

however, the width of the expression possibilities is too small to enable to conclude a distribution modus of controlling elements. A 1 : 1 segregation of the Bc-hybrids in a weak and strong expression class cannot be observed. There is, to this point, no hint as to the existance of *Y* chromosome-linked controlling genes.

Linked controlling genes. Gene dosage compensation is present in *Li* as it is in *Sp* and *Sd* and, therefore, indicates the corresponding compensator elements.

Further indications to the *Li*-linked controlling genes are produced through a crossing over, which appeared after X-irradiation of *P. maculatus/P. variatus*-F_1-hybrids. Hereby *Sd* from the *maculatus X* chromosome was replaced by *Li* from the *variatus X* chromosome. In this position *Li* shows, not only in the genome of *P. variatus* but also in the genome of *P. maculatus*, a strong rise in expression, which far exceeds those that can be obtained by introducing the normal X^{Li} chromosome into the genome of *X. helleri* (Fig. 4c). It is noteworthy that the X^{Li}-hybrid-chromosome in the genome of *X. helleri* shows no additional rise in expression. We conclude from these results that *Li* on the *X* chromosome of *P. variatus* is equipped with a controlling gene which exerts a stronger regulation than all others combined. By the crossing over, *Li* became separated from this controlling gene, or the controlling gene remained linked with *Li*, but was inactive in its new position. Because of the loss or the inactivation of the linked controlling gene the enhanced *Li* expression is transmitted like a Mendelian character.

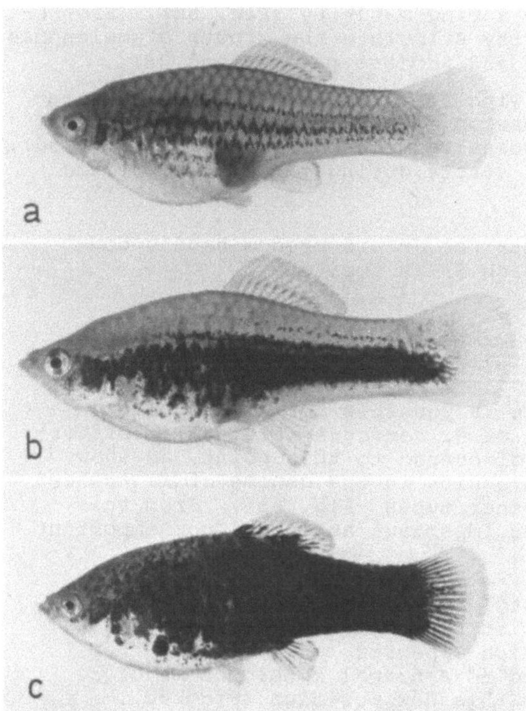

Fig. 4a-c. Expression of a) *Li* (lineatus) in the purebred *P. variatus*, b) *Li* in the genome of *X. helleri*, c) *Li*, which has crossed over to an *X* chromosome of *P. maculatus*, in both the purebred *P. variatus* and the purebred *P. maculatus*. No additional enhancement after crossing with *X. helleri*

Subresults. The non-linked controlling genes of *Li* receed in favor of the linked ones.

Controlling Genes for the *Y* Chromosome-Linked Macromelanophore Locus *Sr* (Stripe Sided) of *Platypoecilus maculatus*

Non-linked controlling genes. *Sr* is responsible for a similar but weaker side striping in *P. maculatus* than *Li* in *P. variatus* (cf. Fig. 4a and 5a). In *P. maculatus*/*X. helleri*-F_1-hybrids appears an insignificant rise in expression by which the character of the stripes is maintained (Fig. 5b). The Bc-hybrids, the backcross parent of which is *X. helleri*, show no additional rise in expression, and segregation of *Sr*-phenotypes is not observed.

It follows from these crossings, that there are only a few weak non-linked controlling genes for *Sr*. It cannot be decided, whether these are distributed over many or few chromosomes.

Linked controlling genes. As it was proved on experimentally produced Y^{Sr}/Y^{Sr}-♂♂(Anders and Anders, 1963; Anders, Anders and Rase, 1969; Anders, Förster, Klinke and Rase, 1969), there is no gene dosage compensation for *Sr* as opposed to *Sp*, *Sd* and *Li*. Thus, there are no Y^{Sr} linked compensator genes present. This is understandable because there are normally no *YY*-animals so that no dosage compensation for *Y* linked genes is necessary.

On the other hand it follows from the expression of three X-ray-induced mutations that very strong controlling elements for *Sr* are located on the *Y* chromosome. In these mutants (*Sr'*, *Sr''* and *Sr'''*) the macromelanophores are already present at birth in great quantity all over the body of the fish, and increase with aging. In addition, the males of the *Sr''* substrain by preference develop extraordinarily rapidly growing melanomas on the gonopodium during maturity (Fig. 5c). Probably, hormonal reactions during maturity stimulate the growth of melanomas. *Sr'* influences the expression of *Sp* and *Sd* (see page 36 and 38).

After hybridization of the mutants with *X. helleri*, a further enhancement of macromelanophore gene expression takes place, which, in the first Bc-generation, proceeds to excessive melanoma formation (Fig. 5d). In these animals melanoma formation starts during the last embryonic stages.

Subresults. The controlling elements for *Sr* are predominantly concentrated on the *Y* chromosome on which *Sr* is located.

Controlling Genes for the *Y* Chromosome-Linked Macromelanophore Locus *Pu* (Punctatus) from *Platypoecilus variatus*

Non-linked controlling genes. *Pu* is responsible for little black, point-like spots on the dorsal part of *P. variatus* (Fig. 6a). Differing from all macromelanophore loci discussed to this point, *Pu* shows not the slightest change of its expression in the genome of *X. helleri* (and neither in the genome of all other types, Fig. 6b). From this we can conclude, that *Pu* has neither important autosomal nor important *X* chromosomal controlling genes.

Linked controlling genes. The control for *Pu* is, therefore, concentrated on the *Y* chromosome on which *Pu* itself is located. This follows from four translocations which appeared independently from each other, by which *Pu* was attached to a normal *X* chromosome of *P. variatus* and causes premelanomas in this new position (Fig. 6c). These premelanomas become benign melanomas in old animals. Since the normal expression of *Pu* results only in very small point-like spots,

Fig. 5a-d. Expression of a) *Sr* (stripe sided) in the purebred *P. maculatus* male (*Y* chromosome-linked). Note the *X* chromosome-linked *Sd*, the expression of which was shown in the female in Fig. 2a. Note also the expression of the autosomal gene *Dot* at the center of the base of the caudal fin, b) *Sr* in the genome of *X. helleri*, c) *Sr"* in the purebred *P. maculatus*. Melanoma on the gonopodium is two months old. Note also the enhancement in expression of *Dot*, d) *Sr"* in the genome of *X. helleri*

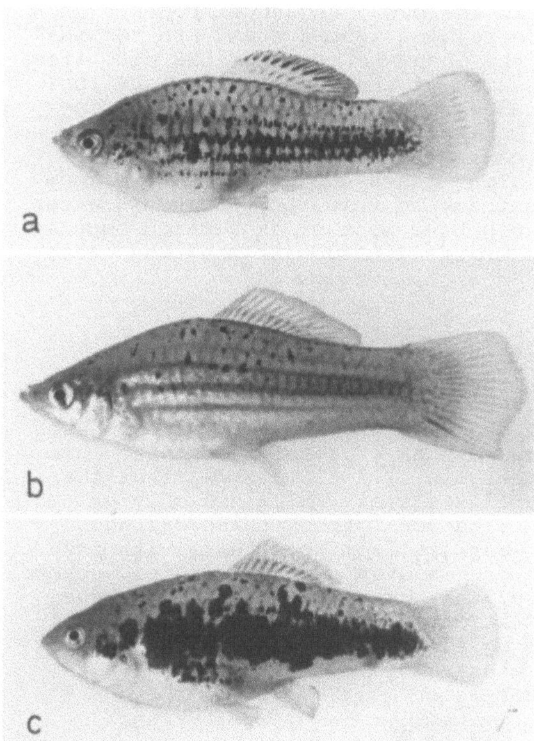

Fig. 6a-c. Expression of a) *Pu* (punctatus) in the purebred *P. variatus* male (*Y* chromosome-linked). Note the *X* chromosome-linked *Li* the expression of which was shown in the female in Fig. 4a. b) *Pu* in the genome of *X. helleri*, c) the *Pu* translocation in the purebred *P. variatus*

this mutation-conditioned increase of gene expression up to benign melanomas in the purebred species is a very strong one. Obviously, the deciding controlling gene was lost during the translocation, or it became ineffective through the position change. The introduction of the translocation chromosome into the genome of *X. helleri* has no marked additional effect on the expression of the *Pu* locus.

Subresults. The essential controlling elements for the expression of *Pu* are linked to *Pu*.

4. Discussion

Hybridization-Conditioned Melanoma Formation

In the platyfish, *Platypoecilus maculatus* and *Platypoecilus variatus*, there is a series of different spot patterns composed of macromelanophores whose cells are determined by a corresponding series of co-dominant sex chromosome-linked loci. To this belong the five loci studied which can be arranged in the following order on the basis of their expression in the genome of the swordtail, *Xiphophorus helleri*: *Sp* (spotted), *Sd* (spotted dorsal), *Li* (lineatus), *Sr* (stripe sided) and *Pu* (punctatus). *Sp* and *Sd* exhibit a strong enhancement of gene expression which results in melanoma formation; on the other hand, *Li* and *Sr* show only a slight, and *Pu* absolutely no enhancement of expression. The result of this paper allows an interpretation of this graduated capability of reaction of the platyfish loci in the swordtail genome (cf. Table 2).

It has been shown that each locus has its own controlling gene system whose single elements are distributed very diversely within the platyfish genome. The controlling elements of *Sp* are distributed over many if not all chromosomes, and from this follows a sequence from *Sd*, *Li*, *Sr* to *Pu* by which the respective controlling gene systems are concentrated on that chromosome on which the locus to be regulated also lies. It follows from this that the expression of the macromelanophore loci in the swordtail genome is determined according to the crossing-conditioned absence of the non-linked controlling genes. Therefore, *Sp* and *Sd* in the swordtail genome can cause the formation of melanomas because their most important controlling genes are not linked to them and are eliminated by crossing. Contrary to this, *Li*, *Sr* and *Pu* after hybridization show only a weak or no enhancement of expression because their most important controlling genes are linked with them and cannot be eliminated by crossings.

Mutation-Conditioned Melanoma Formation

The same, although exactly the reverse, arrangement of the macromelanophore loci is seen when, instead of upon the cross-caused, one focuses upon the mutation-conditioned enhancement of expression of these loci in the species-specific genome (cf. Table 2). It is shown that the mutation or loss of a linked controlling element which has only a weak enhancement of expression for *Sp* becomes stronger and stronger progressing from *Sd*, *Li*, *Sr* to *Pu*. We take from this that, with progressive linkage of the controlling gene system to the macromelanophore locus which has been discussed above, the number of controlling genes is decreased, until the entire regulation of the expression is finally taken over by one or a few controlling elements as in the case of *Pu*. When such a controlling element is missing, defective or inactive, respectively, the enhancement of expression must be very large.

Table 2. Regulation of expression of five macromelanophore loci from platyfish in the species-specific genome and in the genome of the swordtail

	Strength and distribution of controlling elements in the platyfish genome		Macromelanophore gene expression in the platyfish genome			in the swordtail genome	
Macro-melano-phore locus	linked controlling elements	non-linked controlling elements	due to standard conditions	due to mutations in linked controlling elements	due to mutations in non-linked controlling elements	due to hybridi-zation	due to both hybridization and mutations in linked controlling elements
Sp	+	+ + + +	spots	enlarged[a] spots	premelanomas	very large melanomas	not tested
Sd	+	+ + +	spots	additional spots	premelanomas	large melanomas	excessive melanomas
Li	+	+ +	lines	premelanomas	no alteration observed	confluent lines	melanomas
Sr	+	+	stripes	premelanomas melanomas in males	"	enhanced stripes	excessive melanomas
Pu	+	(+)	dots	large premelanomas	"	dots	large premelanomas

[a]Result of Kallman and Schreibman (1971)

Susceptibility of Somatic Cells for Mutation-Conditioned Tumor Formation

The findings on the distribution of the controlling elements in the genome are tightly correlated with the earlier results that certain platyfish genotypes always react to X-irradiation with enhancement of macromelanophore gene expression, while others do not (Anders, Anders and Pursglove, 1971; Pursglove, Anders, Döll and Anders, 1971; Pursglove, 1972). This touches the general problem of the origins of the susceptibility to tumor induction by cancerogenes. The Gordon-Kosswig melanoma system turns out some facilities for discussing this problem (for further discussion see Graffi and Bielka, 1959; Huxley, 1960; Bauer, 1963; Lynch, 1967; Braun, 1969).

It is known that X-irradiation (500 - 1 500 R) of *Sp* and *Sd* platyfish induces enhancement of macromelanophore gene expression in somatic cells. The degree of enhancement corresponds to that exhibited by the *X. helleri/P. maculatus*-F_1-generation; by irradiating *Sp* embryos this enhancement can be induced in all individuals. On the other hand, *Li*, *Sr* and *Pu* do not react, or extremely seldom, to treatment by X-irradiation (unpublished). It is significant that *Sp* and *Sd* also do not react to X-irradiation when they are in the swordtail genome (Pursglove, 1972). These seemingly contradictory results can be explained in the following way: if numerous controlling genes are present, as in the *Sp* and *Sd* platyfish, then many controlling gene mutations can also appear, and the enhanced expression will be a frequent occurrence. On the contrary, if there are only a few controlling genes present as in *Sp* and *Sd* hybrids as well as in all *Li*, *Sr* and *Pu* purebred platyfish, then controlling gene mutations and the enhancement of expression will be rare occurrences.

Heredity and Non-Heredity in Tumor Formation

The important question of cancerologie, why some neoplasms are hereditary and some are not (Strong, 1958; Huxley, 1960; Bauer, 1963; Lynch, 1967; Süss, Kinzel, Scribner, 1970), appears in a new light through the present research.

All neoplasms which were released by cancerogenes are by nature not hereditary. Heredity and non-heredity of germline-caused neoplasms are, on the other hand, harder to judge, as is seen from the following.

The same X-ray-induced mutations of the controlling gene system which appear in somatic cells are found in the germ cells also. These mutations in *Sp* and *Sd* platyfish are almost always mutations in the high polygenic part of the controlling system which is not linked with the macromelanophore loci. These mutations are conserved through inbreeding, but can be eliminated through crossings with standard platyfish. The expression of *Sp* and *Sd* is hereby normalized (Anders, Anders and Pursglove, 1971). These mutation-conditioned enhancements of expression are comparable with those that appear after hybridization of *Sp* and *Sd* platyfish with *X. helleri* and that are cancelled again after backcrossing with a platyfish. They have, to be sure, an heredity basis and are passed on from generation to generation by inbreeding, but they are not inherited like Mendelian characters.

It is different when those controlling genes in germ-line cells mutate or are lost, which are linked to the respective macromelanophore locus. In such cases the enhancement of expression has not only a hereditary basis, but is also inherited like a Mendelian character. The additional enhancement of expression, which appears after hybridization of such mutants with *X. helleri* is, as expected, not hereditary, because it can be cancelled by crossings with *P. maculatus*.

There is no reason to assume that heritability and non-heritability in neoplasms of other animals are determined through fundamentally other biological mechanisms. The genetic bases of the melanomas described here show that it is very difficult to differentiate between heredity and non-heridity of a tumor in single cases. Even the differentiation between tumors that are caused by mutations in somatic cells, and those caused by mutations in germ-line cells is essentially harder than it is generally assumed to be.

Mendelian Segregation of Gene Expression with Respect to Macromelanophore Differentiation

This part of the discussion is concerned with the Mendelian segregation of melanoma, premelanoma, and spot producing *Sp* and *Sd* hybrids, which is due to the presence or absence of the special A_1^{mac}-autosome contributed by *P. maculatus* (see Table 1). A_1^{hell}/A_1^{hell}-hybrids develop melanomas; A_1^{mac}/A_1^{hell}-hybrids, premelanomas; and A_1^{mac}/A_1^{mac}-hybrids, spots. These expression degrees are correlated with the differentiation degree of the chromatophores as was shown by the examination of *Sd* hybrids. The cells of the melanomas are incompletely differentiated and, therefore, continue dividing. The cells of the premelanomas, however, come slowly up to relatively well differentiated macromelanophores which, after differentiation, stop dividing. Finally, the spots consist only of fully differentiated macromelanophores incapable of cell division (U. Vielkind, 1972; Anders, Klinke and Vielkind, 1972; see also Gordon, 1959). These different degrees of differentiation correspond to the different H^3-thymidine labelling indices in autoradiographs which, in the melanomas are high and in the premelanomas and spots are low (Sieger, 1968; Anders, Sieger and Klinke, 1969). This A_1^{mac} chromosome, therefore, occupies an important place in inducing the differentiation of propigment cells to macromelanophores. The controlling elements lying on the A_1^{mac} autosome can be, therefore, perceived as differentiation genes. To these genes of the controlling gene system we can join the standard allele of the albino locus which also has the character of a differentiation gene for macromelanophores because the cells of the albino melanoma are always less differentiated than those of the normal pigmented melanoma (Vielkind, Vielkind and Anders, 1971).

Non-Specific Modifiers which Stimulate Proliferation of Melanoma Cells

In this paper this modifier type has been mentioned only in the case of melanoma growth on the male mating organ (gonopodium) of *Sr* mutants during maturity. In this case the prerequisits for enhancement of expression are fulfilled by the mutation of the controlling element linked to *Sr*. The enhancement of expression is realized already in the embryonic stage. In addition, during maturity of the ♂♂, melanomas grow rapidly on the gonopodium growing equally fast itself. These observations favor the assumption that an additional non-specific stimulation of melanoma growth takes place which is dependent on the hormonal-caused maturing process. In the sense of the two-step hypothesis of carcinogenesis by Berenblum (see Berenblum, 1964, 1967) the working of hormones could be perceived as a "promotion" of the "initiation"-mutation of the regulating element linked to *Sr*. All earlier observations on the influence of the sex on melanoma growth of hybrid fish (Anders, 1967; Siciliano, Perlmutter and Clark, 1971) can be interpreted in this manner if the crossing-conditioned defects on the controlling gene system of the macromelanophore loci are perceived as equivalent to the mutation-conditioned initiations of the cancerogenesis.

Other modifiers which stimulate the expression of the macromelanophore loci in an inspecific way have been known since the disclosure of the Gordon-Kosswig melanoma system (Kosswig, 1927 ff; Gordon, 1927 ff; Breider, 1938; Atz, 1962; Zander, 1962; Öktay, 1964; Kallman, 1965; Anders and Klinke, 1965; Anders, 1967; Anders, Klinke and Vielkind, 1972). Those that originate from *X. helleri* have been especially well studied. They come into view when, for example, *Sd* or *Sp* of *P. maculatus* is crossed into various subspecies of *X. helleri*. It is shown already in the F_1-generation, and more clearly in the backcross generations, that the expression of these macromelanophore loci also depends on the *X. helleri* subspecies used. There are several indications that a non-specific type of modifiers contributed by *X. helleri* controls the free amino acid pools which, in turn, stimulate DNA and RNA synthesis in the melanomas (Anders, Vester, Klinke, and Schumacher, 1961; Anders, Drawert, Klinke and Reuther, 1963; Anders, 1964, 1967; Anders and Klinke, 1965; Anders, Sieger and Klinke, 1969; Sieger, Sieger, Prüssing and Anders, 1969; on stimulation of nucleic acid synthesis by amino acids see also Ley and Tobey, 1970; Tobey and Ley, 1971; Brunner, 1973; Vielkind, Vielkind and Anders, 1973).

The definite degree of macromelanophore gene expression, therefore, depends upon two different modifier types; namely, the macromelanophore locus-specific controlling gene system mainly delt with in this article, and the macromelanophore locus-non-specific stimulating gene system here brought only into the discussion.

Discussion of a Model for Evolution of Macromelanophore Loci and Its Controlling Gene Systems

Melanophores are formed from propigment cells which migrate from the neural crest of the embryo, divide, and populate the whole fish skin (Tavolga, 1949). Only a part of them differentiate to melanophores incapable of cell division while the other part remains as propigment cells capable of division and form, as it were, a reservoir out of which propigment cells are scooped and kept ready for differentiation to melanophores, e.g. in regeneration (Kaeser, 1971; Lueken and Kaeser, 1972).

The macromelanophores are also formed from these propigment cells. They are different from the rest of the melanophores in that they become very big due to several endopolyploidizations. On normal fish skin they, therefore, can be observed already macroscopically as very small black spots.

Since the propigment cells are distributed throughout the whole fish skin, macromelanophores should be produced also throughout the whole fish skin if a gene is present which undertakes an indiscriminate determination to macromelanophores. Further, since the reservoir of propigment cells is always regenerating through cell division, new macromelanophores should continuously be formed in this case.

This paper shows that this is realized in the *Sr* mutations, especially when they are crossed into the genome of *X. helleri* (Fig. 5d). Kosswig (1938) had already described a corresponding mutation of unknown origin and called it *Fu* (fuligonosus = sooty). The *Dr* linked *Li* is, further, *Fu*-like (Fig. 4c) and also *Sd* mutations, which, in the genome of *X. helleri*, are responsible for a soot-like covering over the caudal part of the fish body, can be joined on here (Fig. 3c). The phenotype of the *Fu*-like mutations represents, therefore, the largely unregulated expression of an originally macromelanophore-determining gene which undertakes indiscriminate endoploidizations in propigment cells. The unregulated formation of macromelanophores would then be such an easy process which actually could be brought about by a gene or an oncogene of a virus in the sense of the oncogene theory (Todaro and Huebner, 1972).

In principle it is all the same whether a *Fu*-like factor is introduced into the host as an oncogene or as a gene; it is necessary for the host genome to check the unlimited expression. A relatively simple step could be the limiting of the macromelanophore production to crude spots over the whole body surface, as it is realized in the *Sd* mutants in the genome of *P. maculatus* (Fig. 3a). All other patterns are, then, further restrictions of macromelanophore formation of certain body surfaces which are undertaken through the evolution of more or less specialized controlling gene systems. The present paper has uncovered five such controlling gene systems of varying evolutionary accomplishments. That, for *Sp*, appears to be primitive, and the one for *Pu*, the most highly evolved. It remains to be seen whether further hints as to a common, original macromelanophore gene can be found and whether there are further signs that such a gene corresponds to an oncogene (see Part II, Anders and Klinke, 1973, this issue).

5. Summary

The platyfish, *Platypoecilus maculatus* and *Platypoecilus variatus*, exhibit a series of different macromelanophore spot patterns, the cells of which are determined by a series of codominant loci including *Sp* (spotted), *Sd* (spotted dorsal), *Li* (lineatus), *Sr* (stripe sided) and *Pu* (punctatus). In the case of *Sp* and *Sd*, both hybridization with the swordtail, *Xiphophorus helleri*, and X-irradiation of platyfish embryos induce premelanomas or melanomas. This effect is assumed to be the result of a marked increase in gene expression. In the case of *Li* and *Sr*, however, hybridization induces only a slight enhancement of gene expression, while X-irradiation is ineffective. Finally, *Pu* remains unaffected by both hybridization and X-irradiation.

Extensive crossing experiments as well as studies on crossing over, translocations, hybrid-chromosomes, and mutations revealed that this different susceptibility depends upon the distribution and quantity of controlling genes. *Sp* and *Sd* are controlled by a highly polygenic system distributed all over the chromosomes. Hence, X-irradiation can induce numerous mutations within this large set of controlling genes, while hybridization acts by replacing platyfish chromosomes with the corresponding swordtail chromosomes which lack these controlling genes. In contrast, *Li* and *Sr* may be controlled by other polygenic systems each of which includes a most powerful regulating gene closely linked to the appropriate macromelanophore gene locus, while the regulation of *Pu* is probably restricted to a single gene linked to the *Pu* locus. In the case of *Li*, *Sr* and *Pu*, the prerequisites for susceptibility of the macromelanophore gene expression to hybridization or X-irradiation are, therefore, small or lacking, respectively. - The mutants, from which we derived our knowledge of the linkage of powerful regulating genes to the *Li*, *Sr* and *Pu* loci, exhibit premelanomas or melanomas.

References

Anders, A., Anders, F. (1963): Genetisch bedingte *XX*- und *XY*-Weibchen und *XY*- und *YY*-Männchen beim wilden *Platypoecilus maculatus* aus Mexiko. Z. Vererbungslehre 94, 1-18.
Anders, A., Anders, F., Förster, W., Klinke, K., Rase, S. (1970): *XX*-, *XY*-, *YY*-Weibchen und *XX*-, *XY*-, *YY*-Männchen bei *Platypoecilus maculatus* (Poeciliidae). Verh. Zool. Ges. 1969 in Würzburg, 333-339. Zool. Anz. 33 Suppl. Bd.
Anders, A., Anders, F., Rase, S. (1969): *XY* females caused by X-irradiation. Experientia 25, 871.
Anders, A., Anders, F., Pursglove, D.L. (1971): X-ray-induced mutations of the genetically-determined melanoma system of Xiphophorin fish. Experientia 27, 931-932.
Anders, F. (1964): Aminosäurengehalt, Farbgenmanifestation und Tumorbildung bei lebendgebärenden Zahnkarpfen (Poeciliidae). Verh. Dtsch. Zool. Ges. Kiel 1964, 102-109.
Anders, F. (1967): Über genetische Mechanismen der Regulation niederer und höherer Systeme. Zool. Anz. 179, 1-79.
Anders, F. (1967): Tumor formation in platyfish-swordtail hybrids as a problem of gene regulation. Experientia 23, 1-10.
Anders, F. (1968): Genetische Faktoren bei der Entstehung von Neoplasmen. Zbl. Vet. Med. 15, 29-46.
Anders, F., Vester, F., Klinke, K., Schumacher, H. (1961): Genetisch bedingte Tumoren und der Gehalt an freien Aminosäuren bei lebend gebärenden Zahnkarpfen (Poeciliidae). Experientia 17, 549-551.
Anders, F., Drawert, F., Klinke, K., Reuther, K.H. (1963): Genetische und biochemische Untersuchungen über die Bedeutung der Amino- und Nucleinsäuren im Ursachengefüge von Neoplasmen (Tumoren und Gallen). Experientia 19, 219-224.
Anders, F., Klinke, K. (1965): Untersuchungen über die erbbedingte Aminosäurenkonzentration, Farbgenmanifestation und Tumorbildung bei lebend gebärenden Zahnkarpfen (Poeciliidae). Z. Vererbungsl. 96, 49-65.
Anders, F., Sieger, M., Klinke, K. (1969): Amino acids as stimulating agents of DNA-replication in melanomas. I. Stimulation in Explants. Experientia 25, 871-874.
Anders, F., Klinke, K., Vielkind, U. (1972): Genregulation und Differenzierung im Melanom-System der Zahnkärpflinge. Biologie in unserer Zeit 2, 35-66.
Atz, J.W. (1962): Effects of hybridization on pigmentation in fishes of the genus *Xiphophorus*. Zoologica 47, 153-181.
Bauer, K.H. (1963): Das Krebsproblem. Berlin, Göttingen, Heidelberg: Springer.
Bentvelzen, P. (1968): Genetical control of the vertical transmission of the Mühlbock mammary tumor virus in the GR mouse strain. Hollandia, Amsterdam.
Berenblum, I. (1964): The two-stage mechanism of carcinogenesis as an analytical tool. In: Cellular control mechanisms and cancer (ed. P. Emmelot and O. Mühlbock). Amsterdam-London-New York.
Berenblum, I. (1967): Cancer Research today. Oxford.
Braun, A.C. (1969): The cancer problem. New York and London.
Breider, H. (1938): Die genetischen, histologischen und zytologischen Grundlagen der Geschwulstbildung nach Kreuzung verschiedener Rassen und Arten lebend gebärender Zahnkarpfen. Z. Zellforschung mikrosk. Anatomie 28, 784-828.
Breider, H. (1952): Über Melanosarkome, Melaninbildung und homologe Zellmechanismen. Strahlentherapie 88, 619-639.
Brunner, M. (1973): Regulation of DNA synthesis by amino acids limitation. Cancer Research 33, 29-32.

Emmelot, P., Bentvelzen, P. (Editors)(1972): RNA viruses and host genome in oncogenesis. Amsterdam-London.
Gordon, M. (1927): The genetics of a viviparus top-minnow *Platypoecilus*; the inheritance of two kinds of melanophores. Genetics 12, 253-283.
Gordon, M. (1931): Morphology of the heritable color patterns in the mexican killifish, *Platypoecilus*. Am. J. Cancer 15, 732-787.
Gordon, M. (1931): Hereditary basis of melanosis in hybrid fishes. Am. J. Cancer 15, 1495-1523.
Gordon, M. (1942): Frequencies of seven dominant genes in natural populations of *Platypoecilus maculatus* and experimental evidence of multiple allelic series. Genet. Soc. Amer. 11, 76.
Gordon, M. (1947): Speciation in fishes. Distribution in time and space of seven dominant multiple alleles in *Platypoecilus maculatus*. Adv. in Genetics 1, 95-132.
Gordon, M. (1951): *Platypoecilus* now becomes *Xiphophorus*. Aquarium 20, 277-279.
Gordon, M. (1959): The melanoma cell as an incompletely differentiated pigment cell. Pigment cell biology (ed. M. Gordon), 215-236.
Gordon, M. (1963): Fishes as laboratory animals. The care and breeding of laboratory animals (ed. E.J. Farris), 345-449, New York.
Gordon, H., Gordon, M. (1950): Color patterns and gene frequencies in natural populations of a platyfish. Heredity 4, 61-73.
Gordon, H., Gordon, M. (1957): Maintenance of polymorphism by potentially injurious genes in eight natural populations of the platyfish, *Xiphophorus maculatus*. J. Genetics 55, 1-44.
Graffi, A., Bielka, H. (1959): Probleme der experimentellen Krebsforschung. Leipzig.
Häussler, G. (1928): Über Melanombildung bei Bastarden von *Xiphophorus helleri* und *Platypoecilus maculatus* var. *Rubra*. Klin. Wschr. 7, 1561-1562.
Huebner, J., Gilden, R.V. (1972): Inherited RNA viral genomes (virogenes and oncogenes) in the etiology of cancer. In: RNA viruses and host genome in oncogenesis (ed. Emmelot and Bentvelzen). Amsterdam-London.
Huxley, I. (1960): Krebs in biologischer Sicht. Stuttgart.
Kaeser, U. (1971): Melanoblastenverteilung bei verschiedenen *Xiphophorus* und *Platypoecilus*-Arten (Pisces, Poeciliidae) und deren Bastarden. Diplomarbeit, Universität Giessen.
Kallman, K.D. (1965): Genetics and geography of sex determination in the poeciliid fish, *Xiphophorus maculatus*. Zoologica 50, 151-190.
Kallman, K.D., Atz, J.W. (1966): Gene and chromosome homology in fishes of the genus *Xiphophorus*. Zoologica 51, 107-135.
Kallman, K.D., Schreibman, M.P. (1971): The origin and possible genetic control of new, stable pigment patterns in the poeciliid fish *Xiphophorus maculatus*. J. Exp. Zool. 176, 147-168.
Kosswig, C. (1927): Über Bastarde der Teleostier *Platypoecilus* und *Xiphophorus*. Z. indukt. Abstamm.- und Vererb.- L. 44, 253.
Kosswig, C. (1929): Das Gen in fremder Erbmasse. Nach Kreuzungsversuchen mit Zahnkarpfen. Züchter 1, 152-157.
Kosswig, C. (1929): Melanotische Geschwulstbildung bei Fischbastarden. Verh. Dtsch. Zool. Ges. Marburg, 90-98.
Kosswig, C. (1938): Über einen neuen Farbcharakter des *Platypoecilus maculatus*. Istanbul. Univ. Fen Fak. Mecmuasi 3, 395-402.
Ley, K.D., Tobey, R.A. (1970): Regulation of initiation of DNA synthesis in chinese hamster cells. II. Induction of DNA synthesis and cell division by isoleucine and glutamine in G_1-arrested cells in suspension culture. J. Cell Biol. 47, 453-459.
Lueken, W., Kaeser, U. (1972): The role of melanoblasts in melanophore pattern polymorphism of *Xiphophorus* (Pisces, Poeciliidae). Experientia 28, 1340.

Lynch, H.T. (1967): Hereditary factors in carcinoma. In: Recent results in cancer research. Berlin, Heidelberg, New York: Springer
Öktay, M. (1964): Über genbedingte rote Farbmuster bei *Xiphophorus maculatus*. Ein Beispiel für Nicht-Autonomie der Merkmalsbildung. Mitt. Hamburg. Zool. Mus. Inst.; Kosswig-Festschrift; 133-157.
Payne, L.N., Chubb, R.C. (1968): J. Gen. Virol. 3.
Pursglove, D.L. (1972): Änderungen in der Farbgenmanifestation und Entstehung von Melanomen nach Röntgenbestrahlung bei lebendgebärenden Zahnkarpfen (Teleostei, Poeciliidae). Dissertation, Universität Giessen.
Pursglove, D.L., Anders, A., Döll, G., Anders, F. (1971): Effects of X-irradiation on the genetically-determined melanoma system of Xiphophorin fish. Experientia 27, 695-697.
Rosen, D.E. (1960): Middle-american poeciliid fishes of the genus *Xiphophorus*. Bull. Florida State Mus. 5, 57-242.
Siciliano, M.J., Perlmutter, A., Clark, E. (1971): Effect of sex on the development of melanoma in hybrid fish of the genus *Xiphophorus*. Cancer Research 31, 725-729.
Sieger, M. (1968): Stimulation der DNS-Replikation durch Aminosäuren. Hildesheim.
Sieger, M., Sieger, F., Anders, F. (1968): Stimulation des Tumorwachstums durch Aminosäuren. Untersuchungen an genetisch bedingten Melanomen von lebendgebärenden Zahnkarpfen (Poeciliidae). Verh. Dtsch. Zool. Ges. Innsbruck, 210-225.
Sieger, F., Sieger, M., Prüssing, R., Anders, F. (1969): Amino acids as stimulating agents of DNA-replication in melanomas. II. Stimulation in in-situ melanomas. Experientia 25, 778-779.
Strong, L.C. (1958): Genetic concept for the origin of cancer. Historical review. Ann. New York Acad. Sci. 71, 807-1241.
Süss, R., Kinzel V., Scribner, J.D. (1970): Krebs, Experimente und Denkmodelle. Berlin, Heidelberg, New York: Springer.
Tavolga, W.N. (1949): Embryonic development of the platyfish (*Platypoecilus*), the swordtail (*Xiphophorus*) and their hybrids. Bull. Am. Mus. Nat. Hist. 94, 167-229.
Tobey, R.A., Ley, K.D. (1971): Isoleucine-mediated regulation of genome replication in various mammalian cell lines. Cancer Res. 31, 46-51.
Todaro, G.J., Huebner, R.C. (1972): The viral oncogene hypothesis: New evidence. Proc. Nat. Acad. Sci. USA 69, 1009-1015.
Vielkind, J., Vielkind, U., Anders, F. (1971): Melanotic and amelanotic melanomas in Xiphophorin fish. Cancer Res. 31, 868-875.
Vielkind, J., Vielkind, U., Anders, F. (1973): DNA synthesis in genetically-determined melanomas of platyfish-swordtail hybrids. Z. Krebsforsch. in press.
Vielkind, U. (1972): Tumorwachstum und Differenzierungsgrad der Tumorzellen in erblich bedingten Melanomen von lebendgebärenden Zahnkarpfen (Poeciliidae). Dissertation, Universität Giessen.
Wolf, B., Anders, F. (1973): *Xiphophorus*. Farbmuster, Farbgene, Bibliographie. Giessen.
Zander, C.D. (1961): Künstliche Befruchtung bei lebend gebärenden Zahnkarpfen. Zool. Anz. 166, 81-87.
Zander, C.D. (1962): Untersuchungen über einen arttrennenden Mechanismus bei lebendgebärenden Zahnkarpfen aus der Tribus Xiphophorini. Mitt. Hamburg. Mus. Inst. 60, 205-264.
Zander, C.D. (1964): Physiologische und genetische Untersuchungen zur Systematik xiphophoriner Zahnkarpfen. Mitt. Hamburg. Zool. Mus. Inst. (Kosswig-Festschrift), 333-348.
Zander, C.D. (1967): Ökologische und morphologische Beiträge zur Systematik und geographischen Verbreitung der Gattung *Xiphophorus* (Pisces). Mitt. Hamburg. Zool. Mus. Inst. 64, 87-125.

Regulation of Gene Expression in the Gordon-Kosswig Melanoma System

II. The Arrangement of Chromatophore Determining Loci and Regulating Elements in the Sex Chromosomes of Xiphophorin Fish, *Platypoecilus maculatus* and *Platypoecilus variatus*

A. Anders, F. Anders, and K. Klinke

1. Introduction

The platyfish, *Platypoecilus maculatus* and *Platypoecilus variatus*, exhibit species-specific black spot patterns, the cells of which consist of macromelanophores. In addition these platyfish exhibit species-specific yellowish or reddish patterns consisting of pterinophores. Both macromelanophores and pterinophores (chromatophores) are determined by sex chromosome-linked loci. Up to now no other sex chromosome-linked markers are known.

The preceding part of our studies on regulation of gene expression in the Gordon-Kosswig melanoma system has focussed on the distribution of the controlling genes for the expression of five different macromelanophore loci in the genome (Part I, Anders and Klinke, 1973). As the result of this research one can distinguish between macromelanophore locus-linked and macromelanophore locus-non-linked controlling genes. Especially, the macromelanophore locus-linked controlling genes have proved to be very important regulating elements for the expression of the macromelanophore determining loci since mutation or deletion of these elements may result in an excessive melanoma formation.

As a preparatory work for further studies of these special regulating elements the present investigation is concerned with the arrangement of the sex chromosome-linked pterinophore and macromelanophore determining loci and the regulating elements for the expression of the latter ones. As a subsidiary result we expect new information about the question whether the macromelanophore determining loci may be interpreted as oncogenes with respect to the viral oncogene theory of cancer (Huebner and Gilden, 1972).

2. Materials and Methods

Part I of this research contains detailed instructions on materials and methods (Anders and Klinke, 1973). Only the sex chromosomes which were studied and those linked loci which are responsible for the determination of pterinophores (left) and macromelanophores (right) will be enumerated:

X	*Dr Sd*	(dorsal red and spotted dorsal	Part I, Fig. 2a	*Platypoecilus maculatus*
X	*Dy Sp*	(diluted yellow and spotted)	Part I, Fig. 1a	
Y	*Ar Sr*	(anale red and stripe sided)	Part I, Fig. 5a	

X	*Ye Li*	(yellow and lineatus)	Part I, Fig. 4a	*Platypoecilus variatus*
Y	*Or'Pu*	(orange and punctatus)	Part I, Fig. 6a	

3. Results

The Arrangement of Chromatophore Determining Loci

Strains of platyfish, which naturally have no chromatophore loci linked to the sex chromosomes and those which had lost them after being X-irradiated, exhibit a normal sex determination. From this one may conclude that the sex determining region of the *X* and *Y* chromosomes occupies a position which is nearer to the centromere region than are the macromelanophore and pterinophore loci. We agree in this conclusion with Kallman and Schreibman (1971). Since the above mentioned deficient mutants are unrestricted in vitality even in the homozygous condition, we further conclude that the pterinophore and macromelanophore loci are unimportant for the vitality of the fish and that no essential genes are located distal of these loci. The resulting constitution of the sex chromosomes of *Platypoecilus maculatus* that have no chromatophore loci is accordingly for the *X* chromosome

(a) m O——*F*——

and for the *Y* chromosome

(b) m O——*M*——

whereby m symbolizes " *maculatus*" (the species from which the chromosome originates), O the centromere region, *F* and *M* the female and male sex determining regions respectively, and the line, the linkage group. For the spotted strains, the chromatophore loci would join on behind *F* and *M*.

Certain deficiences, among others, have provided information about the order of pterinophore and macromelanophore loci. A deficiency of the *X* chromosome-linked *Sd* which maintains the *Dr* otherwise linked with the *Sd* shows that *Sd* occupies a region distal of *Dr* on the very end of the chromosome. Accordingly, the standard *X* has the constitution

(c) m O——*F*——*Dr*——*Sd*——

and the *X* chromosome with the *Sd* deficiency the constitution

(d) m O——*F*——*Dr*——

The presence of the pterinophore locus *Dr* and the deficiency of macromelanophore locus *Sd* become particularly clear, when the deficient *X* chromosome (d) of *P. maculatus* is introduced into the genome of *X. helleri*. Now *Dr* shows the expected rise in expression (Anders, 1967), which results in an overproduction of drosopterin and sepiapterin synthesizing pterinophores (Henze, 1973), while not a single macromelanophore is formed. These animals, therefore, exhibit an intense red pigmentation. If *Sd* were present, *Sd* melanomas would have had to have arisen.

A series of X-ray-induced crossovers has confirmed this arrangement and beyond that has provided information about the sequence of other pterinophore and macromelanophore loci. In *P. maculatus* the *Sr* crossed over from *Y* to *X*:

(e) m O—*F* *Dr* *Sr*

Hence it follows, that *Sr* occupies the end of the standard *Y*, also, so that this has the arrangement

(f) m O—*M* *Ar* *Sr*

Kallman and Schreibman (1971) found a *Y* chromosome by which the *Dr* originating from *X* follows on the *M* region. The *Sr* originating from *Y* joins this:

(g) m O—*M* *Dr* *Sr*

If (f) is taken as a basis, then the order of the loci in this aberated *Y* (g) occurred by two crossovers.

A hint that *Sd* occupies the end of the *X* chromosome is also given by a hybrid chromosome which originated at a *P. maculatus/X. helleri* hybrid through translocation of *Sd* from *P. maculatus* to an autosome[1] of *X. helleri*:

(c) m O—*F* *Dr* *Sd* → (d) m O—*F* *Dr*

(h) h O·········· → (i) h/m O·········*Sd*

Since this hybrid chromosome (i) causes neither in the genome of *P. maculatus* nor in the genome of *X. helleri*, a red coloring of the fish, it is certain that *Dr* is missing; if it were present, a *X. helleri* which had this *Sd* hybrid chromosome in its genome would have to be intensively red colored.

In a further case the terminal *Sr*, which had already crossed over from the standard *Y* (f) to the *X* (e) as by (i) had translocated to an autosome[1] of *X. helleri*:

(j) h/m O·········*Sr*

Evidence and Arrangement of Sex Chromosome-Linked Controlling Elements for Macromelanophore Gene Expression

By all chromosome presented to this point neither in inbred strains nor in interspecific hybrids has any indication been given that could point to position effects of the chromatophore loci and, thereby, to a linkage of those loci with regulating elements.

---

[1]*X. helleri* has no sex chromosomes, however, an autosome pair is homologous to the sex chromosomes of *P. maculatus* (Kosswig, 1933, 1939; Anders and Anders, 1963). In (h), (i), and (j), h symbolizes "*helleri*" (the species from which the chromosome originates).

Kallman and Schreibman (1971) were the first who reported an unequal crossover, by which a *Sd* of an *X* chromosome is translocated to a *Y* (g):

(k) m O——*M* *Dr* *Sr* *Sd*——

In the new position, *Sd* causes a rise in expression up to the formation of premelanomas.

Moreover, Kallman and Schreibman mention a translocation of the *Sp* of the *X* to a *Y* that corresponds to our standard *Y* (f):

(l) m O——*M* *Ar* *Sr* *Sp*——

In this position *Sp* shows also a risen expression. Moreover, one may conclude from the terminal position, that *Sp* occupies the distal end of the standard X^{Sp} chromosome:

(m) m O——*F* *Dy* *Sp*——

Kallman and Schreibman conclude from the rise in expression of both *Sd* and *Sp*, at times attached to a complete sex chromosome, that normally a "regulator gene" is closely linked to the macromelanophore locus, which, in the case of the translocation of *Sd* in (k) and of *Sp* in (l), remained in its original site. The enhancement of the expression of *Sd* and *Sp* in (k) and (l), therefore, is due to the lack of the regulator gene. The gene arrangement on both the standard *X* chromosomes (c) and (m) can, therefore, be supplemented with the regulator element:

(n) m O——*F* *Dr* *R* *Sd*——

(o) m O——*F* *Dy* *R* *Sp*——

The following aberrant chromosomes, which appeared in our material, partly from X-irradiation, partly spontaniously, allow further details of expression control to be recognized.

In the case of a F_1 hybrid *Li* crossed over from an *X* chromosome of *P. variatus* to a standard *X* of the *P. maculatus*:

(p)[2] m/v O——*F* *Dr*—— - -*Li*- -

Therefore, one may conclude that *Li* is located on the end of the standard *X* of *P. variatus*, also. In the new position *Li* shows, in all genomes tested, a rise in expression, unknown up to this time (cf. Part I, Fig. 4c). From this result we can conclude that an important controlling element for *Li* is localized on the standard *X* of *P. variatus* which, in the inter-specific crossover, either stayed back on the *X* of *P. variatus* or became inactive in its new position on the *X* chromosome of *P. maculatus*. Crossings with *X. helleri* which normally cause a rise in expression in *Li* did not bring about the expected, additional rise in expression (cf. Part I). The combination of the *DrLi* chromosome (p) with the standard *X* chromosome of *P. variatus* (*DrLi/YeLi*) did not result in the expected reduction of the *Li* linked to the *Dr*. *Li*, in its new position on the *X* of *P. maculatus* (p), was not capable of reacting to regulation signals which normally control its expression. These results hint that the missing or defective controlling element of *Li* corresponds to either only an operator or an operator inclusive regulator element. Under the assumption that the standard *X* chromo-

[2] symbolizes "*variatus*"

some of *P. variatus* in analogy to the *X* chromosomes (n) and (o) of *P. maculatus* actually contains a regulator element for *Li*, we formulate the arrangement of its genes in the following way:

(q) v O---*F*---*Ye*---*R oLi*

The situation is something different in the translocation of a *Or′Pu* fragment from the *Y* chromosome of *P. variatus* to the *X* chromosome of the same species. In this position (*Ye Li Or′Pu*) *Pu* undergoes a large rise in expression which results in the formation of strong premelanomas (Part I, Fig. 6c). This translocation appeared 4 times in 3 years in our inbred wild type strain of *P. variatus*, and Kallman reports (personal communication) that the phenotype of this translocation is also found in wild populations. Since *Or′* and *Pu* have, up to this time, always turned up linked, we do not know their order on the *Y* chromosome. We suppose, however, that the macromelanophore locus *Pu*, as well as all the other macromelanophore loci studied, occupies the end of the chromosome.

As it was pointed out in Part I (Anders and Klinke, 1973), the whole detectable regulation of the expression of the standard *Pu* proceeds from the chromosome where *Pu* itself is located. Crossbreedings with *X. helleri* which would eliminate the non-linked controlling elements, do not influence the *Pu* expression. Such controlling elements are, then, either non-existent or inactive. The enormous rise in expression that *Pu* undergoes through the translocation is however, strengthened somewhat more through crossbreeding with *X. helleri*. This points to the fact that there are some autosomal controlling signals for *Pu* in *P. variatus*. These, however, are only detectably received from the translocated *Pu*. For that reason it may be asked if the translocated *Pu* can also receive control signals from the intact standard *Y* chromosome. To answer this question, we crossed ♀♀ which carried the translocated *Pu* with normal *Pu* ♂♂ of our wild type strain, and we found that the formation of premelanomas is repressed to almost the normal *Pu* expression, when the *Ye Li Or′Pu* - *X* chromosome is combined with an *Or′Pu* - standard-*Y* chromosome. Premelanomas occur only when the translocated *Pu* is homozygous, or hemizygous in combination with the *Ye Li* - standard-*X* chromosome. Hereby it is also clear, that up to now only had females with *Pu* premelanomas been observed. These premelanomas, therefore, are inherited as a sex-limited character. From these results it can be concluded that for *Pu* at least 2 controlling elements belong to the controlling system lying on the *Y* chromosome; namely a regulator element that emits signals and an operator element that receives them. In the *Pu* translocation of the *X* chromosome the regulator element is apparently missing, or it is defect. On the other hand, the operator element is intact because it receives and processes the *Pu* specific controlling signals which are transmitted from the standard *Y* chromosome. It is suggestive to place the regulator element between *M* and *Or′* and the operator element before or in the *Pu* locus. Therefore, we formulate the gene arrangement for the standard *Y* of *P. variatus*

(r) v O---*M*---*R Or′*---*oPu*

and for the translocation chromosome

(s) v O---*F*---*Ye*---*R oLi Or′*---*oPu*

In an individual that has both (r) and (s) chromosomes, a controlling signal emanates from the regulator element *R* of the *Y* chromosome (r) which is not only received and processed by the linked *oPu*, but also by the translocated *oPu* of the *X* chromosome.

It should now be reported about mutations in controlling gene systems in which no chromosomal rearrangements are to be shown. From the wild type strain of *P. maculatus* in which the *X* chromosome has the constitution (n) and the *Y* has the constitution (f), after X-irradiation of pregnant females, we isolated 5 *Sd* and 5 *Sr* mutants, and bred them as substrains. In the *Sd* mutants the rise in expression is relatively small (Part I, Fig. 3a), and in the *Sr* mutants it remains at least within sublethal limits (Part I, Fig. 5c). 3 *Sd* and 2 *Sr* substrains were studied up to the present (in Part I they were referred to as *Sd'*, *Sd"* etc.).

It was shown that the macromelanophore loci of the mutants receive controlling signals from the non-linked controlling genes, because the elimination of these controlling genes through hybridization results in the additional expression of the macromelanophore genes (excessive formation of *Sd* and *Sr* melanomas; Part I, Fig. 3c and 5d). In these mutants, then, the linked regulator elements must be defective while the operator elements are intact. Anyway, we can distinguish between the regulator and operator elements, and can complete the gene arrangement for the standard chromosome (n) and (f) in the following manner:

(t) m O——*F* *Dr* *R oSd*——

(u) m O——*M* *Ar* *R oSr*——

The *Sd* and *Sr* mutants have, then, the same gene arrangement. Solely the *R* has mutated to *R'*, *R"*, etc. Looking back, it should again here be brought to ones attention, that the regulator and operator elements of (t) and (u) must be unchanged in the translocation fragments of (i) and (j), because *Sd* and *Sr* are regulated normally in these positions. Contrary to this, the regulator element for *Sd* at least in the translocation (k) observed by Kallman and Schreibman was lost or it got inactive.

The following experiments show that the regulation of *Sd* and *Sr* can be analyzed still further. Through crossings of animals from the *R'oSd* and *R"oSd* mutant substrains with animals of the original wild type strain, *R'oSd/RoSd* and *R"oSd/RoSd* heterozygotes were produced. In these animals the expression of *R'oSd* and *R"oSd* is repressed to the normal state. Contrary to this the risen macromelanophore gene expression of the *R'oSr* and the *R"oSr* mutant substrains is not repressed in the corresponding *R'oSr/RoSr* and *R"oSr/RoSr* heterozygotes[3]. We must conclude from these differing results, that the operator elements of *R'oSd* and *R"oSd* receive and process controlling signals from the regulator element of the *RoSd*, while the operator elements of *R'oSr* and *R"oSr* receive no signals from the regulator element *RoSr*. This result can be interpreted on the basis of gene dosage compensation and gene dosage effect. For the *RoSd* *X* chromosome (t) it has been known for a long time, that the gene dosage is compensated in the homozygote condition. The controlling signals (compensator signals) must be mutually transmitted, received, and processed (Anders and Klinke, 1966). The normalization of the *Sd* expression in *R'oSd/RoSd*

[3] *YY*-♂♂ can only be experimentally produced (Anders, Anders and Rase, 1969; Anders, Anders, Förster, Klinke and Rase, 1970).

and *R"oSd/RoSd* heterozygotes, therefore, are supposed to be based on the reception and processing of compensator signals. It cannot be determined whether these compensator signals emanate from additional compensator elements or from the regulator element *R*. Neither can it be determined if a supplementary operator element for the reception of compensator signals is present.

On the other hand, it is known for the *RoSr Y* chromosome (u) that it has not developed any gene dosage compensation mechanism (Anders and Klinke, 1966), because there normally are no *YY*-♂♂ which would require a gene dosage compensation. The non-repressed *Sr* expression in *R'oSr/RoSr* and *R"oSr/RoSr* heterozygotes can be interpreted by the lack of special compensator elements, or special compensator constituents of the regulator element *R* of the *Y* chromosome (u). The question of a special operator element is, then, superfluous.

3. Discussion

Part I of this research (Anders and Klinke, 1973) was concerned with the function and distribution of the controlling genes for the expression of the macromelanophore genes *Sp* (spotted), *Sd* (spotted dorsal), *Li* (lineatus), *Sr* (stripe sided) and *Pu* (punctatus) in the genome of the platyfish, *Platypoecilus maculatus* and *Platypoecilus variatus*. As the result we found that the platyfish have developed macromelanophore gene-specific controlling gene systems which can be ranged as the macromelanophore genes were above. *Sp* has the highest, and *Pu* the lowest quantity of non-linked controlling elements. Reversely, *Pu* has the strongest and *Sp* the weakest linked controlling element. It was the intention of the present research to get more information about the macromelanophore loci and the controlling elements which were linked to them.

The only one that was well established up to now is the sex chromosome linkage of macromelanophore loci and that in some cases, the macromelanophore loci are very closely linked to the pterinophore loci (Anders and Anders, 1963; Kallman and Atz, 1967; Kallman, 1970). This latter linkage was found to be true not only in the macromelanophore genes of this research but also in all others so far tested (unpublished).

There could exist some reasons for this linkage. Since we know that the precursor cells of macromelanophores synthesize pterins before starting melanin synthesis, we assume that both macromelanophores and pterinophores are derived from the same precursor cells, which originate from the neural crest of the embryo (Henze, 1973; see DuShane, 1934; Tavolga, 1949). In addition it should be mentioned that both macromelanophores and pterinophores represent special pattern chromatophores which differ from the other chromatophores by their very big shape. In spite of their color they resemble each other. We, therefore, favor the concept that the linkage of pterinophore and macromelanophore genes reflects the joint of macromelanophores and pterinophores.

Mutations, crossovers and other rearrangements of chromosomes which could have served for gene mapping were very rare. The only ones known up to now are those reported by Gordon (1937), MacIntyre (1961), Kallman (1970) and Kallman and Schreibman (1971):

3 different crossovers between the sex determining region and the pterinophore locus

2 crossovers between pterinophore and macromelanophore loci

2 translocations of *Sd* to an *Sr* chromosome

1 translocation of *Sp* to an *Sr* chromosome

To these crossovers and translocations one can join those sex chromosomes which originally do not bear chromatophore loci or have already lost them, respectively, in the wild populations.

In addition we found within the last four years of our research:

1 deficiency	(b): — —	(no chromatophore loci left)
2 deficiencies	(d): *Dr* —	
1 crossover	(e): *Dr Sr*	
1 interspecific translocation	(i): — *Sd*	
1 interspecific translocation	(j): — *Sr*	
1 interspecific crossover	(p): *Dr Li*	
4 *Y/X*-translocations	(s): *Ye Li Or´ Pu*	
1 translocation	*Dr* to (b)	(not taken into consideration in this research)
1 translocation	*Ar Sr Sd*	(not taken into consideration in this research)
5 mutations	*Sd'* to *Sd'''''*	(only 3 were taken into consideration)
5 mutations	*Sr'* to *Sr'''''*	(only 2 were taken into consideration)

To these rearrangements and mutations we can join the fuligonosus mutation which occurred in the breeds of Kosswig (1938; see Part I).

Through these 32 structural and mutational changes it was possible to determine the gene arrangement, as it is demonstrated in the chromosomes (a) to (u). It has been shown that all chromosomes tested are presumably structured on the same principle (v):

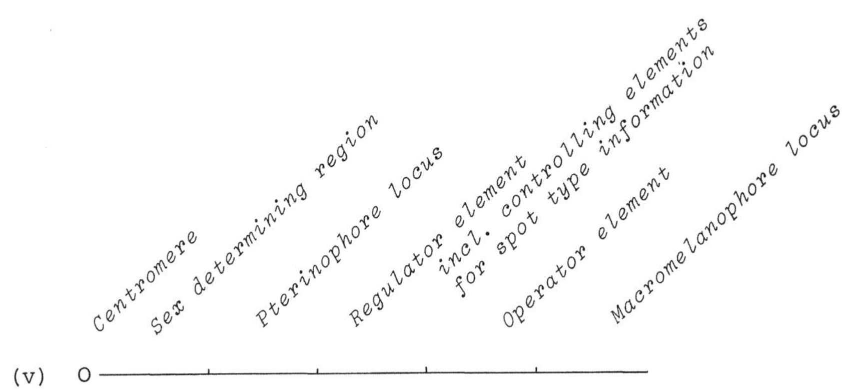

The region adjacent to the centromere is occupied by the sex determining genes. As deficiencies have shown the rest of the chromosome is redundant with respect to the vitality of the fish. The region next to the sex determining part is occupied by the pterinophore loci. There is no hint for special controlling elements associated with them. Since changes in the expression of the macromelanophore loci mostly are paralleled with those of the pterinophore loci, we suppose both loci to be regulated by the same controlling elements.

Distal, adjacent to the pterinophore locus is located the regulator element for the respective macromelanophore locus. The only case in which it seems to be located proximal to the pterinophore locus represents the *Y* chromosome of *P. variatus* (r). Since mutation, inactivity or deletion, respective, of this region may result in an overproduction of macromelanophores on the whole body surface without exhibiting any resemblance to the original spot (see discussion Part I), we conclude that this region includes further controlling genes which are responsible for the typical macromelanophore pattern.

The very end of the chromosome is occupied by the macromelanophore locus, the operator element of which, at least in the chromosome (q), is supposed to be located between the regulator element and the macromelanophore locus. As was discussed in Part I, we assume that this macromelanophore locus is identical in all chromosomes tested so far. It contains the crude information to redirect normal melanophore differentiation into the diverging macromelanophore differentiation, possibly by endopolyploidization only. Macromelanophore locus, operator element, and regulator element including the spot type elements together result in those genetic units that are called *Sp*, *Sd*, *Li*, *Sr* and *Pu*. These symbols, therefore, really represent complex loci or somewhat like operons. Except for *Pu*, additional non-linked controlling genes are responsible for the final stability in the normal macromelanophore gene expression. For the intention of this research the results are important that the chromatophore loci, after having been X-irradiated, can be lost as complete entities or can change their position even to species non-specific chromosomes, some of which never before had possessed macromelanophores (see chromosomes i,j and p). It has never been observed that loci in their new position exhibit a reduced expression. This favors the idea that these loci are more or less independent from the original genome of the species. This is clearest with respect to the *Y*/*X*-translocation, because in this case the *Or'Pu*-region of the *Y* chromosome has been translocated already 4 times as a unit.

These observations allow one to think that these chromosome regions, possibly both pterinophore and macromelanophore loci, originally have been foreign genetic units which were picked up relatively late in evolution and were attached to the very end of a chromosome. Since they originally exhibit a deleterious effect, the platyfish genome was forced to develop a controlling gene system in which non-linked and linked controlling elements were involved. Some of them may act as regulatory genes and operator genes in the meaning of molecular biology. After the controlling gene systems have stabilized the expression of the attached extraneous genetic units, derepression by mutation, inactivation, deletion or elimination (hybridization conditioned; see Part I) of the controlling elements in combination with stimulating factors results in melanoma formation.

The results of this paper may be in accordance with the viral oncogene theory of cancer favored by Huebner.

Further experiments shall clarify whether the macromelanophore loci originally are genes of the platyfish or of a virus.

4. Summary

The platyfish, *Platypoecilus maculatus* and *Platypoecilus variatus*, exhibit species specific patterns each consisting of pterinophores and macromelanophores. The pterinophore and macromelanophore loci are sex chromosome-linked. They are located distal to the sex determining region and can be lost without influencing the vitality of the fish, even as homozygous deletions. The macromelanophore loci are located on the very distal end of a sex chromosome (see the generalized chromosome (v) in the discussion of this paper).

Extensive crossing experiments as well as studies on crossing over, translocations, hybrid-chromosomes, and mutations revealed that at least the macromelanophore loci are regulated by non-linked controling elements and, especially, by powerful regulator and operator elements which may resemble those well known in molecular biology. There are some hints that the macromelanophore loci may be equated with the oncogenes of the tumor viruses. An experimentally induced break-down of macromelanophore gene regulation results in melanoma formation.

5. Acknowledgments

We should like to thank Dr. Ursula Vielkind and Dr. Christina Krauskopf for their stimulating suggestions and constructive criticisms during the preparation of the manuscript, and Thomas Stamey for preparing the text in English.

The experiments have been supported by grants from the Deutsche Forschungsgemeinschaft for many years.

References

Anders, A., Anders, F. (1963): Genetisch bedingte *XX*- und *XY*-Weibchen und *XY*- und *YY*-Männchen beim wilden *Platypoecilus maculatus* aus Mexiko. Z. Vererbungslehre 94, 1-18.

Anders, A., Anders, F., Förster, W., Klinke, K., Rase, S. (1970): *XX*-, *XY*-, *YY*-Weibchen und *XX*-, *XY*-, *YY*-Männchen bei *Platypoecilus maculatus* (Poeciliidae). Verh. Zool. Ges. 1969 in Würzburg. Zool. Anz. Suppl. Bd. 33, 333-339.

Anders, A., Anders, F., Rase, S. (1969): *XY* females caused by X-irradiation. Experientia 25, 871.

Anders, A., Anders, F., Klinke, K. (1973): Regulation of gene expression in the Gordon-Kosswig melanoma system. I. The distribution of controlling genes in the genome of the xiphophorin fish, *Platypoecilus maculatus* and *Platypoecilus variatus*. This issue.

Anders, F. (1967): Tumor formation in platyfish-swordtail hybrids as a problem of gene regulation. Experientia 23, 1-10.

Anders, F., Klinke, K. (1966): Über Gen-Dosis, Gen-Dosiseffekt und Gen-Dosiskompensation. Verh. Dtsch. Zool. Ges. Göttingen, 391-401.

DuShane, G.P. (1934): The origin of pigment cells in amphibia. Science 80, 620-621.

Gordon, M. (1937): Genetics of *Platypoecilus*. III. Inheritance of sex and crossing over of the sex chromosomes in the platyfish. Genetics 22, 376-392.

Henze, M. (1973): Über die Verteilung der Pterine bei lebend gebärenden Zahnkarpfen (Poeciliidae). Untersuchungen an verschiedenen Organen, Melanomen und Zelltypen. Dissertation, Universität Giessen.

Huebner, J., Gilden, R.V. (1972): Inherited RNA viral genomes (virogenes and oncogenes) in the etiology of cancer. In: RNA viruses and host genome in oncogenesis (ed. Emmelot and Bentvelzen). Amsterdam-London.

Kallman, K.D. (1970): Different genetic basis of identical pigment patterns in two populations of platyfish, *Xiphophorus maculatus*. Copeia 3, 472-487.

Kallman, K.D., Atz, J.W. (1967): Gene and chromosome homology in fishes of the genus *Xiphophorus*. Zoologica 51, 107-135.

Kallman, K.D., Schreibman, M.P. (1971): The origin and possible genetic control of new, stable pigment patterns in the poeciliid fish *Xiphophorus maculatus*. J. Exp. Zool. 176, 147-168.

Kosswig, C. (1933): Die Geschlechtsbestimmungsanalyse bei Zahnkarpfen. Z. indukt. Abstamm.- und Vererb.-L. 67, 200-205.

Kosswig, C. (1939): Die Geschlechtsbestimmung in Kreuzungen zwischen *Xiphophorus* und *Platypoecilus*. Istanbul. Univ. Fen Fak. Yeni Seri 4, 91-144.

Kosswig, C. (1939): Die Geschlechtsbestimmungsanalyse bei Zahnkarpfen. Istanbul. Univ. Fen Fak. Yeni Seri 4, 239-270.

MacIntyre, P.A. (1961): Crossing over within the macromelanophore gene in the platyfish (*Xiphophorus maculatus*). Amer. Nat. 95, 323-324.

Tavolga, W.N. (1949): Embryonic development of the platyfish (*Platypoecilus*), the swordtail (*Xiphophorus*) and their hybrids. Bull. Am. Mus. Nat. Hist. 94, 167-229.

3. Mutagenesis

Radiation Damage in Fish

C. E. Purdom and D. S. Woodhead

From a purely practical point of view it has seemed unlikely that radiation damage in fish from environmental contamination could ever prove to be a cause for public concern. Standards for permissible contamination of the environment are decided on the grounds of human welfare and these are far more stringent than any considerations which would be applied to fish or any other wildlife.

Nevertheless, it was necessary to establish a scientific basis for this a priori assumption, and a considerable body of information now exists on the effects of ionizing radiation on fish (Polikarpov, 1966a; Templeton et al., 1971). The lethal and sterilizing effects of acute exposures have been described for a variety of species and so also has the effect of a given dose at different stages during the development from fertilized eggs onwards (Allen and Mulkay, 1960). These results show that radiation damage in fish parallels that found in mammals, although with a much lower overall sensitivity. Similar conclusions came from studies of the effects of chronic irradiation, particularly of fish eggs, arising either from discrete sources or from dissolved radionuclides. The one exception to this, the controversial findings of very high sensitivity of fish eggs by Polikarpov and his colleagues (Polikarpov, 1966b), has not been accepted by other workers in the field.

Although radiation damage to gonads following acute exposure has been described by several authors (Foster et al., 1949; Welander et al., 1949), the effects of chronic irradiation on fertility are not known. Similarly, very little is known of the genetic effects of ionizing radiation in fish. The present paper describes experiments with the guppy, *Lebistes reticulatus*, designed to give information on the induction of mutations and on the effects of chronic irradiation on fecundity.

1. Production of Mutation

The estimation of mutation rates in any organism is simply a matter of having appropriate genetic systems within which the induction of a mutation is easily recognized. The guppy provides two possible systems of this type. The first comprises the complex of genes which controls the colour polymorphisms normally expressed only in males. These genes act as dominants in males and are also expressed in masculinized females. They are believed to comprise many gene loci which are mostly situated on the X and Y chromosome (Winge, 1927). Mutation at such loci can be recognized in F_1 males and in masculinized F_1 females.

The second system includes recessive genes which may be used in the conventional 'specific locus' technique. Two such genes are *gold* and *blond* (Goodrich et al., 1944), both of which reduce the number of melanophores in the skin. Mutation from wild-type to mutant at either of these loci may be recognized in the F_1 when a homozygous wild-type fish is crossed with a homozygous *gold-blond* fish.

In the present experiments, males with the Y-linked genes *Pauper* or *Maculatus* were irradiated and crossed with females lacking colour genes on the X chromosomes but homozygous for *gold* and *blond*. F_1 offspring were scored for colour polymorphism characters and for *gold* or *blond*. On occasion, irradiated males were crossed with wild-type females from an *Armatus* stock. F_1 offspring of these crosses were scored for colour polymorphism characters only.

Males were irradiated with 250 kVp X-rays or ^{60}Co gamma rays at doses ranging from 1 000 to 2 000 rad. They were individually or mass-mated with F_0 females for approximately one month, on the assumption that germ cells used during this time were irradiated as mature or near-mature spermatozoa.

The results of tests for dominant colour pattern mutants are shown in Table 1. These data were in part reported at the IAEA Conference on the Disposal of Radioactive Wastes into Seas, Oceans and Surface Waters (Purdom, 1966). There were three clear-cut cases which appeared to represent mutations:

1. A change from *Pauper* to a pattern which resembled *Ferrugineus*, described by Winge (1927). This pattern appeared to replace *Pauper* and was inherited in a Y-linked manner.

2. A yellow flush which appeared on the caudal fin of a masculinized female. This pattern was not identical to any known pattern of the *Lebistes* colour polymorphisms but it appears to be a component of several described patterns (e.g. *Coccineus* and *Luteus*). Its inheritance could not be checked, because the masculinized female was sterile.

3. A striped tail pattern in addition to *Maculatus*; this resembled *Variabilis* but was not confirmed by breeding tests.

Other anomalous patterns included two colourless males, one of which was confirmed as an XX female; the other was sterile but presumably also XX. An apparent change from *Pauper* to *Maculatus* possibly arose from contamination and was therefore disregarded.

No colour deviants were observed in the controls. There were, therefore, 3 mutants out of a total of 852 irradiated gametes.

In mouse spermatozoa, the mean induced mutation rate of the 7 loci used in the normal specific-locus technique is 6×10^{-7} mutation/locus/gamete. A similar rate for the colour polymorphism loci, on the assumption that 20 loci are involved, would give an expectation of 13.8 mutations and the observed number is clearly lower than this. However, it would be wrong to assume that this implies a lower mutagenic effect of radiation in fish, since the nature of the colour polymorphism characters is quite different from that of the loci studied in the mouse. Mutation at the latter loci comprises disadvantageous alleles which probably represent some loss of gene action or even deletion of the locus itself, whereas mutation within the colour polymorphism system of *Lebistes* would appear to be a more subtile change, possibly of the forward mutation type.

A report on similar studies by Samokhvalova (1938) was discovered after the completion of the present work. Samokhvalova examined F_1 males and the offspring of F_1 females following X-irradiation of F_0 males. One possible mutation was observed in 77 offspring and concerned the loss of black spotting from the *Bimaculatus* pattern in about half the male offspring of an F_1 female. This was interpreted as an X-chromosome change which inhibited the expression of the black-spots component of the Y-linked gene *Bimaculatus*, but no further tests of the inheritance of the modified character were made.

Table 1. Colour polymorphism mutants in F_1 offspring

	F_1 Males	Mutants	Masculinized F_1 Females	Mutants	Total	Mutants	Expected[a]
Control	230	-	244	-	474	-	-
1 000 rad	107	1	105	1	212	2	2.5
1 200 rad	211	-	-	-	211	-	3.0
1 500 rad	147	-	176	-	323	-	5.8
2 000 rad	38	1	68	-	106	1	2.5
Total Irradiated	503	2	349	1	852	3	13.8

[a]Expected on mutation frequency 6×10^{-7} mutation/locus/rad/gamete with 20 loci involved.

Although no mutations were observed in the control series of the present experiments - nor indeed in the case of any controlled matings other than after irradiation - one case has been observed in a stock tank. This occurred in a stock of *Maculatus* and was observed as an additional colour pattern in a single male. The extra coloration comprised a white tail pattern with a prominent black spot at the base of the caudal fin. On crossing with an X_0X_0 female from a *Pauper* stock, all male offspring to date have shown the new pattern. It therefore appears to be Y-linked.

Thus it appears that mutation within the colour genes of *Lebistes* can occur spontaneously and can be induced by ionizing radiation, although at a fairly low rate.

Within the *gold-blond* system, one mutation was observed in 765 F_1 offspring, which is almost exactly the number expected on the basis of mouse specific locus rates. Many more data are required to establish the magnitude of mutation rates at these two loci.

2. Effect of Chronic Irradiation on Fecundity

This experiment was performed to investigate the effect of comparatively low-level chronic irradiation on the continuing fecundity of the guppy. Single pairs of fish were irradiated at dose rates of 0.25, 0.5 and 1.7 rad/h (levels 1 to 3) in small (2 l) polythene containers at different distances surrounding a caesium-137 source. The fish used were F_1 hybrids between males of a *Maculatus* stock and females of an *Armatus* stock and they were placed in pairs in the irradiation and control aquaria at about 3 days of age. The temperature of the aquaria was maintained at 23.5 ± 1.0°C and the fish were fed daily with newly hatched brine shrimp nauplii and a commercially available dried flake food. At the onset of maturity (at approximately 2 months of age), some reshuffling was done at each dose level to ensure one male and one female in each aquarium. Offspring were removed from the aquaria within 24 hours of birth and grown on in separate aquaria outside the irradiation unit. The offspring were counted at birth and recounted and sexed at maturity. Neonatal deaths, abnormalities and post-natal mortalities were also recorded.

Fig. 1 shows the overall fecundity of the fish throughout the duration of the experiment, which is still in progress after 2 years. It can be seen that all three dose rates produced a marked reduction in fecundity. At 1.7 rad/h complete sterility occurred after about 4 months, during which time the fish had accumulated doses of 5 000 rad. All the fish at the highest dose rate survived for 8 months and accumulated a total dose of 9 000 rad. At this time they were killed and histological examination of the gonads showed, not surprisingly, that they were entirely devoid of germ cells. The two lower dose rates both reduced fecundity by about 50% and there was no significant difference between them. This lack of a dose effect is difficult to explain, but there is some evidence that the fish receiving 0.5 rad/h are approaching complete sterility more rapidly than those receiving 0.25 rad/h. Table 2 shows the numbers of fertile and sterile pairs at the present time and whilst all nine surviving control pairs are still producing broods, at 0.25 rad/h one pair out of six is sterile and at 0.5 rad/h five pairs out of seven are sterile. The final column of Table 2 lists the doses received to date at the lower two dose rates and the accumulated dose at the time the fish at the highest dose rate became sterile. There appears to be a marked dose rate effect, in that the fish were completely sterile after 5 000 rad at 1.7 rad/h but not all sterile after 8000 rad at 0.5 rad/h.

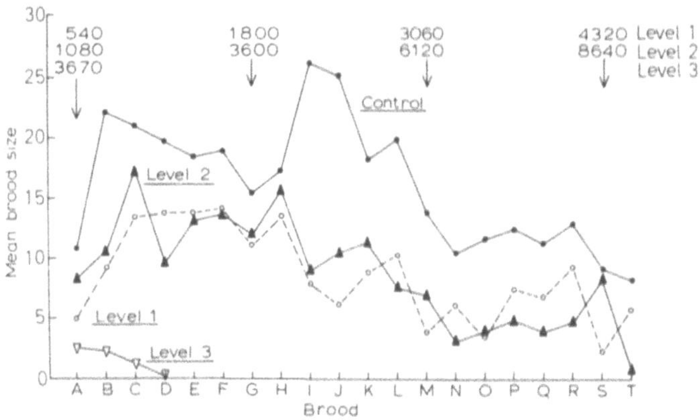

Fig. 1. Fecundity in successive broods from *Lebistes* populations maintained under continuous irradiation

Table 2. Sterilizing effect of continuous gamma-irradiation

	Dose rate (rad/h)	No. of pairs	Male or female deaths	No. fertile at 23 mths.	No. sterile at finish	Approx. total dose rad
Control	-	12	3	9	-	-
Level 1	0·25	10	5	4	1	4 000
Level 2	0·5	10	3	2	5	8 000
Level 3	1·7	6	-	-	6	5 000[a]

[a]Dose to time of last brood. Fish were killed for histology after 8 000 rad.

Returning to the reduced fecundity at the lower two dose rates, there may be three basic mechanisms by which this could arise:

1. physiological effects in the female parent,
2. lethal effect on the developing embryo,
3. lethal or mutagenic effects on the germ cells.

There is no relevant information available on the effect of radiation on the physiology of reproduction in the guppy. However, any such effect would presumably be progressive and there is no clear evidence of this (Fig. 1).

Relatively low doses of radiation delivered acutely at certain precise stages in the early development of fish embryos have been shown to have severe effects. Silver salmon embryos irradiated at 0.17% of the incubation period (i.e. early in the single cell stage) were found to have an LD_{50} at hatching (approximately 60 days post-fertilization) of 30R, and at 90 days after hatching the LD_{50} had decreased to 16R (Bonham and Welander 1963). Plaice eggs irradiated at the blastodisc stage (22 h after fertilization or approximately 5% of incubation) were found to have an LD_{50} for survival of the larvae to metamorphosis of 90 rad (Ward, Beach and Dyson, 1971). At later stages of embryogenesis the radiosensitivity markedly decreased. Assuming a 22-day gestation period, the total doses received by the developing guppy embryos at levels 1 to 3 during the first 5% of the incubation period are 6.5, 13 and 44 rad respectively. Although these doses are of the order of those mentioned above, the fact that they are chronic as opposed to acute leads to the expectation that somatic damage to the embryos would be unlikely to be apparent except possibly at the highest dose rate. This expectation is supported by the data given in Table 3. There was no evidence whatsoever of radiation effects on the number of neonatal deaths or survivors to maturity; also no abnormal offspring have been observed.

Table 3. Survival and sex of offspring from continuously irradiated parents

	Total young	Neonatal death	Survivors to maturity	%	Number of males	%
Control	2100	3	1977	94.1	653	33.0
Level 1	1075	6	1007	93.7	257	25.5
Level 2	1187	0	1145	96.5	376	32.8
Level 3	21	0	21	100.0	6	28.6

Thus, the reduction in fecundity seems most likely to be due to the killing of germ cells or to the induction of dominant lethal mutations. Destruction of germ cells is demonstrably complete at level 3 after a total dose of 5 000 rad, but the surviving fish at levels 1 and 2 have not yet been examined in this respect. Dominant lethal mutations are by far the most common forms of mutation in mice and *Drosophila*, and their effects are so drastic that they normally result in the death of the embryos at an early stage of development and are therefore not expressed in post-embryonic life.

Attempts were made to score dominant lethals in gravid females which had been mated to males exposed to acute radiation doses of 600 or 1 000 rad, but the macroscopic appearance of the contents of the Mullerian ducts was too confused to allow interpretation. Live embryos and immature eggs were found together with large eggs which could conceivably have included eggs for the subsequent brood and also dead early embryos of the current brood. A histological examination of these large eggs might be a worthwhile approach.

It is clear that chronic irradiation at dose rates equal to or greater than 0.25 rad/h have a marked effect on the fecundity of *Lebistes*. Unfortunately the absence of a simple relationship between dose rate and effect in the present experiment precludes any extrapolation to the very much lower irradiation regimes encountered in contaminated environments.

3. Summary

Mutagenic and sterilizing effects of ionizing radiations were observed in the guppy, *Lebistes reticulatus*. Mutations were detected within the colour-pattern polymorphism complex and at an autosomal locus with recessive alleles. Mutation rates were low in the colour-pattern system in comparison with conventional specific-locus rates in the mouse, but the data were too sparse for quantitative evaluation of mutation rates at autosomal loci. Chronic exposure at dose rates of 1.7, 0.5 and 0.25 rad/h greatly reduced fecundity over a 2-year period. Fish were sterile within 4 months at the highest rate, and fecundity was reduced overall by about 50% at the two lower dose rates. Germ cell killing and the induction of dominant lethal mutations appeared to be the most likely causes of the reduced fecundity.

References

Allen, A.L., Mulkay, L.M. (1960): X-ray effects on embryos of the paradise fish, with notes on normal stages. Growth 24, 131-168.

Bonham, K., Welander, A.D. (1963): Increase in radio resistance of fish to lethal doses with advancing embryonic development. In: Radioecology: proceedings of the first national symposium on radioecology (ed. V. Schultz and A.W. Klement), pp. 353-358. New York: Reinhold Publishing Corporation and Washington D.C.: The American Institute of Biological Sciences.

Foster, R.F., Donaldson, L.R., Welander, A.D., Bonham, K., Seymour, A.H. (1949): The effect on embryos and young of rainbow trout from exposing the parent fish to X-rays. Growth 13, 119-142.

Goodrich, H.B., Josephson, H.D., Trinkaus, J.P., Slate, J.M. (1944): The cellular expression and genetics of two new genes in *Lebistes reticulatus*. Genetics 29, 584-592.

Ivanov, V.N. (1967): Effect of radioactive substances on the embryonic development of fish. In: Problems of Biological Oceanography. (Trans. 2nd Int. Oceanogr. Congr., 30 May-9 June 1966). Acad. Sci. Ukr. S.S.R., Inst. Biol. Sthn Seas. A.O. Kovaleskii. Naukova Dumka, Kiev. A.E.C.-tr-6940 (1968), pp. 47-51.

Polikarpov, G.G. (1966a): Radioecology of aquatic organisms. Amsterdam: North Holland Publishing Co.

Polikarpov, G.G. (1966b): Certain biological aspects of radioactive contamination of the seas and oceans. In: Radioactive Contamination of the Sea. (eds. V.I. Baranov and L.M. Khitrov), pp. 78-104. Translated from the Russian and published by the Israel program for scientific translations, Jerusalem.

Purdom, C.E. (1966): Radiation and mutation in fish. "Disposal of radioactive wastes into seas, oceans and surface waters". IAEA, Vienna, 861-867

Samokhvalova, G.V. (1938): Effect of X-rays on fishes (*Lebistes reticulatus, Xiphophorus hellerii* and *Carassius vulgaris*). Biol. Zhur. 7, 1023-1034.

Templeton, W.L., Nakatani, R.E., Held, E.E. (1971): Chapter 9, Radiation Effects. In: Radioactivity in the Marine Environment. Washington, D.C.: National Academy of Sciences, 223-239.

Ward, E., Beach, S.A., Dyson, E.D. (1971): The effect of acute X-irradiation on the development of the plaice *Pleuronectes platessa* L. J. Fish Biol. 3 (2), 251-260.

Welander, A.D., Donaldson, L.R., Foster, R.F., Bonham, K., Seymour, A.H., Lowman, F.G. (1949): The effect of Röntgen rays on adult rainbow trout. U.S.A.E.C. Doc. No. UWFL-17, 20 pp.

Winge, O. (1927): The location of eighteen genes in *Lebistes reticulatus*. J. Genetics 18, 1-43.

Dominant Lethal Mutation Rates in the Fish, *Oryzias latipes*, Irradiated at Various Stage of Gametogenesis

N. Egami and Y. Hyodo-Taguchi

Except for two series of studies by Schröder with the guppy, and Newcombe and his co-workers with trout, the number of reports of fish radiation genetics is still small (cf. Schröder, 1969; Newcombe and McGregor, 1968). In a series of papers, the present authors have reported the cell killing effects of X-rays on germ cells at different stages in the freshwater fish, *Oryzias latipes* (Egami, 1963; Egami and Hyodo, 1965 a,b; Konno and Egami, 1966). Since a preliminary experiment designed to study the relative sensitivity of various germ cell stages of the fish to radiation-induced dominant lethal mutations was carried out, the results will be briefly presented in this report.

1. Materials

Materials used were the orange-red variety of the Medaka, *Oryzias latipes*. A pair of this species lay eggs almost every morning during the breeding season. Fertilization takes place immediately after the oviposition. Oogenesis and rate of yolk deposition of this fish have been studied by many workers (cf. Yamamoto and Yoshioka, 1964). Frequency distribution of oocytes at different diameters in the ovary of laying female, and effects on oviposition of whole body irradiation of female fish have also been reported (Egami and Hyodo, 1965b). If male fish are irradiated with 2 000-8 000 R of X-rays and kept with intact females, the fertilization rate will decrease (Konno and Egami, 1966). From the autoradiographic examination of the testis, the rate of spermatogenesis has also been demonstrated (Egami and Hyodo-Taguchi, 1967; Egami, Hyodo-Taguchi and Konno, 1967). On the basis of this fundamental knowledge on the fish, the experiments were performed.

2. Methods

Several pairs of adults *Oryzias* were kept in a laboratory tank under favorable conditions at 23°C. For the experiment laying pairs were selected and three groups were made: in the first group, both male and female fish were irradiated with 0, 2 000, 4 000, 8 000 and 16 000 R of X-rays, respectively. Each single pair given the same dose were kept together in the same vessel. In the second group only males were irradiated with 0, 2 000, 4 000, 8 000 and 16 000 R and kept with non-irradiated females. Males of the third group were not irradiated but the female partners were exposed to the radiation. X-rays were generated at 200 kVP and 20 mA with 0.5 mm Cu and 0.5 mm Al filters. The target-object distance was 25 cm, dose-rate in air being 400 R/min.

After the irradiation, each pair was transferred to a glass vessel containing about 1.5 liters of water and given fresh water oligochaetes. Eggs laid by the pairs were collected every morning and incubated at 23°C. In the non-irradiated control, the majority of fry hatched 10-11 days after fertilization under the present conditions. In the irradiated groups hatching was delayed. Survival of embryos and fry was recorded every day for a period of 30 days. For convenience sake, the mortality rates at 10 (embryos) and 30 days (fry) after fertilization were carefully examined and the relative mortality rate (R M R) was calculated.

$$R\ M\ R = \frac{\text{Mortality percentage of the given group} - \text{Mortality percentage of the control group}}{\text{Survival percentage of the control group}} \times 100$$

3. Results

Results of the experiments are summarized in Tables 1, 2 and 3. In these tables the mortality data of embryos and fry developed from the eggs laid 1 and 2, 3 and 4, 5 and 6, 7 and 8 and 9 and 10 days, respectively, after irradiation of fish are pooled. After 11 days and later no fertilized eggs were obtained if the dose was 4 000 R or more.

When both parents were irradiated (Group 1, Table 1), the mortality rate was very high, and exposure doses of 4 000 R or more killed all embryos before hatching. However, if the parents were given 2 000 R, some of the offsprings could survive for more than 30 days. The mortality rate was higher in eggs collected shortly after the irradiation. Most eggs collected 9 or 10 days after irradiation could develop into fry and survive for 30 days.

Table 1. Effects of X-irradiation of male and female on mortality of the following generation (1st group)

Dose of Irradiation (R)	Duration between irradiation and fertilization (Day)	Number of eggs examined	Relative mortality rate	
			10 days (embryo)	30 days (fry)
0	control	379	0	0
2 000	1,2	111	92	97
	3,4	4	-	-
	5,6	72	71	92
	7,8	95	44	68
	9,10	18	0	10
4 000	1,2	96	100	100
	3,4	20	100	100
	5,6	127	100	100
	7,8	13	100	100
8 000	1,2	35	100	100
	3,4	9	100	100
	4,5	86	100	100
	6,7	2	100	100
16 000	1,2	5	100	100
	3,4	20	100	100
	5,6	5	100	100
32 000	3,4	5	100	100

In general, mortality rates of fish in the second group were slightly lower than the corresponding ones in the first group (Table 2, Fig. 1). No embryos could survive if the facher had been irradiated with 8 000 R or more. Small number of fry survived even if their father were exposed to 4 000 R of X-rays. Mortality of the zygotes was significantly low, if the interval between irradiation and fertilization was long (9-10 days).

Table 2. Effects of X-irradiation of male on mortality of the following generation (2nd group)

Dose of Irradiation (R)	Duration between irradiation and fertilization (Day)	Number of eggs examined	Relative mortality rate 10 days (embryo)	Relative mortality rate 30 days (fry)
0	control	561	0	0
2 000	1,2	144	68	99
	3,4	147	73	95
	5,6	167	47	96
	7,8	127	64	90
	9,10	17	33	20
4 000	1,2	180	98	100
	3,4	103	87	98
	5,6	134	59	97
	7,8	46	84	100
	9,10	31	66	100
8 000	1,2	230	100	100
	3,4	168	100	100
	5,6	88	100	100
16 000	1,2	128	100	100
	3,4	121	100	100
	5,6	109	100	100
32 000	1,2	79	100	100
	3,4	114	100	100
	5,6	23	100	100

The results of the third group (Table 3, Fig. 2) indicate that mortality rates were much lower than those in the second group. Furthermore, in the case of mother irradiation, lethal effects of X-rays markedly decreased within 3 days. In other words, radiation sensitivity of oocytes clearly increased during a few pre-oviposition days.

In order to compare radiosensitivities of male and female gametes Fig. 3 is drawn. In this figure the mortalities of zygotes obtained on the second and third days after irradiation are used. Perusal of this figure indicated that the relationship between the relative mortality rate and radiation dose is not always proportional and that male gametes are much more radiosensitive than female gametes in this respect.

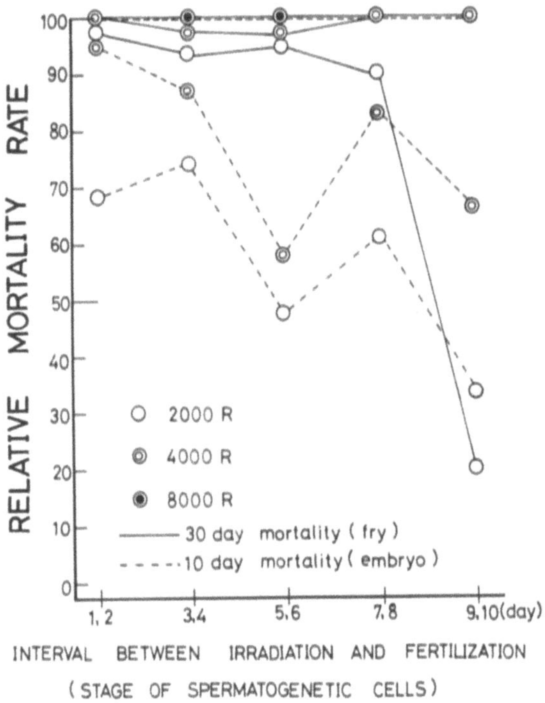

Fig. 1. Relative mortality rate of embryos (-----) and fry (———) fertilized with sperm irradiated at different days (second group)

Table 3. Effects of X-irradiation of female on mortality of the following generation (3rd group)

Dose of Irradiation (R)	Duration between irradiation and fertilization (Day)	Number of eggs examined	Relative mortality rate	
			10 days (embryo)	30 days (fry)
0	control	195	0	0
2 000	1,2	51	25	48
	3,4	53	0	19
	5,6	56	0	27
	7,8	114	0	2
	9,10	36	0	2
4 000	1,2	88	30	65
	3,4	124	20	29
	5,6	75	0	35
	7,8	83	6	42
	9,10	10	0	12
8 000	1,2	53	43	80
	3,4	68	22	47
	5,6	22	20	68
16 000	1,2	47	60	91
	3,4	60	30	67

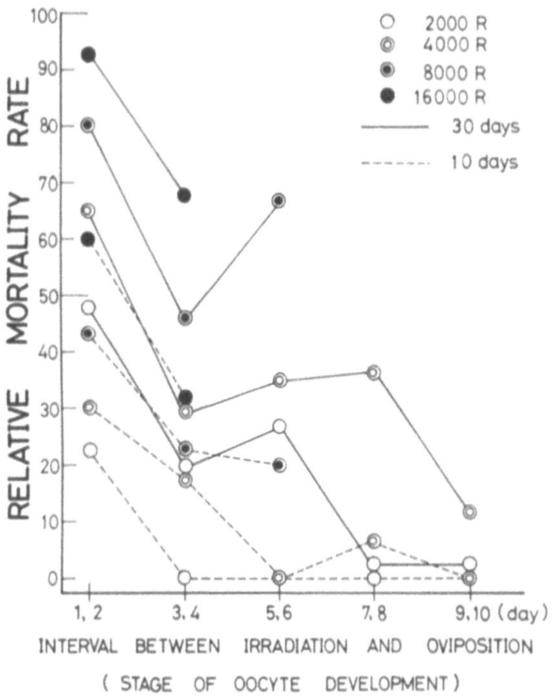

Fig. 2. Relative mortality rate of embryos (-----) and fry (———) of different brood (third group)

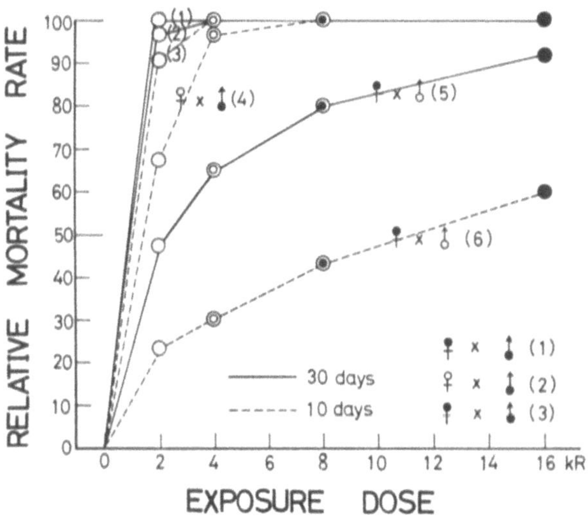

Fig. 3. Relationship between radiation dose and relative mortality rate of different groups. Sex symbols with closed circle indicate irradiated parents

4. Discussion

Although the present experiments are of a preliminary nature, the following conclusions may be drawn:

1. Spermatozoa are more sensitive to X-rays in dominal lethal mutation induction than oocytes. In the present experimental system, the stage of oocytes irradiated is easily defined. In the case of male irradiation, however, the stage of the irradiated spermatogenetic cells is not uniform, since autoradiographic examination of the rate of spermatogenesis at 25°C indicated that the interval from the spermatocyte DNA synthesis to the early spermatid is at least 5 days, that it takes about 7 days for a spermatid to differentiate into a spermatozoon and that mature sperm are stored for 1-7 days in the testis. Therefore, in most cases eggs were fertilized with spermatozoa which were irradiated at the stage of mature sperm. Some sperm ejaculated at 7-10 days after male irradiation might be exposed at spermatid stage.

2. It is highly probable that spermatids are more radioresistant than the sperm since if the interval between irradiation of the male and fertilization is 9-10 days, the relative mortality rate of the zygote is low (2nd group). It is evident that the younger oocytes are more resistant to X-rays than older ones in the induction of dominal lethal mutations (3rd group). These facts are in good agreement with the results obtained by Solberg (1938) in *Oryzias latipes*. It is likely that the younger oocytes are more potent in recovery from radiation induced injury than older oocytes and mature sperm.

3. Death date frequency distribution and kinds of malformation observed in the present experiment were somewhat different from those found in embryos irradiated after fertilization (Hyodo-Taguchi and Egami, 1969). This fact may suggest a role of lethal genes in fish development. The detailed comparison will be reported elsewhere.

4. At any rate this fish is very suitable vertebrate material for the experiments of such nature.

5. However, since 1. the dose-effect relationship curves presented in Fig. 3 were not linear, and 2. fertilized eggs were obtained only within 10 days after the irradiation, more quantitative analyses with lower doses of X-rays are necessary before final conclusions.

References

Egami, N. (1963): Radiation induced malformation in fishes (in Japanese). Iden (The Heredity) 17 (2), 13-16.

Egami, N., Hyodo, Y. (1965a): Inhibitory effect of X-irradiation on the development of the ovaries of the fish, *Oryzias latipes*, in sexually inactive seasons. Annot. Zool. Japon. 38, 8-11.

Egami, N., Hyodo, Y. (1965b): Effect of X-irradiation on the oviposition of the teleost, *Oryzias latipes*. Annot. Zool. Japon. 38, 171-181.

Egami, N., Hyodo-Taguchi, Y. (1967): An autoradiographic examination of rate of spermatogenesis at different temperatures in the fish, *Oryzias latipes*. Experimental Cell Res. 47, 665-667.

Egami, N., Hyodo-Taguchi, Y., Konno, K. (1967): Autoradiographical studies on spermatogenesis in the fish, *Oryzias latipes*, with special reference to radiation effects on fertility of fish. Gunma Symposia Endocrinol 4, 147-159.

Hyodo-Taguchi, Y., Egami, N. (1969): Change in dose-survival time relationship after X-irradiation during embryonic development in the fish, *Oryzias latipes*. Jour. Radiation Res. 10, 121-125.

Konno, K., Egami, N. (1966): Notes on effects of X-irradiation on the fertility of the male of *Oryzias latipes* (Teleostei, Cyprinodontidae). Annot. Zool. Japon. 39, 63-70.
Newcombe, H.B., McGregor, J.F. (1967): Major congenital malformation from irradiation of sperm and eggs. Mutation Res. 4, 663-673.
Schröder, J.H. (1969): X-ray induced mutations in the poeciliid fish, *Lebistes reticulatus* Peters. Mutation Res. 7, 75-90.
Solberg, A.N. (1938): The susceptibility of the germ cells of *Oryzias latipes* to X-radiation and recovery after treatment. Jour. Exp. Zool. 78, 417-440.
Yamamoto, K., Yoshioka, H. (1964): Rhythm of development in the oocytes of the medaka, *Oryzias latipes*. Bull. Fac. Fisch. Hokkaido Univ. 15, 5-19.

Chromosome Manipulation in Fish

C. E. Purdom and R. F. Lincoln

Manipulation of chromosomes or chromosome sets is not used in animal breeding although it is a powerful tool in plant genetics. There seems no reason why such techniques should not be applied to fish breeding where, in contrast to the situation in most animals, the breeder can handle the gametes separately and has some control of the developing zygote after fertilization.

The application of temperature shocks during cell division, so as to destroy the functioning of the metaphase spindle, is a standard technique for controlling the number of chromosome sets. This results in the doubling of chromosome numbers during one cell cycle. This technique has been used widely in plants; amongst animals it has been used extensively only in amphibia (Fankhauser, 1945), for the production of diploid embryos from potential haploids or polyploids from normal diploids. The present paper discusses the use of post-fertilization cold shocks as an aid to inbreeding in fish and in the production of polyploids.

1. Inbreeding by Gynogenesis

One of the principal difficulties in genetic analysis in fish, particularly for quantitative characteristics which might be of commercial value, is the lack of inbred lines. To produce inbred lines by conventional methods requires the maintenance of several lines with close inbreeding for up to 20 generations. For commercially important fish this would entail much labour over a long time, and inbred lines might be lost through inbreeding depression before the required degree of inbreeding had been achieved. Theoretically, the rate of inbreeding can be increased by using parthenogenesis instead of conventional sib mating. Artificial parthenogenesis has been described for a number of animal groups, but particularly amongst the amphibia (Beatty, 1964; Fankhauser, 1945), and one particular form which is suitable for fish eggs is gynogenesis - sperm-activated parthenogenesis.

Gynogenesis was first described by Hertwig (1911) in connection with experiments on the effect of ionizing radiation on frog sperm. Hertwig showed that with an increasing radiation dose to sperm the survival of embryos decreased to a zero point, after which a further increase in dose led to an improvement in survival rate. This paradoxical result, which is now known as the 'Hertwig effect', was attributed to the complete destruction of sperm chromosomes at the highest doses, permitting parthenogenetic development of the haploid embryo without the deleterious effects of radiation-damaged chromosomes. These gynogenetic embryos were basically haploid, but later work (Subtelney, 1958) on gynogenesis and other forms of parthenogenesis in amphibia showed that amongst the haploids there was a low frequency - about 1% - of diploid individuals. Furthermore, the incidence of diploids could be greatly increased by applying temperature shocks to parthenogenetic eggs prior to cleavage.

Studies by Soviet scientists at the Institute of Pond Fish Culture in Moscow have shown that these features of gynogenesis also occur in fish (Romashov et al., 1961; Golovinskaia, 1968). Work with carp *(Cyprinus carpio)*, sturgeon *(Acipenser ruthenus)* and loach *(Misgurnus fossilis)* demonstrated the existence of a typical 'Hertwig effect' following the irradiation of sperm prior to fertilization. The gynogenetic fish embryos were mostly haploid with a low frequency of diploids, which could be increased by applying a cold shock just after fertilization.

Similar studies with plaice and flounder have confirmed these results in most respects (Purdom, 1969). An irradiation dose of 100 000 rad of ^{60}Co gamma-rays was used to de-activate the sperm genetic material, 50 000 rad having proved insufficient. After fertilization with irradiated sperm, the majority of the embryos showed a typical haploid syndrome, with a very short body and gross abnormalities of the head and gut. When the gynogenetic eggs were cold-shocked by immersion in sea water at about 0°C for 4 hours, from 15 minutes after fertilization, a majority of embryos were of the diploid type. The timing and the duration of the cold shock are both important factors in the success of the diploidization process.

Table 1 shows the effect of the time of application of the shock. The frequency of diploids was highest over the range from 5 minutes to 25 minutes after fertilization. Further delay resulted in a decrease in the frequency of diploids, and by 40 minutes after fertilization the diploid frequency was down to the spontaneous level. The duration of the cold treatment was less critical (Table 2). Cold treatment for one hour was not very efficient, but high frequencies of diploid gynogenomes were produced with cold shocks of from 2 to 5 hours duration.

Table 1. Effect of time of application of 4 hour cold shock on the frequency of gynogenetic diploids. Scored 5 days after fertilization

Time of application of cold shock (minutes after fertilization)	Number of haploids	Number of diploids	% diploid
5	2	198	99.0
10	9	191	95.5
15	5	195	97.5
20	3	197	98.5
25	9	191	95.5
30	41	159	79.5
35	172	28	14.0
40	198	2	1.0
45	197	3	1.5
50	192	8	4.0

The important feature of gynogenesis, from the point of view of inbreeding, is the way in which diploidy is re-established. There are three basic routes by which this may occur:

1. failure of meiosis I;
2. failure of meiosis II;
3. failure of first mitosis of the zygote.

Table 2. Effect of duration of cold shock on the frequency of gynogenetic diploids. Scored 5 days after fertilization

Duration of cold shock in hours	Number of haploids	Number of diploids	% diploid
0.5	304	31	9.3
1	184	27	12.8
2	110	703	86.4
3	168	766	82.1
4	74	438	85.5
5	230	534	69.9

The degree of inbreeding produced by gynogenesis will depend on which mechanism prevails. Thus, failure of meiosis I, which is equivalent to self-fertilization, will produce 50% inbreeding. Fusion of the products of first mitosis is equivalent to 100% inbreeding, while failure of meiosis II will lead to a level of homozygosity by descent somewhere between 50 and 100%, depending on the degree of crossing-over during meiosis I.

The first possibility can be ruled out, since teleost eggs at ovulation are in the prophase or early metaphase of meiosis II (Ginzburg, 1968). In amphibia, the diploidization process has been attributed to the third possibility, the failure of first mitosis (Subtelney, 1958). Under these conditions, the first observed cleavage division is, in fact, the second mitosis and thus arises with a delay equal to the duration of one cell cycle. Measurement of the onset of first cleavage in plaice eggs - which is simple because of their transparency - showed that cleavage times were not delayed by the process of diploidization. Two groups of eggs were compared, each of which had been fertilized with irradiated sperm. The first group was cold shocked from 20 minutes after fertilization, the second from 40 minutes after fertilization. No differences were observed, although the first group developed into diploids and the second into haploids. Thus diploids induced in gynogenetic fish eggs by cold shocks soon after fertilization must arise by the failure of meiosis II in the eggs.

Each individual gynogenetic fish will be inbred in the sense that it will be homozygous at many loci, due to the inheritance of replicate genes from the female parent. The actual degree of inbreeding will depend on the frequency of crossing-over. On average, one cross-over per chromosome will reduce the level of inbreeding to 50% and, in general, the degree of inbreeding will be $1-1/2$, where α is the cross-over frequency. For fish, cross-over values are small and on the assumption of $\alpha = 0.1$, the degree of inbreeding will be 0.95, which is equivalent to the degree of inbreeding achieved by 14 generations of full sib mating. Each individual gynogenome will, of course, be unique, but a second generation would produce an inbred line from each female gynogenome.

Diploid gynogenetic plaice were not normal. The most obvious abnormality was their low viability, which is perhaps not surprising if they are 95% inbred. So far about 60 gynogenetic plaice have been reared through metamorphosis from many thousands of eggs. Plaice are normally rather difficult to rear, but 50% survival to metamorphosis can be achieved with care. Of the 60 metamorphosed fish reared in 1970, 12 survive now, approaching maturity at 3 years of age. Some of the remainder were deliberately killed for examination and some died by accident or disease.

A further feature of the gynogenetic fish was that they included both males and females (Table 3). This observation implies that sex-determination in plaice is not of the XX, XY form, since this would ensure that all gynogenetic fish were female. Heterogametic systems have been described in fish (Hickling, 1960; Kallman, 1965) and it is possible that a WZ female, ZZ male system operates in plaice. In this case, one would expect that WW and WZ genotypes were female in view of the preponderance of this sex in the gynogenetic plaice.

Table 3. Gynogenetic plaice sexed at 1 year of age

Female parent	Female offspring	Male offspring
1	7	1
2	7	7
3	5	3
4	1	2
5	4	1
Total	24	14

2. Polyploidy

Since diploid individuals can be produced by cold shocking haploid eggs, it seemed probable that triploids could be produced in a similar way, by cold shocking diploid eggs. A further possibility was the production of tetraploids by applying cold shocks to diploid zygotes during the first mitotic division.

Triploid fish should be sterile and this might be of great practical value in fish cultivation, where maturation can affect both growth rate and food conversion efficiency. Tetraploidy is potentially useful as an aid to producing fertile hybrids.

Production of triploids in plaice was accomplished by applying a cold shock to eggs as in gynogenesis, but after fertilization with normal spermatozoa (Purdom, 1972). To test the effectiveness of the treatment, hybrids between plaice and flounder were used in the critical experiments. The larval pigmentation patterns of plaice and hybrids are very different and it was expected that the triploids, with two sets of plaice chromosomes, would be intermediate - this proved to be the case. Subsequently, triploidy was confirmed by measurement of red blood cell nuclei and also by observation of chromosomes in cultured leucocytes from juvenile fish. Triploid production by heat treatment methods in amphibia is well established and has also been demonstrated in fish by Swarup (1956), who produced triploid sticklebacks. Our observations with plaice differ from these only in respect to the frequency with which triploids were produced. In the amphibian studies and in Swarup's work temperature shocks produced an array of heteroploids, diploids and mosaics. In plaice, the induction of triploids has been almost 100% successful in several experiments but a complete failure in others. This variation between experiments is difficult to explain, but it is possible that physical features of the temperature shock and differences in the degree of maturity of eggs may be important factors.

The triploid fish themselves appear perfectly normal and as viable as the diploid controls. The sex ratio seems undisturbed by triploidy but the triploids were sterile; the ovaries contained only a few abnormal oocytes and the testes remained very small, although they have not yet been examined critically for evidence of spermiogenesis. A batch of 3-year-old female triploids and their hybrid sibs are currently being examined for their food conversion efficiency. The ovaries of the diploids are now visibly filling whilst the triploids are not, but no clear-cut differences in food conversion have been observed (Table 4). Following the successful induction of triploids in plaice, similar treatments were given to eggs of trout. There is an immediate practical use for such triploids, since the trout which are stocked in still-water reservoirs cannot spawn effectively and sexual maturation or attempts to spawn are often associated with a high mortality rate. Similarly, the current practice of salmon farming is bedevilled by the tendency of male smolts to mature precociously and females to mature early as grilse. In both cases growth rate is seriously diminished, and in the case of maturing male smolts the fish are worthless. The possible utilization of sterile triploids, therefore, promises to be economically advantageous.

Table 4. Condition factor, $W/L^3 \times 100$, and food conversion efficiency in diploid and triploid plaice/flounder hybrid females

	Diploid			Triploid		
Date	W/L^3 x 100	% food conversion	n	W/L^3 x 100	% food conversion	n
11 September	12.85	-	7	13.84	-	6
25 September	13.03	16.85	7	14.33	19.74	6
9 October	13.65	20.68	7	14.57	14.28	6

One of the problems with salmonids is the very long period between fertilization and first cleavage (12 hours). In addition, little is known of the timing of the final stages of meiosis in the fertilized egg, and observation is difficult because of the yolky, translucent nature of salmonid eggs. According to Böhm (in Svardson, 1945), in trout eggs the first polar body is extruded after fertilization. Ginzburg (1972), however, describes the second meiotic metaphase as present prior to fertilization. Experiments were therefore performed in which cold shocks were applied to eggs at different times throughout the period from fertilization to first cleavage. Two experiments were run. In the first, successive batches of eggs were cold shocked at 15-minute intervals up to 2 hours after fertilization. Nuclei were measured in 28-hour-old blastomeres, but were too variable to give reliable data. Measurement of nuclei in embryos fixed 7 days after fertilization, however, suggested that all were diploid. A second experiment was run in which cold shocks were applied to successive batches of eggs at 30-minute intervals, starting 2 1/2 hours after fertilization and continuing up to 10 hours after fertilization. Chromosome configurations were examined in squash preparations of blastodiscs fixed 3 1/2 days after fertilization (Svardson, 1945) and stained in aceto-orcein. No triploids were found, but very high levels of heteroploidy were observed.

In general, normal diploid complements were observed in blastodiscs cold shocked up to 5 1/2 hours after fertilization. Eggs cold shocked from 6 hours after fertilization and later showed gross levels of aneuploidy, multipolar spindles and anaphase bridges, whilst eggs cold shocked at 10 hours after fertilization showed many highly polyploid cells. The simple interpretation of this is that the cold treatment was effective after 6 hours but not sufficient to disrupt completely the metaphase of meiosis II or first mitosis. This partial disruption of metaphase may have produced unbalanced chromosome and centrosome sets which later led to mitotic disturbances. On balance, it seems more probable that these effects were consequences of disrupted first mitotic metaphase. If, as according to Ginzburg (1972), the mitotic metaphase is in existence prior to fertilization, cold treatment may be more effective if applied at, or even prior to, fertilization.

Attempts to produce tetraploids by cold shocks applied during the first mitotic divisions in plaice have so far failed, although this type of phenomenon is reported to be the prime factor in diploid parthenogenesis in frogs. Plaice eggs were fertilized with flounder sperm and successive batches were cold shocked at intervals from fertilization up to first cleavage. The suppression of first mitosis would result in an apparent delay in cleavage and in three experiments such a delay was in fact observed, when cold shocks were applied from 4 hours after fertilization. The first cleavage normally occurs at about 5 hours after fertilization in eggs incubated at 7°C. However, despite this apparent suppression of first mitosis, all surviving fish proved to be diploid. The third experiment supplied the answer to this paradox. First cleavage was suppressed and instead of a two-cell stage, the egg at this time contained what appeared to be a single-cell stage. At the time of the next division, however, two cleavages occurred almost simultaneously. Later histological studies showed that the single-cell stage of the egg contained two nuclei which were very close together. Thus the cold shock again appeared to be partially effective and the question remains whether more drastic shocks might be more successful.

3. Summary

The manipulation of chromosome sets in fish by radiation treatment of spermatozoa and by cold treatment of eggs after fertilization is shown to be feasible in some circumstances.

Diploid gynogenetic plaice were produced by applying cold shocks to eggs 15 minutes after fertilization with sperm made genetically inert by a dose of 100 000 rad of gamma-rays. The mechanism by which diploidy was induced is discussed in relation to the use of gynogenesis for inbreeding.

Triploids plaice were produced by cold shocking eggs after normal fertilization. Attempts to produce plaice tetraploids by the suppression of first mitotic metaphase were unsuccessful. Trout eggs did not respond in the same way as plaice eggs to post-fertilization cold shocks, although considerable chromosome imbalance was induced when the cold treatment began 6 hours after fertilization, or later. The timing of meiotic metaphase, relative to fertilization, may differ between salmonids and pleuronectids and this may explain the different responses to cold treatment.

References

Beatty, R.A. (1964): Gynogenesis in vertebrates: fertilization by genetically inactivated spermatozoa. Proc. Int. Symp. on Effects of Ionizing Radiation on the Reproductive System, Fort Collins, Colorado, 1962 (eds. W.D. Carlson and F.X. Gassner), 229-238.
Fankhauser, G. (1945): The effects of changes in chromosome number on amphibian development. Cl. Rev. Biol. 20, 20-78.
Ginzburg, A.S. (1968): Fertilization in fishes and the problem of polyspermy. Moscow, "Nauka" Pub. House. Jerusalem, Israel Program for Scientific Translations, for U.S. National Marine Fisheries Service, 1972. Russian Translation.
Golovinskaia, K.A. (1968): Genetics and selection of fish and artificial gynogenesis of the carp (*Cyprinus carpio*). F.A.O. Fish. Rep. No. 44, Vol. 4, Proc. World Symp. on warm water pond fish culture, Rome, 1966 (ed. T.V.R. Pillay), 215-222.
Hertwig, O. (1911): Die Radiumkrankheit tierischer Keimzellen. Arch. Mikr. Anat. 77, 1-97.
Hickling, C.F. (1960): The Malacca *Tilapia* hybrids. J. Genet. 57, 1-10.
Kallman, K.D. (1965): Genetics and geography of sex determination in the poeciliid fish, *Xiphophorus maculatus*. Zoologica, N.Y. 50, 151-190.
Purdom, C.E. (1969): Radiation-induced gynogenesis and androgenesis in fish. Heredity, Vol. 24, Pt. 3, 431-444.
Purdom, C.E. (1972): Induced polyploidy in plaice (*Pleuronectes platessa*) and its hybrid with the flounder (*Platichthys flesus*). Heredity, Vol. 29, Pt. I, 11-24.
Romashov, D.D., Belyaeva, V.N., Golovinskaia, K.A., Prokof'eva-Bel' govskaya, A.A. (1961): Radiation disease in fish. (In Russian) Radiatsionnaya Genetika, Acad. of Science, USSR, Moscow, 1961, 247-266.
Subtelney, S. (1958): The development of haploid and homozygous diploid frog embryos obtained from transplantation of haploid nuclei. J. Exp. Zool. 139, 263-305.
Svardson, G. (1945): Chromosome studies on Salmonidae. Meddn St. Unders.o. FörsAust. SöttvattFisk. 23, 1-151.
Swarup, H. (1956): Production of heteroploidy in the three-spined stickleback (*Gasterosteus ciculeatus L.*). Nature 178, 1124-1125.

Teleosts as a Tool in Mutation Research

J. H. Schröder

1. Introduction

The purpose of this paper is to review what has been done concerning mutation research on teleostean fish. Although earlier mutagenesis studies have been restricted to the effects of radiation, teleosts would probably be very helpful in elucidating the mutational response of vertebrate germ cells to chemical mutagens[1]. This paper is confined to radiation-induced mutations in fish. Accordingly, all the results dealing with somatic radiation damage to the gonads at the cellular and sub-cellular level are excluded. Table 1 summarizes the following survey.

2. Presumed Mutations at Dominant *X*- and *Y*-Linked Color Genes

Teleosts were first used in mutation research in 1938 by Samakhvalova. She found a presumed dominant mutation of the gene complex "Bimaculatus" in the guppy, *Lebistes reticulatus (= Poecilia reticulata)*, following 750 R of X-rays to spermatozoa. However, because she used only a very small sample size, it was impossible to reliably estimate the mutation rate. Since 1964 this species has also been studied by C.E. Purdom and his co-workers (1966, 1973) who checked F_1 males and masculinized F_1 females for dominant visible mutations on the sex chromosomes following 1 000, 1 500 and 2 000 rads of X-rays to spermatozoa. He compared the observed frequency of mutants with that expected if it were assumed that: 1. the spermatozoa of *Lebistes* are as sensitive as mouse spermatozoa and 2. 20 different loci are involved in the inheritance of color patterns in *Lebistes*. It was shown that the gonosomal loci of *Lebistes* were less sensitive than the autosomal loci of the house mouse as often used by mouse geneticists. Purdom (1966) also used the specific-locus technique itself in attempting to estimate the mutation rate of the guppy. However, because the number of loci tested was too small, this method was not yet successful. In the meantime, however, guppy strains were bred which contained recessive mutations at four autosomal loci (albino, blond, gold and blue).

3. Dominant Lethal Mutations and Changes of Viability

Egami and Hyodo-Taguchi (1973) determined the frequency of dominant lethal mutations in the egg-laying tooth carp, medaka (*Oryzias latipes*). They either irradiated both parents or only one of them with different doses of X-rays (ranging from 2 000 to 32 000 R) between 1 and 10 days

[1]Bevilacqua and Giovenzana (1971; "Zierfische", Südwest-Verlag München) mention a mutational shortening of the vertebral column of the guppy induced by HCHO.

Table 1. Survey on the results of mutation research in fish

Mutational response to irradiation	Species treated
Presumed mutations of dominant gene complexes on the sex-chromosomes	*Lebistes reticulatus*
Dominant lethal mutations	*Oryzias latipes*
Reduction in litter size, change of sex-ratio	*Lebistes reticulatus*
Reduction of litter size, increase of still-births, postnatal mortality and anomalies	*Lebistes reticulatus*
Major congenital malformations of embryos; "beneficial" effect of low doses (25 and 50 rads) on hatching rate	*Salmo gairdnerii*
Inheritance of radiation-induced curvatures of the vertebral column	*Xiphophorus helleri; Lebistes reticulatus*
Induced exchange between the sex-chromosomes	*Lebistes reticulatus*
Synergistic interaction of newly induced mutations with certain recessive mutant genotypes	*Lebistes reticulatus*
Formation of pre-melanomas in pure-breeding platies	*Xiphophorus maculatus*
Occurrence of functional *XY*-females	*Xiphophorus maculatus*
Unidirectional shift of the mean values of quantifiable traits	*Lebistes reticulatus*
Reduction of male aggressiveness to subadult male conspecifics	*Cichlasoma nigro-fasciatum*
Induced gynogenesis by fertilization of eggs with genetically inactivated sperm	*Cyprinus carpio; Pleuronectes platessa; Acipenser ruthenus; Misgurnus fossilis*

Table 1. (continued)

Irradiated stage	Radiation dose	Authors
Spermatozoa	750 R; X-rays	Samokhvalova, 1938
	1 000 - 2 000 rads; X-rays	Purdom, 1966; Purdom and Woodhead, 1973
Oocytes or spermatozoa or spermatids or both female and male gametes	2 000 - 32 000 R; X-rays	Egami and Hyodo-Taguchi, 1973
All stages of gametogenesis	4 000, 5 000 and 8 000 rads of chronic γ-rays at 0.25, 0.5 and 1.7 rad/h	Purdom, 1973
Various stages of gametogenesis	500, 1 000 and 2 000 R; X-rays	Schröder, 1969a
Sperm and eggs	25 - 400 rads; 200, 2 000 and 20 000 rads; X-rays	McGregor and Newcombe, 1967; 1972
Embryos; various stages of gametogenesis	400 - 1 000 R	Penners, 1957, 1962; Schröder, 1969b
Spermatogonial stem-cells	1 000 R; X-rays	Schröder, 1969c
Oogonia, spermatogonia or spermatozoa	1 000 R; X-rays 2 x 500 R (24 h apart	Schröder and Holzberg, 1972
Embryos or germ-cells of adult females	500 - 2 000 R; X-rays	Anders et al., 1971
Embryos	1 000 - 2 500 R; X-rays	Anders et al., 1969
Oogonia or spermatogonia in neonatal fish	1 000 R; X-rays	Schröder, 1969d, 1969e
Oogonia and spermatogonia in immature fish	2 x 500 R (24 h apart); X-rays	Holzberg, 1973; Holzberg and Schröder, 1972
Sperm	100 000 rads; ^{60}Co γ-rays	Golovinskaia, 1968; Purdom, 1969; Purdom and Lincoln, 1973; Romashov et al., 1961

prior to oviposition. They found a non-linear dose-mutation relationship for several stages of spermatids and spermatozoa. Furthermore, male gametes exhibited a higher radiosensitivity to cell killing than oocytes.

Because the estimation of living and dead embryos in the ovaries of viviparous tooth carps was found to be rather difficult, the reduction of the litter size after irradiation as compared to that of control lines was used to determine the rate of dominant lethal mutations. Purdom and Woodhead (1973) demonstrated a pronounced reduction of litter size after chronic gamma irradiation with 4 000, 5 000, and 8 000 rads at dose rates of 0.25, 0.5 and 1.7 rad/h, respectively. The observed decrease in litter size was probably due to the induction of dominant lethal mutations in all stages of gametogenesis of the guppy. The present author also used *Lebistes* for somewhat similar experiments. He, however, used X-irradiation delivered at 200 R/min instead of chronic gamma irradiation at doses ranging from 500 to 2 000 R. The results were as follows: 1. irradiation of primordial germ cells did not lead to a significant change in the number of living young born per litter in the first four litters, although in the inbred lines a trend towards increasing litter size was seen in the F_2 to F_4 generations after irradiation; 2. the frequency of stillborn fish (expressed as percent of living neonatal ones) was higher only in the F_1 and F_2 offspring after spermatogonial irradiation; 3. in experiments in which gonial stages were irradiated, postnatal mortality (between birth and 90 days of age) was enhanced only in the F_2; 4. viability effects in post-irradiation generations were more pronounced in inbred than in hybrid lines; 5. the incidence of skeletal abnormalities (curvatures of the vertebral column) and of pigmentation defects was higher in irradiated than in control lines; and 6. the overall sensitivity to the induction of mutations by X-irradiation of the guppy apparently lies between that of the mouse and *Drosophila* (Schröder, 1969a and b; Schröder and Holzberg, 1972).

4. Major Congenital Malformations

McGregor and Newcombe (1967) studied the incidence of major congenital malformations such as eye, head, back and tail defects following irradiation of either mature eggs or sperm of the rainbow trout, *Salmo gairdnerii*. The frequency of major eye malformations in the first generation offspring followed a linear relationship in the low dose range (25 - 400 rads) of ^{60}Co-gamma exposure at a dose rate of approximately 64 rad/min (McGregor and Newcombe, 1972).

After *in vitro* irradiation of mature sperm and eggs with acute X-ray doses of 200, 2 000 and 20 000 rads at high dose rates (delivered in 2.1 min, 2.4 min, and 13.4 min, respectively) the response per unit dose fell at high doses. At 200 rads the yield was about 3×10^{-4} malformations per embryo per rad for irradiation of both sperm and eggs. Approximately 26 rads would be required to double the rate of malformations observed in the controls. The authors believe that these major malformations may be associated predominantly with gross chromosome changes, although this was not proved.

Analysis of the survival data in the low dose range (25 - 400 rads) at 64 rad/min revealed a significant increase over controls in the proportion of eggs with embryos following an exposure of sperm to 25 and 50 rads. After 400 rads, however, the proportion of eggs with embryos was greatly reduced. The apparently "beneficial" effect of the lower doses at gamma-irradiation was seen during early and intermediate stages of embryonic development, while the harmful effect of higher doses predominated during the intermediate stages.

5. Inheritance of Radiation-Induced Spinal Curvatures

Penners (1959, 1962) was the first author who showed that radiation-induced malformations of the vertebral column of the swordtail, *Xiphophorus helleri*, are hereditary. These curvatures of the spine proved to be recessive, but the number of genes involved in the formation of the abnormalities could not be assessed.

The spinal curvatures and shortenings which occurred in post-irradiation generations of the guppy also proved to be hereditary (Schröder, 1969 a and b). The abnormal shapes of the spine could be expressed in clinical terms of scolioses, kyphoses, or lordoses. After outcrossing of lordotic fish with non-related normal mates the mutant to wild-type ratios in F_2 fitted a di- and tri-genic segregation ratio, respectively. This was the first demonstration of the manner of inheritance of radiation-induced recessive mutations affecting the vertebral column. Thus, generally speaking, the curvatures of the spine in poeciliid fish are caused by recessive oligo-genes located on the autosomes.

6. Radiation-Induced Exchange between the Sex Chromosomes

After a dose of 1 000 R of acute X-rays to guppy spermatogonia, an increase of crossovers between the *X* and *Y* chromosomes was observed. This excess of exchanges in the irradiated series occurred only after irradiation of spermatogonia and not after exposure of later spermatogenic stages (Schröder, 1969c). Recombinant offspring were found in clusters, presumably because mitotic crossing over occurred in relatively few spermatogonia that then repopulated a significant fraction of the testis.

7. Synergistic Action of Newly Induced Recessive Mutations

The irradiation of wild-type guppies with 1 000 R of X-rays and their subsequent mating to fish homozygous at three loci (albino, blond and blue) resulted in a significant change in the phenotypic segregation ratio in the F_2 in favor of wild type and blue offspring. This unidirectional effect was greatest after exposure of spermatozoa to 500 + 500 R (24 hours apart) followed by single-dose exposure of 1 000 R to either spermatogonia or oogonia. It was concluded that radiation-induced recessive mutations may interact synergistically with other gene loci, thereby affecting the viability of certain mutant genotypes of the guppy (Schröder and Holzberg, 1972). Since both the albino and blond loci are concerned with melanin formation, it appears that abnormal tyrosinase synthesis decreases viability in comparison to that of guppies with wild type and blue phenotypes.

8. Formation of Pre-Melanomas after Irradiation

Anders and co-workers using the platyfish, *Xiphophorus (Platypoecilus) maculatus*, found an increase in macromelanophore gene expression after irradiation of either embryos in pregnant females or germ cells in adult females. The enhancement in the production of macromelanophore spots resulted in the formation of pre-melanomas similar to those produced by F_1 hybrids between *X. maculatus* (from Mexico) with *X. helleri*. Since this increase of the gene expression (e.g., *Sp* of *X. maculatus*) proved to be inherited, the authors concluded that the genetic repression system of the macromelanophore gene, *Sp*, was altered both in somatic and germ cells in the same direction through X-irradiation with 500 - 2 000 R (Anders et al., 1971; Pursglove et al., 1971). In other words, any unspecific mutation of the apparently polygenic repression system may result in the same phenotypic effect.

9. Radiation-Induced Sex-Reversal

Functional *XY*-females occurred with high frequencies after embryonic irradiation of platyfish with 1 000 - 2 000 R of X-rays (Anders et al., 1969). Because these functional *XY*-females were fertile, the production of *YY*-males was possible. Thus, an *XX* ♀ - *XY* ♂ sex-determining mechanism could be changed experimentally into an *XY* ♀ - *YY* ♂ mechanism, i.e. homogametic females were transformed into heterogametic females and heterogametic males were transformed into homogametic males. Because this radiation effect was relatively unspecific (in the same sense as was mentioned above for the post-irradiation formation of premelanomas), it was suggested that polygenic sex-determining genes located on the autosomes perhaps dominate over those located on the sex chromosomes as the consequence of radiation-induced mutations in the autosomal system or gonosomal system or both. These radiation-induced alterations may be associated with chromosomal rearrangements which are known to be very frequent after irradiation of mouse gametes with corresponding doses.

10. Mutability of Quantifiable Characters

After 1 000 R of X-rays to either oogonia or spermatogonia of neonatal guppies, the mutability of polygenically determined traits such as body proportions and number of vertebrae was determined in post-irradiation F_1 and F_2 generations (Schröder, 1969d and e). An unidirectional shift of the mean values towards a more compact average fish was found in the irradiated line as compared to the unirradiated control line. This tendency to become shorter and broader after irradiation was also reflected by a slight decrease of the mean vertebral number. There was an increase of the degree of variability in post-irradiation generations as expected by a radiation-induced increase of heterozygosity. These findings completely correspond to similar results in higher plants.

11. Radiation-Induced Changes of Behavioural Traits

A comparison was made between the aggressive behaviour of groups of convict cichlid males (*Cichlasoma nigrofasciatum*) derived either from irradiated (500 + 500 R of X-rays, 24 h apart, both to oogonia and spermatogonia) parents or from sham-treated control parents (Holzberg, 1973; Holzberg and Schröder, 1972). The male level of aggression was measured as bites delivered to subadult male conspecifics per 15 min observation time. For the same individual F_1 males the activity was determined by counting the number of 90° turns within 15 min. The aggressivity of F_1 males was also examined during the first 9 days of the breeding cycle starting with the day of spawning.

The results of this study were as follows:

a) the aggressiveness of the nonmated control F_1 males was 3.5 times as high as that of the nonmated F_1 males of the irradiated series;

b) the aggressiveness of the mated control F_1 males was 2.1 times as high as that of the mated F_1 males of the irradiated series;

c) the activity of the F_1 males of both series did not differ in any way;

d) the variability of aggressiveness of nonmated F_2 males of the irradiated series was higher than that of the nonmated F_2 control males;

e) the mean values of aggressiveness were lower in post-irradiation F_2 than in control F_2; and

f) the mean value of aggression of F_2 males belonging to the same brood derived from a given F_1 male roughly corresponded to the mean value of aggressiveness of their father.

From these findings on quantifiable characters as well as on formation of pre-melanomas, sex-reversal, and reduction of male aggressiveness, one can conclude that unidirectional changes of the average phenotype that may be of evolutionary significance tend to occur in mutated polygenic systems.

12. Radiation-Induced Gynogenesis

Already in 1911 Hertwig showed that with an increasing radiation dose to sperm survival of embryos decreased to a zero value, after which a further increase in radiation dose led to an improvement in survival rate. This so-called Hertwig effect was due to the appearance of gynogenetic animals following extremely high doses of ionizing radiation to sperm prior to fertilization. Gynogenetic fish occurred only when doses of at least 100 000 rads of ^{60}Co-gamma irradiation were applied to mature sperm (Purdom, 1969; Purdom and Lincoln, 1973). After this dose, spermatozoa were still able to inseminate eggs, but fusion of the haploid parental nuclei did not occur because the paternal genome was inactivated due to irradiation. Thus, haploid zygotes were produced which could be diploidized by a cold shock just after fertilization. This method of obtaining gynogenetic fish is of practical importance in fish cultivation because it saves on many generations of inbreeding to get homozygous strains. Thus far gynogenetic fish have been produced this way in the following species: the carp (*Cyprinus carpio*) (Golovinskaia, 1968), the sturgeon (*Acipenser ruthenus*), the loach (*Misgurnus fossilis*) (Romashov et al., 1961), and the plaice (*Pleuronectes platessa*) (Purdom, 1969; Purdom and Lincoln, 1973).

13. Conclusions

The knowledge obtained in mutation research on fish is obviously important both from the theoretical and practical point of view. Since intensive fish farming seems to be necessary to provide either directly or indirectly the base for the protein food of mankind in the future, chromosome manipulation of both marine and fresh-water fish would help to increase the productivity of yields in cultured fish. To improve the variability of highly inbred strains and to search for new mutants resistant to various infectious diseases (e.g., dropsy of the carp), the treatment of germ cells with ionizing radiations or chemical mutagens or both might be the way to reach this goal. Further elucidation of the role of new mutant genes associated with the formation of melanomas and pre-melanomas in xiphophorin fish should help to understand carcinogenesis, this work being obviously justifiable because of its application to humans. The results of mutation research in teleosts have already contributed to the estimation of the genetic risk of ionizing radiation. Chemical mutagens, therefore, should also be studied for their effects on fish. Besides this extension to chemical compounds, more sophisticated methods are desirable to screen the genetic material for mutational changes. For example, the response of biochemical traits (such as isozyme patterns) should be investigated after mutagenic treatment. Mutation research on fish may also be helpful in answering the question of how vertebrates have evolved phylogenetically since their beginning.

14. Acknowledgment

The author would like to express his sincere appreciation to Dr. Paul Selby for improving the English.

References

Anders, A., Anders, F., Förster, W., Klinke, K., Rase, S. (1969): *XX-*, *XY-*, *YY*-Weibchen und *XX-*, *XY-*, *YY*-Männchen bei *Platypoecilus maculatus* (Poeciliidae). Zool. Anz. Suppl. 33, Verh. Zool. Ges. 1969, 333-339.

Anders, A., Anders, F., Pursglove, D.L. (1971): X-ray-induced mutations of the genetically-determined melanoma system of xiphophorin fish. Experientia 27, 931-932.

Anders, A., Anders, F., Rase, S. (1969): *XY*-females caused by X-irradiation. Experientia 25, 871.

Egami, N., Hyodo-Taguchi, Y. (1973): Dominant lethal mutation rates in the fish, *Oryzias latipes*, irradiated at various stages of gametogenesis. This volume.

Golovinskaia, K.A. (1968): Genetics and selection of fish and artificial gynogenesis of the carp (*Cyprinus carpio*). F.A.O. Fish Rep. No. 44, Vol. 4, Proc. World Symp. on warm water pond fish culture, Rome, 1966 (ed. Pillay, T.V.R.), 215-222.

Hertwig, O. (1911): Die Radiumkrankheit tierischer Keimzellen. Arch. Mikr. Anat. 77, 1-97.

Holzberg, S. (1973): Change of aggressive readiness in post-irradiation generations of the convict cichlid, *Cichlasoma nigrofasciatum*. This volume.

Holzberg, S., Schröder, J.H. (1972): Behavioural mutagenesis in the convict cichlid fish, *Cichlasoma nigrofasciatum* Guenther. I. The reduction of male aggressiveness in the first post-irradiation generation. Mutation Res. 16, 289-296.

McGregor, J.F., Newcombe, H.B. (1967): Major congenital malformations from irradiation of sperm and eggs. Mutation Res. 4, 663-673.

McGregor, J.F., Newcombe, H.B. (1972): Dose-response relationships for yields of major eye malformations following low doses of radiation to trout sperm. Radiat. Res. 49, 155-169.

Penners, R. (1959): Durch Röntgenstrahlen verursachte biologische Schäden. Untersuchungen mit und an dem Schwertträger. Z. f. Natur u. Technik 1959, 5, 403-407.

Penners, R. (1962): Experimentell erzeugte Mißbildung bei Fischen. Jahrbuch der Fürsorge für Körperbehinderte 1962, 35-38.

Purdom, C.E. (1966): Radiation and mutation in fish. In: Disposal of radio-active wastes into seas, oceans and surface waters. IAEA, Vienna, 861-867.

Purdom, C.E. (1969): Radiation-induced gynogenesis and androgenesis in fish. Heredity 24, 431-444.

Purdom, C.E., Lincoln, R.F. (1973): Chromosome manipulation in fish. This volume.

Purdom, C.E., Woodhead, D.S. (1973): Radiation damage in fish. This volume.

Pursglove, D.L., Anders, A., Döll, G., Anders, F. (1971): Effects of X-irradiation on the genetically-determined melanoma system of xiphophorin fish. Experienta 27, 695-697.

Romashov, D.D., Belyaeva, V.N., Golovinskaia, K.A., Prof'eva-Bel'govskaya, A.A. (1961): Radiation disease in fish (in Russian). Ratiatsionnaya Genetika, Acad. of Sciences, USSR, Moscow 1961, 247-266.
Samokhvalova, G.V. (1938): Effect of X-rays on fishes (*Lebistes reticulatus*, *Xiphophorus helleri* and *Carassius vulgaris*) (in Russian). Biol. Zhur. 7, 1023-1034.
Sankaranarayanan, K. (1972): Ionizing radiations, genetic loads and population fitness. Panel on radiation effects on population dynamics in ecosystems. IAEA Report PL-515/2, October 2-5, 1972, Reykjavik.
Schröder, J.H. (1969a): X-ray-induced mutations in the poeciliid fish, *Lebistes reticulatus* Peters. Mutation Res. 7, 75-90.
Schröder, J.H. (1969b): Inheritance of radiation-induced spinal curvatures in the guppy, *Lebistes reticulatus*. Can. J. Genet. Cytol. 11, 937-947.
Schröder, J.H. (1969c): Radiation-induced spermatogonial exchange between the *X* and *Y* chromosomes in the guppy. Can. J. Genet. Cytol. 11, 948-954.
Schröder, J.H. (1969d): Quantitative changes in breeding groups of *Lebistes* after irradiation. Can. J. Genet. Cytol. 11, 955-960.
Schröder, J.H. (1969e): Die Variabilität quantitativer Merkmale bei *Lebistes reticulatus* Peters nach ancestraler Röntgenbestrahlung. Zool. Beitr. N.F. 15, 237-265.
Schröder, J.H., Holzberg, S. (1972): Population genetics of *Lebistes (Poecilia) reticulatus* Peters (Poeciliidae; Pisces). I. Effects of radiation-induced mutations on the segregation ratio in post-irradiation F_2. Genetics 70, 621-630.

4. Chromosomes and Cytology

Chromosomes of Two Fish-Species of the Genus Diretmus (Osteichthyes, Beryciformes: Diretmidae)

A. Post

1. Introduction

In addition to an ichthyological program cytological studies were carried out during the 36th cruise of the FRS "Walther Herwig". The smear preparations of gonads dissected from adult males of 47 species of oceanic fishes furnished the material for these investigations.

With these preparations the chromosome cycle can be observed during spermatogenesis and the chromosome number can be ascertained in meiotic metaphase. In the metaphase of the meiosis the bivalents usually are arranged in the equatorial plane as shown in Fig. 1.

In species with the abundant haploid chromosome number of n = 24 (as in Fig. 1), the diameter of the metaphase plane measures to about 7 μ. Single bivalents are usually very small and equal in size. Chromosomes differing from this pattern in size and shape rarely have been observed before (for example in *Notobranchius rachowi* Ahl) (Post 1965).

Among the species prepared on board "Walther Herwig", there are two specimens belonging to the family Diretmidae. The shape and, moreover, the size and number of their chromosomes are conspicuously

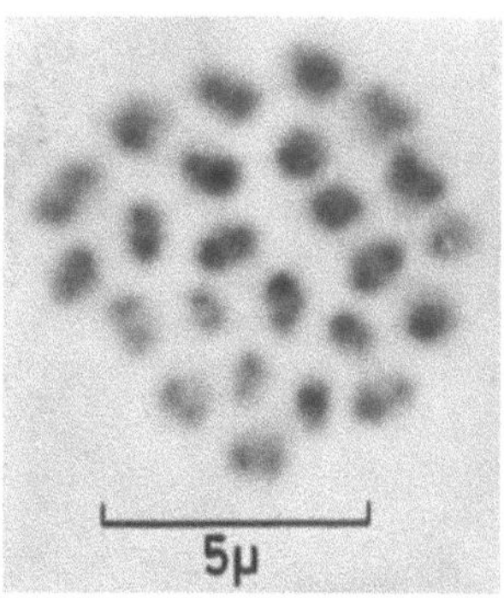

Fig. 1

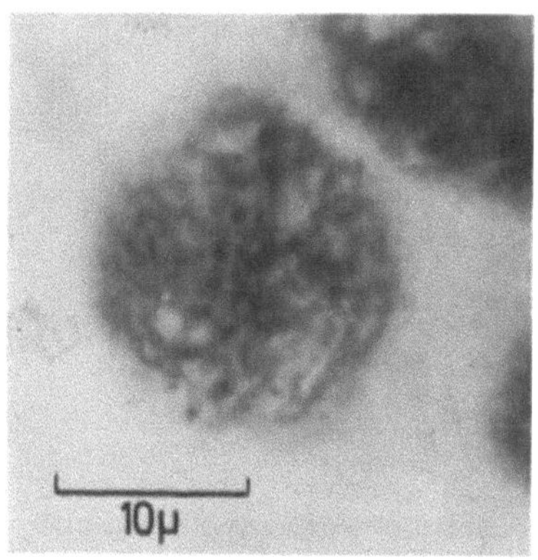

Fig. 2

Fig. 1. *Notoscopelus resplendens* (Myctophidae). Example of a meiotic metaphase-I plane showing n = 24 chromosomes. The diameter of the plane is about 7 μm

Fig. 2. *Diretmus argenteus*: late pachynema

different from all known fish-chromosomes. The basic structures of the two chromosome sets in question are quite similar. But there are some considerable differences in the number and shape of the chromosomes, indicating taxonomic differences between the two *Diretmus*-specimens.

2. Taxonomic Remarks

Until now the genus *Diretmus* was regarded to be monotypic, with its only species *Diretmus argenteus* Johnson, 1863. However, during the past several years some ichthyologists have suspected that at least two sympatrically living *Diretmus*-species exist, which have been confused for many years.

L.P. Woods, taking an interest in this problem, confirmed this suspicion to be justified.

I am indebted to Dr. Woods for his kindness to allow me the use of the results of his revision on the genus *Diretmus*, in advance of the publication on his paper in "Fishes of the Western North Atlantic 6".

According to the key given therein, most of the *Diretmus*-specimens of the ISH[1]-collection can easily be identified as either *Diretmus argenteus* or the other species which will be described in Woods' paper. Some specimens, however, do not belong either to one or the other species, but look like multiple dominant-recessive F_1-polyhybrids.

They combine alternate characters of both species and so represent a corresponding link between them.

Of the two specimens of which the chromosomes are described here, one belongs to *Diretmus argenteus*. The other one, however, is of indeterminate systematic position. In this paper it will be signified as *Diretmus* spec. C.

3. Materials and Methods

1. *Diretmus argenteus* stat. 471-III/71 WH, 10.4.1971
02°27'S 19°00'W, 657 - 592 m, SL 72.0 mm

2. *Diretmus* spec. C. stat. 502/71 WH, 18.4.1971
20°27'N 21°58'W, 1900 - 2100 m, SL 82.0 mm

Diretmus spec. C was the only *Diretmus*-specimen caught at station 502. At station 471-III 11 specimens of *Diretmus argenteus* were collected. In preparing and staining the gonad-tissue, I followed the methods of Hsu and Pomerate (1953) and Karbe (1961). Observations of spermatogenesis were carried out under the microscope (250 X and 1250 X). Microphotographs were taken with linear magnification of 500 X and 1250 X.

In *Diretmus argenteus* and in *Diretmus* spec. C two and three types, respectively, of chromosomes of different size were observed. These chromosomes will be indicated in this paper as

a) "giant" chromosomes = supermacro-chromosomes = SM-chromosomes
b) chromosomes of "normal" size = normal-chromosomes = N-chromosomes
c) "tiny" chromosomes = micro-chromosomes = M-chromosomes

4. Results

Diretmus argenteus

The median diameter of the spermatocyte-I-nucleus measures approximately 15 µm, i.e. about double the size of comparable phases in most of the other fish-species.

---

[1]Institut für Seefischerei, Hamburg.

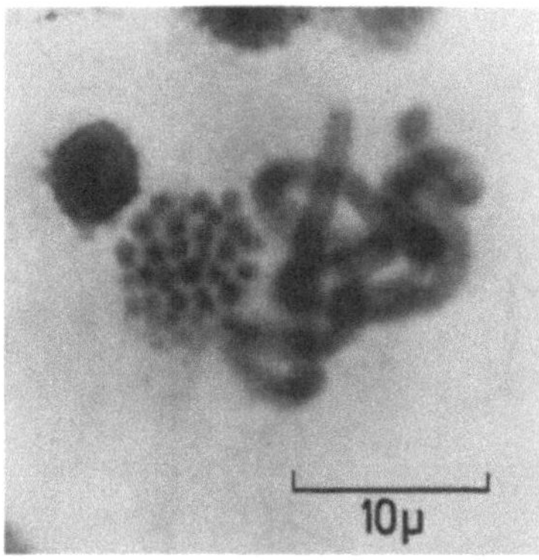

Fig. 3

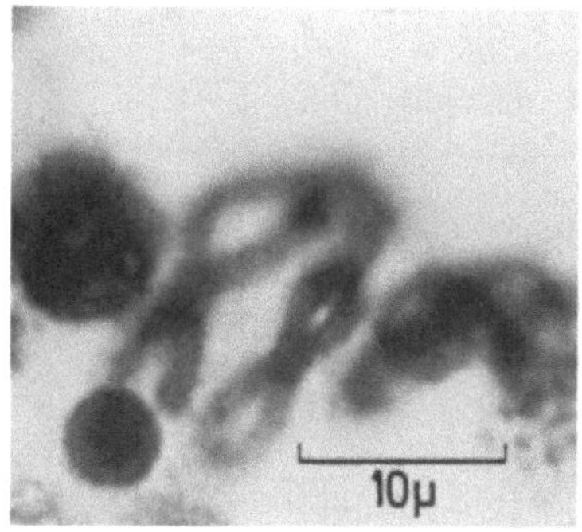

Fig. 4

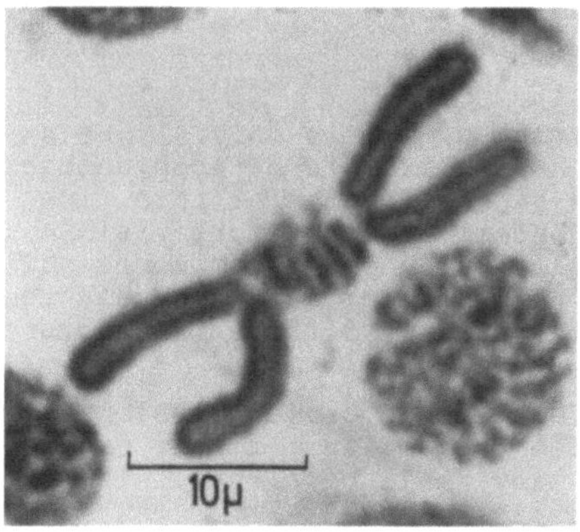

Fig. 5a

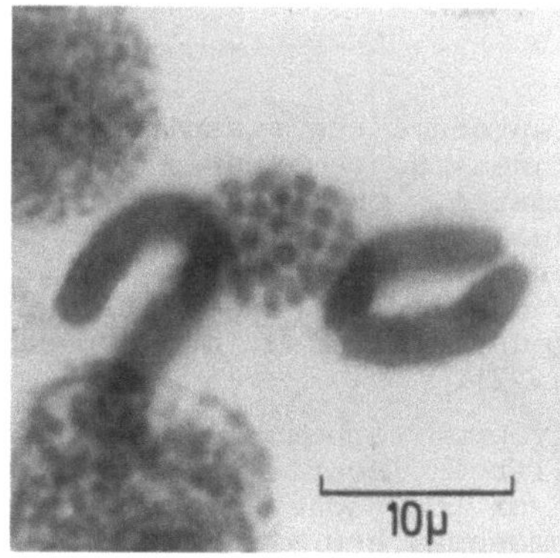

Fig. 5b

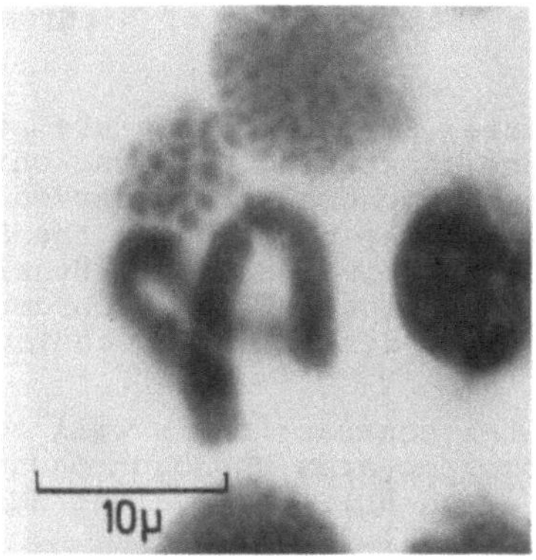

Fig. 6a

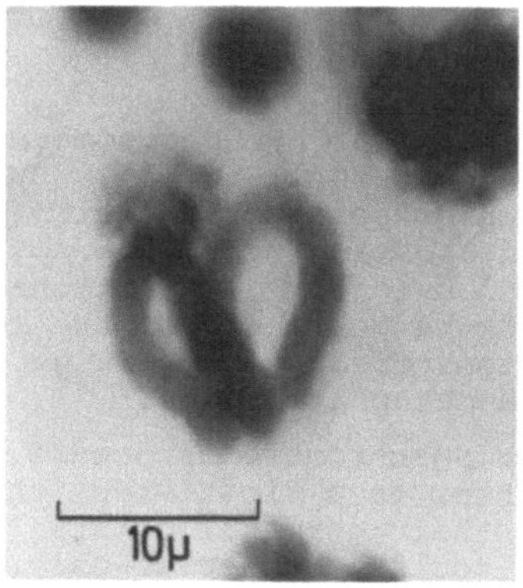

Fig. 6b

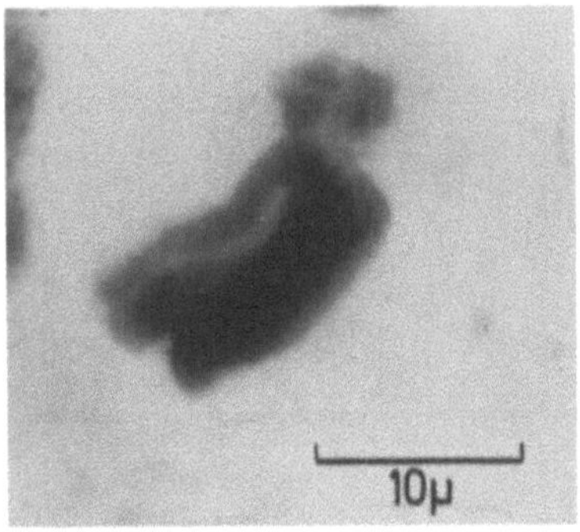

Fig. 7

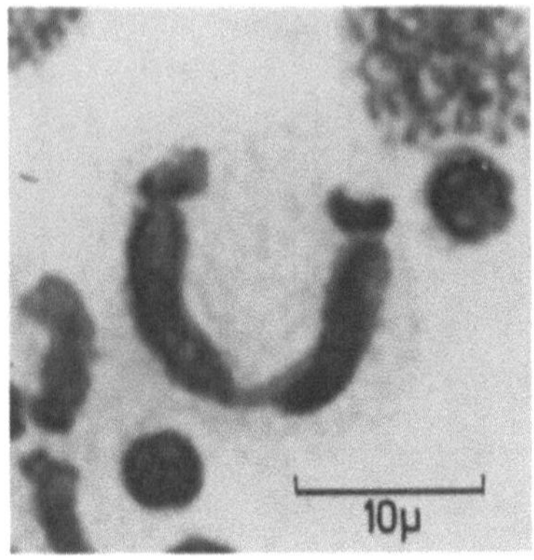

Fig. 8

Fig. 3-8. *Diretmus argenteus*: meiotic cycle in spermatocyte-I from early metaphase to anaphase. Subdivided numbers (a and b) indicate same developmental phases

In leptonema the staining of the spermatocyte-I is only slight and diffuse. No remarkable structures, as for example heteropycnotic chromatin, indicate different forms of chromosomes at this time. By gradual shortening, the chromosomes become more and more visible in zygonema and pachynema. While in early pachynema chromosomes still keep equally dispersed all over the nucleus, in late pachynema the nucleus is divided into two chromatin areas of different structure (Fig. 2).

Early prometaphase-I shows long coiled SM-chromosomes developed from the larger, coarse grained area, while the smaller, finely grained part of the nucleus decomposed into many small bivalents. These small chromosomes are arranged in an equatorial plane.

The equatorial plane and its single bivalents are comparable in size and shape to "normal" fish-chromosomes (Fig. 3).

The further meiosis of N-chromosomes probably is retarded by a diapause, while the SM-chromosomes continue spiralization and finally enter into synaptic phase. As shown in Figs. 3, 5 and 6 differently contracted SM-chromosomes are combined to the metaphase of N-chromosomes.

At the beginning of N-chromosome-metaphases, SM-chromosomes are still loosely coiled cords (Fig. 3). During the shortening process, chiasmata of the SM-chromosomes or the arms of one chromosome were observed (Fig.4). Linkage and crossing-over were not evident. The position of the SM-chromosomes obviously does not change during spiralization. Usually they lie at the periphery of the metaphase-I plane of the N-chromosomes in opposition to each other (Fig. 5a and 5b), directing the centromeres towards the metaphase plane.

Synapsis of SM-chromosomes occurs only when contraction is well advanced. Then, as a first step, the chromosome-arms approach each other from their terminal points (Fig. 6a and 6b). A bivalent SM-chromosome could be observed only once (Fig. 7).

In anaphase the SM-chromosomes detach, taking with them a haploid set of N-chromosomes (Fig. 8). In Fig. 8 the thickness of the chromatin is probably an artifact.

The number of N-chromosomes cannot be determined exactly: n = 18 to n = 38, mostly n = 20-26 chromosomes were counted. The mitotic metaphase of spermatogonia, which somewhat clearer, shows 2n = 42-44 chromosomes (Fig. 9a and 9b).

The most probable number of chromosomes in *Diretmus argenteus* is 2n = II + 42-44; n = I + 21-22.

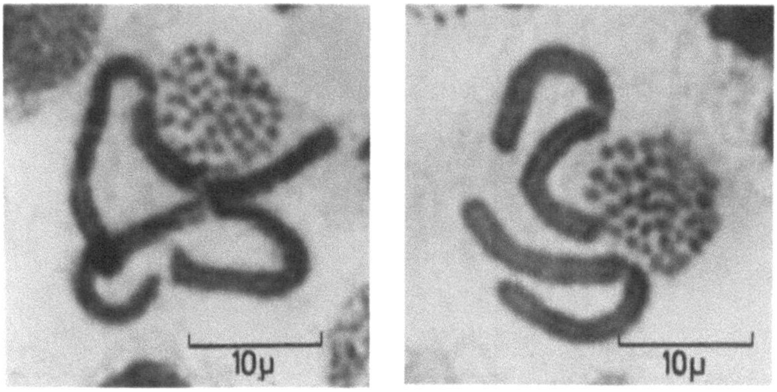

a b

Fig. 9. *Diretmus argenteus*: early mitotic metaphases

The photograph of the mitotic prometaphase in Fig. 9 distinctly shows two chromatids in each SM-chromosome and the centromeres being directed towards the metaphase plane of the N-chromosomes.

Diretmus spec. C

The median diameter of the spermatocyte-I-nucleus is 35 µ. The nuclei of spermatocytes-I in leptonema are not spherical or elliptical as they are in other fishes but unsymmetrically laced up (Fig. 10). The groove is S-shaped as can be seen from a horizontal view. It divides the nucleus into two differently structured and differently staining areas (Fig. 11). The groove extends during the next phase to form a ring around the surface of the nucleus, dividing the chromatin into two parts distinctly different in volume and density (Fig. 12).

From the smaller, slightly stained part of the nucleus, a metaphase plane of N- and M-chromosomes is built up. The larger, deeply stained part of the nucleus is transformed to a coiled ball of SM-chromosomes (Figs. 13 and 14).

The further process of shortening forms four metacentric SM-chromosomes.

The centromeres of these chromosomes are directed towards the metaphase plane of the smaller chromosomes which stay in a diapause until the SM-chromosomes finish the process of spiralization (Figs. 15 to 17).

The total number of chromosomes could not be counted exactly but is approximately n = 70.

No further meiotic phases could be observed, thus nothing can be said about the synapsis and homology of SM-chromosomes.

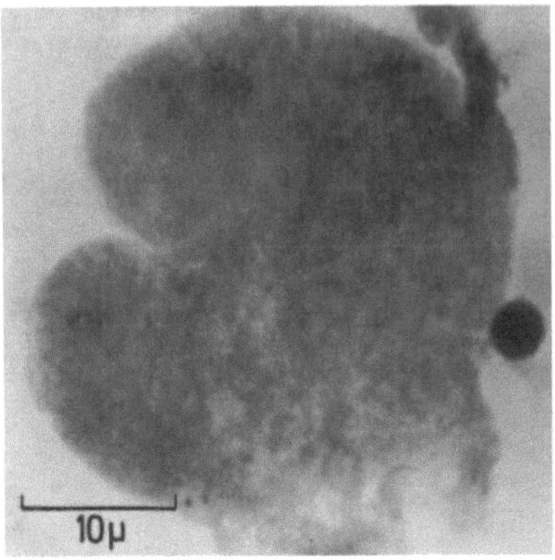

Fig. 10

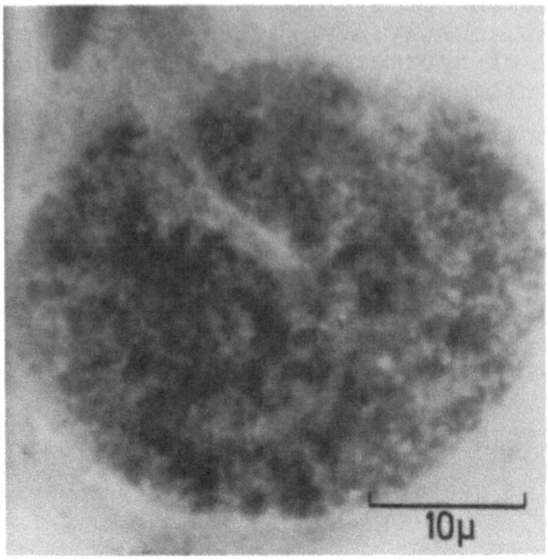

Fig. 11

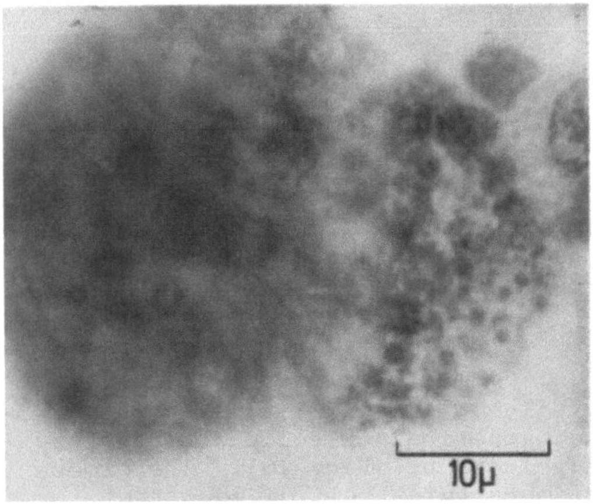

Fig. 12

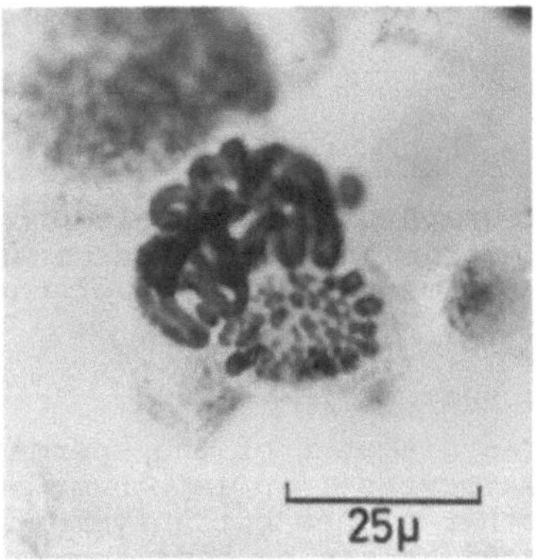

Fig. 13

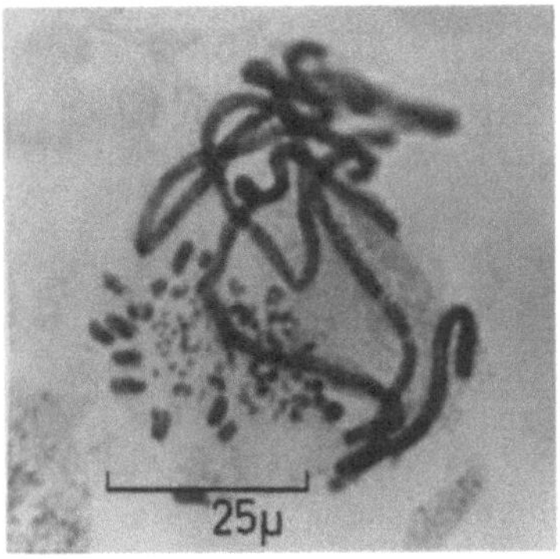

Fig. 14

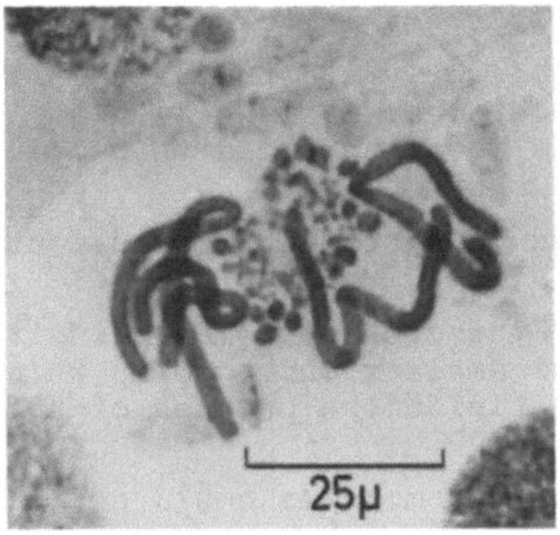

Fig. 15

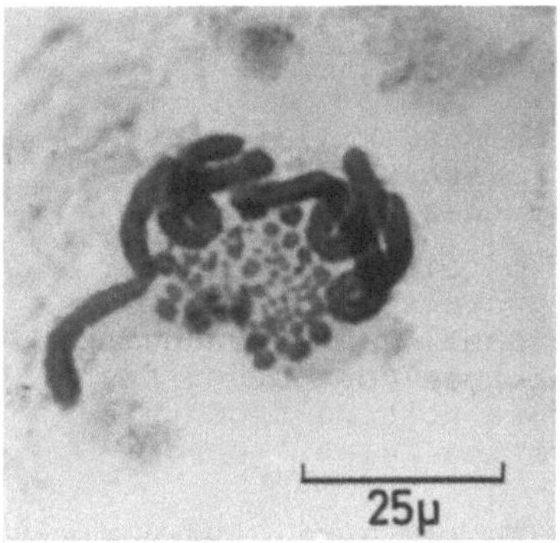

Fig. 16

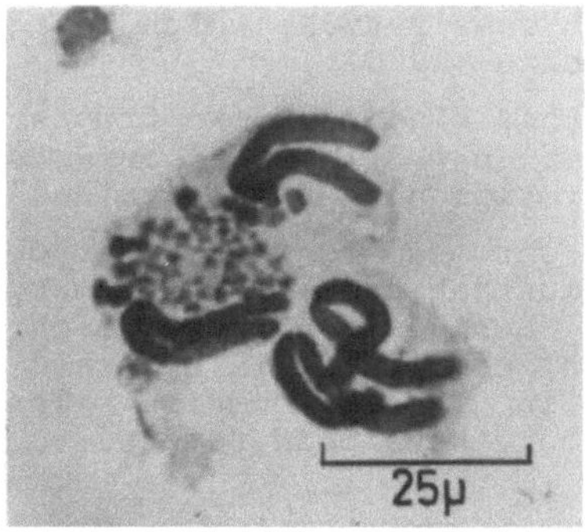

Fig. 17

Fig. 10-17. *Diretmus* spec. C: meiotic cycle from early pachynema to early metaphase

5. Discussion

The approximately 300 to 350 fish-species, of which the chromosomes have been described, include such rather primitive groups as for instance Cyclostomata as well as members of modern phyletic groups, e.g. Perciformes and Pleuronectiformes. So a relatively good survey of this vertebrate class was obtained.

Chromosome-numbers in fishes cover nearly the whole scope of chromosome-numbers in vertebrates, ranging from n = 8 in *Notobranchius rachowi* (Post 1965) to n = 66 in *Corydoras aneus* (Scheel et al. 1972) or n = 82 in *Ichthyomycon gagei* (Hubbs and Trautmann cit. after Howell and Duckett 1971). Within the whole range, peak frequencies of chromosome-numbers were found at n = 22 to 28 and n = 46 to 52.

Apart from microchromosomes in birds and some reptiles, fishes have the smallest chromosomes in vertebrates. All chromosome types rod- J- and V-shaped have been found even in the same single nucleus. Heteromorphic chromosomes were not often revealed in teleostean fishes. Except for a larger component in *Bathylagus wesethi* (Chen and Ebeling 1966) and *Megupsilon aporus* (Miller and Walters 1972) or several big elements in *Notobranchius*- and *Aphyosemium*-species, fish-chromosomes are not considerably different in size.

To my knowledge, in no fishes, differences between smallest and largest chromosomes are so striking as demonstrated herein for the two *Diretmus*-species.

At this moment nothing can be said about evolution and function of the unusual karyotype and its SM-chromosomes, in so far as no somatic cell divisions were obtained and only one male of each species was investigated.

Big chromosomes may result from diverse abnormalities of cell division, for example from centric fusions or translocations. Increase of chromosome volume by addition of chromatin consequently must be followed by decrease of chromosome number or volume of other chromosomes. Extreme diminution of chromosome number, probably caused by multiple fusions, is seen in the Indian muntjak (*Muntiacus muntjak*). This is a species of deer which has only few but very big chromosomes (2n = 6 in female,

2n = 7 in male) is fertile in offspring of cross-breedings with the closely related *Muntiacus reevesi* which has 2n = 46 chromosomes (Wurster and Benirschke 1970).

Another example of this kind is known from some related killifishes, *Notobranchius rachowi* (n = 8) and *Notobranchius guentheri* (n = 19), though no fertile hybrids have been obtained from these species.

The formation of SM-chromosomes in the karyotype of Diretmidae cannot be explained as in muntjak and killifishes by a process of fusions and translocations. Besides the large SM-chromosomes, a certain number of small chromosomes, comparable in size and number to most other fish-chromosomes, exists in *Diretmus*-species. Therefore nothing from this part of chromatin could have been removed to form the "giant" chromosomes.

Analogous to the karyotype of *Bathylagus wesethi* or *Megupsilon aporus*, the SM-chromosomes may be regarded as gonosomes which are mainly composed of heterochromatin. The staining used here is not suitable to differentiate euchromatin and heterochromatin, so as to give information over different substances in the nucleus.

Special heteropycnotic chromatin areas in leptonema, like Barr-bodies, which may indicate sex-chromatin in early meiotic phases, has not been observed.

The chromosome-number and the nuclear diameter in *Diretmus* spec. C is about double of that of *Diretmus argenteus*. Exactly the double number can be stated for the SM-chromosomes. Both the double number of chromosomes and diameter of nucleus support a theory which regards *Diretmus* spec. C to be tetraploid in relation to *Diretmus argenteus*. Contrary to this view stands the different shape of small chromosomes of the two species in question. Even if a process of tetraploidization played a part in the development of the species, further karyological evolution must have taken place to form M- and N-chromosomes in *Diretmus* spec. C. As mentioned above, from morphological aspects *Diretmus* spec. C must be placed systematically between the two *Diretmus*-species. Possibly *Diretmus* spec. C either is a hybrid of the two *Diretmus*-species which is not able to produce fertile gametes, considering that no meiotic phases beyond metaphase were observed, or the two different genomes were stabilized by allopolyploidy and formed a stable hybrid. The last hypothesis is not incompatible with the karyotype of *Diretmus* spec. C but assigns to the second *Diretmus*-species a chromosome pattern formed of two SM-chromosomes and a certain number of M-chromosomes.

To the best of my knowledge, this would be the first case of allopolyploidy in fishes.

To give the true interpretation of the chromosome pattern of *Diretmus* spec. C, it would be very helpful to know the karyotype of Woods' new *Diretmus*-species.

Another important question to be resolved is how the variance in volume of the nuclei, which are eight times as large in *Diretmus* spec. C as in *Diretmus argenteus*, effects the growth of the specimens of the different species.

6. Summary

During the 36th cruise of the FRS "Walther Herwig" to the South Atlantic chromosomes of two *Diretmus*-species have been examined: *Diretmus argenteus* Johnson, 1863, and an undescribed species, here called *Diretmus* spec. C. The latter seems to be intermediate between

D. argenteus and a new *Diretmus* species which L.P. Woods will describe in Vol. 6 of "Fishes of the Western North Atlantic".

The karyotype of the two species turned out to be extremely unusual for fishes and even other vertebrates. Two types of chromosomes appear during meiosis in *D. argenteus* and three types in *D.* spec. C. The chromosomes have been classified and described in this paper.

Some hypothetical aspects have been discussed on the formation of the large chromosomes by fusion, translocation or by the development of heterochromatic substances. Furthermore, aspects of the evolution of the karyotype of *D.* spec. C by tetraploidization or alloploidization in a constant hybrid have also been discussed.

References

Chen, T.R., Ebeling, A.E. (1966): Probable male heterogamety in the deep-sea fish *Bathylagus wesethi* (Teleostei: Bathylagidae). Chromosoma (Berl.) 18, 88-96.

Howell, W.M., Duckett, C.R. (1971): Somatic chromosomes of the Lamprey, *Ichthyomycon gagei* (Agnatha, Petromyzonidae) - Experientia (Basel) 27 (2), 222-223.

Hsu, T.C., Pomerate, C.M. (1953): Mammalian chromosomes in vitro II. A method for spreading the chromosomes in cells in tissue culture. J. Hered. 44, 23-29.

Johnson, J.Y. (1863): Description of three new genera of marine fishes obtained at Madeira. Proc. Zool. Soc. (London), 403-410.

Karbe, L. (1961): Cytologische Untersuchungen der Sterilitätserscheinungen bei anatolischen Zahnkarpfen, ein Beitrag zum Speziationsproblem. Mitt. Hamburg, Zool. Mus. Inst. 59, 73-104.

Miller, R.R., Walters, V. (1972): A new genus of cyprinodontid fish from Nuevo Leon, Mexico. Contr. Sci., Nat. Hist. Mus., Los Angeles County 233, 1-13.

Post, A. (1965): Vergleichende Untersuchungen der Chromosomenzahlen bei Süßwasser-Teleosteern. Z. zool. Syst. Evol.-Forsch. 3, 47-93.

Post, A. (1972): Ergebnisse der Forschungsreisen des FFS "Walther Herwig" nach Südamerika. XXIV. Die Chromosomenzahlen einiger atlantischer Myctophidenarten (Osteichthyes, Myctophoidei, Myctophidae). Arch. Fischereiwiss. 23(2), 89-93.

Scheel, J.J., Simonsen, V., Gyldenholm, A.O. (1972): The karyotype and some electrophoretic patterns of fourteen species of the genus *Corydoras*. Z. zool. Syst. Evol.-Forsch. 10(2), 144-152.

Woods, L.P.: Family Diretmidae. In press. In: Fishes of the Western North Atlantic. Mem. Sears-Found. mar. Res., No. 1, Pt. VI.

Wurster, D., Benirschke, K. (1970): Indian muntjak, *Muntiacus muntjak*: a deer with a low diploid chromosome number. Science 168, 1364-1366.

The Chromosomes of Some African Anuran Species

J. J. Scheel

Rather little information concerning the karyotypes (chromosome complements) of African anuran species has been published. The present note concerns the results of my own study together with karyotypic information published by others.

In the table the dipoid number (2n) is given for all species, together with information concerning the gross structure of the karyotypes. The symmetry or asymmetry of the elements is indicated by the figures under "S1", which give the total length of the short arms of two-armed elements. S1 = 50 (%) indicates the highest grade of symmetry, i.e. all elements have arms of equal length. Increasing asymmetry is indicated by decreasing figures of S1. One-armed or telocentric chromosomes were not seen in these karyotypes, but such elements do exist in other anuran karyotypes.

The symmetry of the complement is indicated by the figures for "S2", which gives the ratio of length of the longest and the shortest element of the complement. Low figures, close to 1.0 indicate symmetrical complements with elements of almost equal size.

In many anuran karyotypes the elements distinctly separate into two groups by length. Such grouping is indicated by high figures for "S3", which gives the ratio of length of the shortest of the group of long elements and the longest of the group of short elements.

1. Discussion

The karyotypes of *Xenopus laevis* and of *X. fraseri* have more elements than the other complements. Also, there is no distinct grouping by length. The karyotype is rather symmetrical, but many of the elements are distinctly asymmetrical. These two karyotypes undoubtedly are rather ancient or basic compared to the other complements of this study. On the basis of these karyotypes or similar basic ones most karyotypes of frogs and toads (?) are easily constructed. First pericentric inversions replace the centromeres of some of the two-armed elements to a terminal position, and next the one-armed chromosomes fuse two and two to produce large, often rather symmetrical, two-armed elements (J.J. Scheel, 1971a).

This type of evolution of karyotypes is characterized by an initial decrease of S1 (by pericentric inversions) followed by a sudden increase of both S3 and S1 (centric fusions). S2 most probably remains almost constant.

The karyotype of *Hymenochirus* is very different from those of *Xenopus* and appears to represent a highly specialized complement. It resembles that of *Bombina bombina*.

The karyotypes of African *Bufo* and of *Nectophrynoides* are very similar and exhibit a distinct grouping of the elements by size. The karyotype of *Nectophryne* has a less marked grouping and seems to be closer to some bufoids of the New World (*Bufo marinus* a.o.).

The karyotype of *Petropedetes* resembles that of European *Rana* (n = 13 group), but differs by some traits. The large elements of the former are more asymmetrical than those of the latter and the long arm of the longest element of the former has a marked secondary constriction. The karyotype of *Hylarana* is also of the ranid type, but in one of two specimens 2n = 27 and not 2n = 26 was discovered. The unmatched element is the smallest of the complement and is a metacentric. It may represent a sex chromosome.

The karyotype of *Ptycadena* is different. The elements make a rather evenly graduated series by length, and some of the smaller chromosomes have secondary constrictions. The character last mentioned is even more marked in the karyotype of *Discoglossus* in which all elements have several secondary constrictions. These two karyotypes may be close, but they are very different from those of *Rana* and its allied.

The karyotype of *Arthroleptis* most probably represents the peak of specialization (of karyotypes) of the present study. It most probably derived from the karyotype of *Rana* or from a similar complement.

The karyotypes of the 21 species of *Leptopelis*, *Hyperolius* and *Afrixalus* are very similar and (by some traits) different from the other karyotypes. All have 12 haploid elements. These karyotypes differ from those of *Rana* and its allied by the ratio of length (S2). The longest elements of the former group are shorter than the corresponding elements of the latter group, and the shortest element of the former is longer than the corresponding element of the latter. Except for the longest element, which differentiates from the other elements by its length, the other elements make a rather evenly graduated series by length. If the basic karyotype of the Hyperolidae has derived from a n = 12 karyotype of *Rana* or from a similar karyotype, an exchange of arms between long and short elements probably took place. If the longest element of the ranid karyotype exchanges arms with the shortest element, two highly asymmetrical elements will develop. There are three to four such elements in the hyperolid karyotype and the highest degree of asymmetry occurs in *Leptopelis*, the lowest degree in *Hyperolius pictus*, *H. parallelus* and *H. nasutus*. Also by other karyotypic features, the complement of *Leptopelis* appears to be closer to *Rana* than the other forms.

The karyotypes of *Afrixalus* and of *Hyperolius* are very similar. A close study of the morphology of the small elements brings about certain differences. In some of the karyotypes one of these elements is subtelocentric whereas in other karyotypes it is metacentric or nearly so. In several karyotypes this particular element has a distinct secondary constriction. This element is subtelocentric in the species nos. 21, 22, 24, 28 and 35 (Table). It has a satellite on the long arm in the nos. 28 and 35. The element is symmetrical and with no secondary constriction in the nos. 23, 25, 31, 32, 34 and 41. The secondary constriction is close to the centromere in the nos. 30 and 40, and not quite so close in the nos. 26, 27 and 33. The constriction is rather terminal in the no. 29 and is represented by a satellite in the nos. 36 and 37. All these replacements of the centromere and of the secondary constriction probably were caused by peri- and paracentric inversions, which consequently were rather numerous during the evolution of the karyotype of the Hyperolidae, at least in this particular element. The absence of secondary constrictions in the other elements hinders the demonstration that inversions also acted frequently on the other chromosomes of the basic karyotype of this family. The different grades of asymmetry of the long elements have been mentioned and also indicate that inversions took place in these elements.

	2n	S1	S2	S3	Author
PIPIDAE:					
1. Xenopus laevis	36	34	2.1	1.2	A.Morescalchi, 1968a J.J.Scheel, 1971b
2. Xenopus fraseri	36	30	2.6	1.3	J.J.Scheel, 1971b
3. Hymenochirus boettgeri	22	34	5.9	1.8	J.J.Scheel, 1971b
- -	24	?	?	?	A.Morescalchi, 1968a
BUFONIDAE:					
4. Bufo brauni	20	SY	H	H	J.P.Bogart, 1968
5. Bufo carens	22	SY	H	H	J.P.Bogart, 1968 A.Morescalchi, G.Gargiulo, 1968
6. Bufo gariepensis	22	SY	H	H	J.P.Bogart, 1968
7. Bufo garmani	20	SY	H	H	J.P.Bogart, 1968 A.Morescalchi, G.Gargiulo, 1968
8. Bufo gutturalis	20	SY	H	H	J.P.Bogart, 1968
9. Bufo mauritanicus	22	SY	H	H	J.P.Bogart, 1968 A.Morescalchi, G.Gargiulo, 1968
10. Bufo preussi	22	44	6.0	1.6	J.J.Scheel, 1971b
11. Bufo rangeri	20	SY	H	H	J.P.Bogart, 1968 A.Morescalchi, G.Gargiulo, 1968
12. Bufo regularis	20	SY	H	H	J.P.Bogart, 1968 A.Morescalchi, G.Gargiulo, 1968
13. Bufo rosei	22	SY	H	H	J.P.Bogart, 1968
14. Nectophryne afra	22	39	5.8	1.4	J.J.Scheel, 1970
15. Nectophrynoides tornieri	22	42	3.8	1.5	J.J.Scheel, 1971b
RANIDAE:					
16. Arthroleptis peocilonotus	14	43	2.5	1.4	J.J.Scheel, 1971a
17. Discoglossus occipitalis	26	SY	H	H	J.J.Scheel, 1971b
18. Hylarana albolabris	27+	40	5.7	1.5	J.J.Scheel, 1971b
19. Petropedetes species	26	35	4.7	1.8	J.J.Scheel, 1971b
20. Ptycadena species	24	43	4.4	1.3	J.J.Scheel, 1971b
HYPEROLIDAE:					
21. Afrixalus congicus	24	39	2.9	1.3	J.J.Scheel, 1971b
22. Afrixalus dorsalis	24	39	3.5	1.3	J.J.Scheel, 1971b
23. Africalus f.fulvovittatus	24	41	3.0	1.2	J.J.Scheel, 1971b
24. A.fulvovittatus leptosoma	24	37	3.1	1.3	J.J.Scheel, 1971b
25. Afrixalus uluguruensis	24	41	3.2	1.2	J.J.Scheel, 1971b
26. Hyperolius concolor	24	38	3.2	1.3	J.J.Scheel, 1971b
27. Hyperolius kivuensis	24	35	3.2	1.3	J.J.Scheel, 1971b
28. Hyperolius nasutus	24	39	2.7	1.1	J.J.Scheel, 1971b
29. H.affinis steindachneri	24	39	3.4	1.3	J.J.Scheel, 1971b
30. Hyperolius ocellatus	24	40	3.1	1.3	J.J.Scheel, 1971b
31. Hyperolius parallelus	24	40	2.6	1.2	J.J.Scheel, 1971b
32. Hyperolius picturatus	24	41	3.1	1.2	J.J.Scheel, 1971b
33. Hyperolius pictus	24	42	3.7	1.3	J.J.Scheel, 1971b
34. Hyperolius platyceps	24	39	2.3	1.3	J.J.Scheel, 1971b
35. Hyperolius aff pusillus	24	36	3.6	1.2	J.J.Scheel, 1971b
36. Hyperolius aff bolifambae	24	38	2.9	1.3	J.J.Scheel, 1971b
37. Hyperolius viridiflavus	24	40	2.9	1.2	J.J.Scheel, 1971b
38. Hyperolius argentovittis++	24	SY	2.5	1.1	A.Morescalchi, 1968b

	2n	S1	S2	S3	Author
39. Kassina senegalensis	24	SY	2.5	1.2	A.Morescalchi, 1968b
40. Leptopelis gramineus	24	35	4.9	1.3	J.J.Scheel, 1971b
41. Leptopelis concolor ssp.indit	24	?	?	?	J.J.Scheel, 1971b

+The unmatched chromosome of one of the specimens of *Hylarana albolabris* is the smallest element and may represent a sex chromosome

++Probably misidentified

SY = symmetrical
H = relatively high figure

The analysis of the karyotypes of Hyperolidae indicates that a certain grouping of the different species on the basis of the structure of the small element above mentioned is possible, and that the genus *Leptopelis* appears to separate from the other two genera. On the basis of the present karyotypic informations *Afrixalus* and *Hyperolius* cannot be separated.

The specimens originated from the following countries:

Cameroon: 2, 10, 16, 17, 18, 19, 20, 21, 24, 29, 30, 34 and 36.
Ethiopia: 28, 37, 40 and 41.
Ferando Poo: 14.
Liberia: 22, 23, 26 and 32.
Tanzania: 15, 25, 33 and 35.
Zambia: 27 and 31.

Specific localities will be presented elsewhere.

References

Bogart, J.P. (1968): Chromosome number difference in the amphibian genus *Bufo*: the Bufo regularis species group. Evolution 22, 42-45.

Morescalchi, A., Gargiulo, G. (1968): Su alcune relazioni cariologiche del genere *Bufo* (Amphibia Salientia). Rc. Accad. Sci. fis. mat. Napoli.(4) 35, 117-120.

Morescalchi, A. (1968a): I chromosomi di alcuni *Pipidae* (Amphibia Salientia). Experientia 24, 81-82.

Morescalchi, A. (1968b): Initial cytotaxonomic data on certain families of amphibious *Anura* (Diplasiocoela, after Noble). Experientia 24, 280-283.

Scheel, J.J. (1970): Notes on the biology of the African Tree-toad *Nectophryne afra* Buchholz and Peters, 1875 (Bufonidae, Anura) from Fernando Poo. Rev. Zool. Bot. Afr. 81, 225-236.

Scheel, J.J. (1971a): The seven-chromosome karyotype of the African frog *Arthroleptis*, a probable derivative of the thirteen-chromosome karyotype of *Rana*. Hereditas 67, 287-290.

Scheel, J.J. (1971b): Unpublished cytological findings.

Incorporation of H^3-Thymidine into the DNA of Melanotic and Amelanotic Melanoma Explants after Whole-Body X-Irradiation of Melanoma-Bearing Platyfish-Swordtail Hybrids[1]

J. Vielkind, D. L. Pursglove, and U. Vielkind

1. Introduction

In higher organisms, two main biological effects of X-rays have been observed which have quite opposite results. In most cases, irradiation with X-rays causes a drastic reduction in cell proliferation and tumor growth, while in other cases this treatment gives rise to an accelerated proliferation of cells which results in tumor formation (for review, see Little, 1968).

The spot patterns of platyfish and the premelanomas and melanomas of platyfish-swordtail hybrids, which had been subject of numerous studies during the last four decades (Gordon, 1959; for last review, see Anders et al., 1972), provide a useful in vivo system for experiments with X-irradiation, since it is possible to study separately tumor induction and tumor regression within a known genetically determined background. Those cases, in which premelanoma or melanoma formation could be induced by X-irradiation, have been interpreted as a result of mutations of a highly polygenic control system which normally represses the expression of certain tumor-inducing macromelanophore genes (Anders et al., 1973); those cases in which temporary regression of melanoma or premelanoma growth occurred were not attributed to genetic changes but to physiological damage of the tumor cells (Pursglove et al., 1971).

It has been observed that regression or discontinuation of melanoma growth occurred in about 50% of the irradiated albino hybrids but only in about 30% of the corresponding irradiated pigmented hybrids (Pursglove, 1972). This indicates that the amelanotic melanomas of albino hybrids, which are known to grow very rapidly, are more injured by X-rays than the melanotic melanomas of pigmented hybrids, which are known to grow also rapidly but not as fast as the amelanotic ones. It seems, therefore, that a subject suffering from a more rapidly growing tumor, e.g. an amelanotic melanoma, has a higher chance to be cured by radiotherapy than a subject suffering from a less rapidly growing tumor, e.g. a melanotic melanoma.

From the studies mentioned above, it is expected that the observed regression of melanoma growth caused by X-irradiation is reflected by a decrease in DNA synthesis, whereby DNA synthesis in amelanotic melanomas should be more affected than that in melanotic ones. The present study on H^3-thymidine incorporation into melanotic and amelanotic melanoma explants from irradiated and non-irradiated fish confirms this statement.

---

[1]This research has been supported by grants from Deutsche Forschungsgemeinschaft given to Dr. F. Anders, Giessen. For radioactive measurement, a liquid scintillation counter was placed to the authors' disposal by the Institut für Biophysik, Strahlenzentrum, Giessen.

2. Materials and Methods

The experiments were carried out with rapidly growing melanotic melanomas and very rapidly growing amelanotic melanomas of platyfish-swordtail hybrids. Only melanomas caused by the gene *Sd* (spotted dorsal), which is responsible for macromelanophore differentiation and tumor formation within the dorsal fin, were used in this study.

Part of the pigmented hybrids carrying melanotic melanomas and part of the albino hybrids carrying amelanotic melanomas were whole-body X-irradiated with 1 000 R (for detail, see Pursglove, 1972). 2 hours later, melanomas of the irradiated fish as well as melanomas of non-irradiated control fish were cut off and incubated at 27°C in 1 ml of a sterile medium for various periods of time[2]. The medium contained 15 μCi methyl-H^3-thymidine per ml (specific activity, 5 Ci/mM) and consisted of M199 plus 20% calf serum, pH 7.4, diluted 8:3 with distilled water and supplemented with 100 units of streptomycin and 100 units of penicillin per ml.

The incorporation of labelled thymidine into the melanoma DNA was followed by determination of radioactivity in the DNA extracted from the melanoma explants after various times of incubation. The explants were homogenized in cold trichloroacetic acid, centrifuged, and the sediment extracted with ethanol and ether. For quantitative extraction of DNA, the sediment was hydrolyzed in perchloric acid for 20 minutes at 70°C. The amount of DNA in the perchloric acid hydrolysate was measured by a modified diphenylamine reaction (Burton, 1956) and the radioactivity by using a Philips liquid scintillation counter.

3. Results

The incorporation of H^3-thymidine into the DNA of melanotic and amelanotic melanoma explants from irradiated and non-irradiated fish is demonstrated in Fig. 1a and b, respectively. It is obvious that the amount of labelled thymidine incorporated into DNA of melanotic melanomas was significantly lower than that found in the corresponding amelanotic melanomas. In addition, melanotic and amelanotic melanoma explants from irradiated fish showed a quite lower incorporation than explants from non-irradiated control fish. Whereas the labelled thymidine incorporated into DNA of melanomas from control fish rose significantly with increasing incubation times, it was only slightly increasing during the first few hours of incubation in melanomas of irradiated fish. After about 10 hours in the case of melanotic melanomas, and after about 16 hours in the case of amelanotic melanomas, there seemed to be no further incorporation of radioactive material into the melanoma DNA of irradiated fish. At the end of the incubation period, the amount of H^3-thymidine incorporated into DNA of melanotic melanomas of irradiated fish was about one third of that obtained for control fish, and in the case of amelanotic melanomas it was only one fourth of the control value.

---

[2]In the case of very large melanomas (melanomas of about more than 50 mg fresh weight), the melanomas were divided into two equal halves; one half was incubated in normal medium, while the other half was incubated in medium supplemented with additional amounts of amino acids. The results of this experiment are published elsewhere (Vielkind et al., 1973).

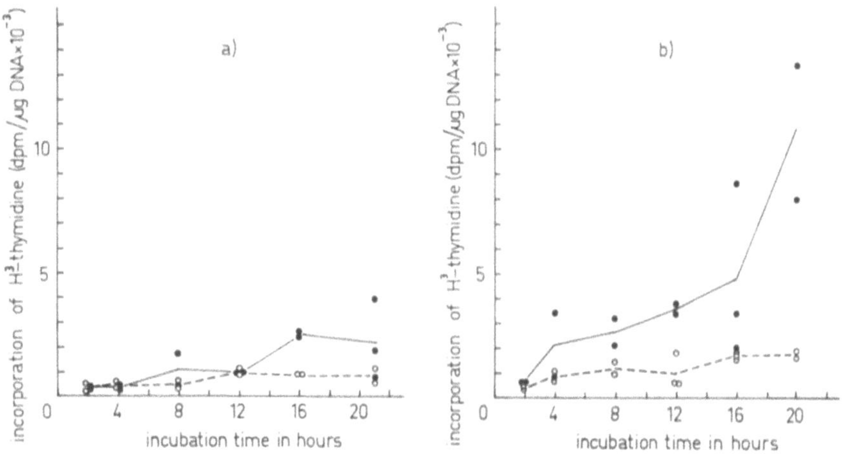

Fig. 1a and b. Effect of whole-body X-irradiation on the incorporation of H^3-thymidine into the DNA of melanotic (a) and amelanotic (b) melanoma explants cultivated in H^3-thymidine-containing medium. •——• melanoma explants from non-irradiated fish. o----o melanoma explants from irradiated fish

4. Discussion

Since there exists a correlation between the growth rate of a tumor and the rate of DNA synthesis within the tumor cells (Lea et al., 1966; Pierce and Wallace, 1971), the rate of H^3-thymidine incorporation into a tumor DNA has been used as a measure to estimate the growth rate of the tumor (Lea et al., 1966). As judged from their higher content of incompletely differentiated cells (Vielkind et al., 1971), the higher DNA to protein ratio (Vielkind et al., 1973), and the increased tendency to metastasize (Greenberg and Kopac, 1963), the amelanotic melanomas of albino platyfish-swordtail hybrids grow more rapidly than melanotic melanomas of pigmented hybrids. They are, therefore, expected to incorporate more H^3-thymidine into their DNA during a given period of time than melanotic melanomas do.

In fact, the present results clearly indicate a higher rate of H^3-thymidine incorporation into amelanotic melanoma DNA than into melanotic melanoma DNA. Thus, the method to determine the amount of labelled thymidine incorporated into the tumor DNA, in order to estimate the growth rate of the tumor, is useful also in the case of platyfish-swordtail melanomas and can be employed to study the effect of X-irradiation on the growth of these melanomas. The drastic suppression of H^3-thymidine incorporation into melanoma DNA caused by whole-body X-irradiation of melanoma-bearing hybrids can be interpreted as an inhibition of melanoma growth. Because of the more serious suppression of label incorporation into amelanotic melanomas, these melanomas can be regarded to be more sensitive to growth inhibition by X-rays than the melanotic ones.

This inhibition might be due to temporary inhibition of DNA replication as well as to irreversible cellular damage and subsequent cell loss. There are numerous investigations which have shown that most cells appear maximally sensitive to X-rays during mitosis and early

S phase of the cell cycle, while they are resistant during G_1 and late S or early G_2 phase (for review, see Little, 1968). Since the radiosensitivity of tissues depends upon the mitotic activity of their cells, the amelanotic melanomas, which are composed of rapidly proliferating cells, are more inhibited in growth by X-irradiation than the melanotic melanomas, which are composed of less proliferating cells. Besides inhibition of DNA synthesis, X-irradiation favors premature differentiation of cells leading also to growth inhibition (Evans, 1965; von Wangenheim, 1970). In addition, X-rays cause inhibition of growth by the induction of irreversible cytoplasmic damage as the destruction of mitochondria (Schäfer and Wendt, 1967) or the activation of lysosomes which results in degradative changes within both cytoplasm and nuclei of irradiated cells (Brandes et al., 1967).

The inhibition of melanoma growth after whole-body X-irradiation of melanoma-bearing fish probably results from a combination of decreased DNA synthesis and cell division with some of the other effects of X-rays described above. In the present experiments, this inhibition is expressed in a decreased incorporation of H^3-thymidine into melanoma DNA. In order to investigate also the morphological changes induced by X-irradiation within the melanoma cells, light and electron microscopic studies are in progress.

The stronger inhibition of thymidine incorporation into DNA of amelanotic melanomas than into melanotic melanomas confirms the earlier observation that there is a difference in radiosensitivity between melanotic and amelanotic melanomas (Pursglove, 1972). Since the amount of pigment present in the cells influences only the resistance against UV-irradiation but not against X-irradiation (Barranco et al., 1971), this result can be best explained by the higher amount of rapidly dividing cells present in amelanotic albino melanomas and by the increased tendency of these cells to undergo autophagy (Vielkind et al., 1971) rather than by a lack of protection by melanin.

5. Summary

Melanoma-bearing platyfish-swordtail hybrids have been whole-body irradiated with X-rays, and the H^3-thymidine incorporation into the DNA of melanoma explants has been determined after various times of cultivation. While melanomas of non-irradiated control fish continued to incorporate labelled thymidine into DNA during the whole cultivation period of 20 hours, the melanomas of X-irradiated fish showed a drastic inhibition of H^3-thymidine incorporation already after few hours of cultivation. This suppression of DNA synthesis, which was more distinct in amelanotic melanomas of albino hybrids than in melanotic melanomas of pigmented hybrids, is consistent with the finding that X-irradiation of melanoma-bearing fish causes a temporary regression of melanoma growth. The data also confirm that the two melanoma types differ in growth rate and radiosensitivity. Due to the higher content of proliferating cells, the rapidly growing amelanotic melanomas incorporated more labelled thymidine into DNA and are more radiosensitive than the less rapidly growing melanotic melanomas.

References

Anders, A., Anders, F., Klinke, K. (1973): Regulation of gene expression in the Gordon-Kosswig melanoma system. I. The distribution of the controlling genes in the genome of the xiphophorin fish, *Platypoecilus maculatus* and *Platypoecilus variatus*. This issue.

Anders, F., Klinke, K., Vielkind, U. (1972): Genregulation und Differenzierung im Melanom-System der Zahnkärpflinge. Biologie in unserer Zeit 2, 35-45.
Barranco, S.C., Romsdahl, M.M., Humphrey, R.M. (1971): The radiation response of human malignant melanoma cells grown in vitro. Cancer Res. 31, 830-833.
Brandes, D., Sloan, K.W., Anton, E., Bloedorn, F. (1967): The effect of X-irradiation on the lysosomes of mouse mammary gland carcinomas. Cancer Res. 27, 731-746.
Burton, K. (1956): A study of the conditions and mechanisms of the diphenylamine reaction for the colorimetric estimation of deoxyribonucleic acid. Biochem. J. 62, 315-323.
Evans, H.J. (1965): Effects of radiations on meristematic cells. Radiation Botany 5, 171-182.
Gordon, M. (1959): The melanoma cell as an incompletely differentiated pigment cell. In: Pigment Cell Biology (ed. M. Gordon), Academic Press, New York, 215-239.
Greenberg, S.S., Kopac, M.J. (1963): Studies of gene action and melanogenic enzyme activity in melanomatous fishes. Ann. N.Y. Acad. Sci. 100, 887-923.
Lea, M.A., Morris, H.P., Weber G. (1966): Comparative biochemistry of hepatomas. VI. Thymidine incorporation into DNA as a measure of hepatoma growth rate. Cancer Res. 26, 465-469.
Little, J.B. (1968): Cellular effects of ionizing radiation. New Engl. J. Med. 278, 308-315 and 369-376.
Pierce, G.B., Wallace, C. (1971): Differentiation of malignant to benign cells. Cancer Res. 31, 127-134.
Pursglove, D.L. (1972): Änderungen in der Farbgenmanifestation und Entstehung von Melanomen nach Röntgenbestrahlung bei lebengebärenden Zahnkarpfen (Teleostei, Poeciliidae). Dissertation, Universität Giessen.
Pursglove, D.L., Anders, A., Döll, G., Anders, F. (1971): Effects of X-irradiation on the genetically-determined melanoma system of xiphophorin fish. Experientia 27, 695-697.
Schäfer, D., Wendt, E. (1967): Das Verhalten der Mitochondrien gezüchteter Zellen nach sublethalen Strahlendosen. Verh. Dtsch. Zool. Ges. in Heidelberg, 722-734.
Vielkind, J., Vielkind, U., Anders, F. (1971): Melanotic and amelanotic melanomas in xiphophorin fish. Cancer Res. 31, 868-875.
Vielkind, J., Vielkind, U., Anders, F. (1973): DNA synthesis in genetically determined melanomas of platyfish-swordtail hybrids. Z. Krebsforsch. 80, in press.
Wangenheim, K.H. von (1970): Genetische Defekte und Strahlenschädigung. Radiation Botany 10, 469-490.

Fate of Bacterial DNA Injected into Embryos of Poeciliid Fish[1]

J. Vielkind, M. Schwab, and F. Anders

1. Introduction

Since Avery, MacLeod and McCarty (1944) and later on other investigators (for review, see Hotchkiss, 1952; Schaeffer, 1964) could successfully demonstrate that genetic transformation in bacteria is caused by DNA externally added to bacterial cultures, many experiments had been done to achieve genetic transformation in higher organisms also. Whereas the results of the first experiments carried out by Benoit et al. (1957) on ducklings were judged to be positive, the findings of further experiments on the same as well as on other subjects could not be interpreted as genetic transformation (for review, see Ledoux, 1965). In the following years, however, it became possible to do genetic transformation of eukaryotic cells at least in tissue culture (Szybalska and Szybalski, 1962) and, only recently, in some cases also in intact higher organisms (for review, see Hess, 1972; collected literature in Ledoux, 1971, and in Raspé, 1972). It had been possible, thereby, to show that the applied DNA was integrated into the host cell genome. But up to now, nobody had been able to show simultaneously that the integrated DNA was transcribed and translated, nor had it been proved that a new phenotype which occurred after DNA treatment was the result of integration, transcription, and translation, of the extraneous DNA. Nevertheless, transformation experiments had become of special interest not only for the development of genetic therapies or genetic change but also for the solution of some basic biological problems as genetic recombination, provirus insertion, and genetic inversion.

Many investigations on genetic transformation in eukaryotes are, therefore, in progress, although one is faced with numerous difficulties. Because of the complexity of higher organisms, the cells to be transformed cannot be brought into contact with the donor DNA as easily as in bacterial or tissue cell cultures. In most cases, the donor DNA has to be transported over relative wide distances underlying thereby the action of nucleases and has also to overcome numerous membraneous barriers. For this reason and because of the low relative concentration of single specific genes within the genome, the frequency of transformation events in higher organisms is expected to be even lower and more difficult to detect than in tissue cell cultures, where the frequency of transformed cells is about 10^{-4} and the transformants can be easily detected on selection media.

It is inevitable, therefore, to look for a system in which the difficulties mentioned can be reduced to a minimum. In the follow-

[1]This investigation was supported by grants from Deutsche Forschungsgemeinschaft and Stiftung Volkswagenwerk given to F. Anders. - Radioactive measurement was carried out either with a Philips liquid scintillation counter of the Institut für Biophysik, Strahlenzentrum, Giessen, or with a Packard Tricarb liquid scintillation counter of the Institut für Humangenetik, Giessen.

ing, the macromelanophore gene system of certain poeciliid fish will be introduced, and its advantages for transformation experiments will be cited.

The life-breeding platyfish, e.g. *Platypoecilus maculatus* and swordtails, e.g. *Xiphophorus helleri*, possess species-specific spot patterns which are composed of pigment cells, i.e. erythrophores and melanophores. Concerning the melanophores, micro- and macromelanophores can be distinguished which originate from the same precursor, the melanoblast. During early embryonic development, melanoblasts differentiate from the neural crest and migrate to various regions of the body where they later differentiate into micro- and macromelanophores. This differentiation is due to the expression of micro- and macromelanophore genes, respectively. Whereas the platyfish have both kinds of these genes, the swordtails possess only micromelanophores. Since platyfish and swordtails can be crossed, it is possible to compare the effect of the platyfish and the swordtail genome on macromelanophore genes. Both hybridization of platyfish with swordtail and backcrossing the resulting F_1-hybrids with swordtail cause an increase in macromelanophore gene expression which results in premelanoma formation within the F_1-generation and in melanoma formation within the backcross generation. Obviously, the replacement of platyfish chromosomes by swordtail chromosomes causes a derepression of the macromelanophore genes (for review, see Anders, 1967).

For the purpose of transformation experiments in this system, macromelanophore genes of a platyfish should be applied to early melanoblasts of a swordtail embryo. If integrated, a macromelanophore gene will potentiate these melanoblasts to differentiate into macromelanophores. After proliferation and migration of the transformed melanoblasts into various areas of the body, melanoma-like spots will be produced because of the derepressed state of the platyfish macromelanophore gene when present in the swordtail genome.

The macromelanophore gene system thus provides at least three advantages for transformation experiments. First, a melanoblast which has incorporated a macromelanophore gene very early in development, will still proliferate and, therefore, intensify the transformation event. Second, when the descendants of the transformed melanoblast differentiate, the swordtail genome will induce an overproduction of macromelanophores, a situation which can be compared to that in bacterial or tissue cell cultures where transformants are visualized and multiplied on a selection medium. Third, the transformed cells are easy to detect because of their high melanin content. Finally, there are presumably two further advantages which are based on recent findings (Anders et al., 1973). It is possible to concentrate within a single platyfish several pigment cell genes which are all increased in gene expression after hybridization with swordtails, so that up to ten of these genes can be applied to swordtail embryos in form of DNA isolated from such a platyfish. In addition, crossover and translocation studies revealed that all these genes studied so far are located at the end of a chromosome and are easily translocated to its homologous chromosome without regard to presence or absence of other pigment cell genes on this chromosome. Since these translocations occur both within a given species as well as in hybrids, for instance between the chromosomes of platyfish and swordtail, also the externally added pigment cell genes may easily attach to a swordtail chromosome.

The present studies were performed in order to find out a suitable mode of DNA application for transformation experiments in the platyfish-swordtail system and, if degradation of donor DNA occurs, how this can be reduced to a minimum so that a good chance exists for intact high-molecular weight DNA to reach the target cells, the melanoblasts. In this study, bacterial DNA was used as donor DNA, and, since platyfish can be bred during the whole year, embryos of *Platypoecilus maculatus* were taken as recipients instead of swordtail embryos.

2. Materials and Methods

Isolation and culture of embryos. As recipients for donor DNA, embryos from the platyfish, *Platypoecilus maculatus*, were used. Eight days after the last brood had hatched, embryos were isolated from dissected gravid females (for detail, see Tavolga, 1949, and Vielkind, 1971). Three day old embryos of about stage 10 were selected according to Tavolga (1949) and transferred each to 1 ml PBS (Dulbecco's phosphate-buffered saline, diluted 8:3 with distilled water) containing 100 units each of penicillin (Calbiochem) and streptomycin (Bayer). The embryos were maintained in the dark at 27°C.

Isolation and characterization of donor H^3-DNA. H^3-DNA was isolated from *Escherichia coli* strain 15 T^- which had been cultivated at 37°C in M 9 medium containing 2 µg thymine and 1 µCi H^3-methyl-thymine (Amersham, specific activity, 23 Ci/mM) per ml. In order to obtain very high molecular weight DNA, the single steps of the following isolation procedure were reduced to a minimum, DNase activities were inhibited by the use of EDTA and ethanol, and shearing forces were avoided by omission of both pipetting and vigorous stirring of DNA-containing solutions (Schwab and Vielkind, 1973). - The bacteria were collected by centrifugation, washed once with NaCl-EDTA (0.15 M NaCl, 0.005 M EDTA, pH 8.0), once with a 1:1 (v:v) mixture of 95% ethanol and NaCl-EDTA, and twice with NaCl-EDTA. The bacteria were suspended in 0.01 M Tris-HCl, pH 8.0 (20 ml per 1 g wet weight). After addition of a lysozyme-EDTA solution (5 ml, see Godson, 1967), the bacteria were incubated for 10 minutes at 37°C and then lysed by the addition of 2% sodium dodecylsulfate (12 ml) at room temperature. The following steps were performed according to Marmur (1961), with a few exceptions. Sodium perchlorate was used in a final concentration of 2 M at 60°C, and only two steps of deproteinization with chloroform/isoamyl alcohol (24:1, v:v) were done prior to the first precipitation of DNA with cold ethanol. Digestion with RNase was not carried out. The isolated H^3-DNA was dissolved in 0.1 M borate buffer, pH 9.2, and further purified by CsCl density gradient centrifugation, each gradient being loaded with about 500 µg of DNA. After centrifugation, the DNA-containing fractions were pooled, diluted with 1 volume of a 1/10 solution of standard saline citrate (SSC, 0.15 M NaCl, 0.015 M Na-citrate) and precipitated with cold ethanol. DNA was dissolved in SSC, precipitated once again, transferred to a SSC-ethanol mixture (1:2, v:v) for 30 minutes and redissolved in sterile SSC. - The DNA content of this solution was determined by UV absorption measurement and a modified diphenylamine reaction (Burton, 1956). RNA and protein content were measured according to Mejbaum (1939) and Lowry et al. (1951), respectively. Molecular weight of the isolated DNA was determined in a sucrose gradient, using P^{32}-labelled phage T_7 DNA as an internal marker (molecular weight, 25 x 10^6 daltons) and the relationship $D_2/D_1 = (M_2/M_1)^{0.35}$ (Burgi and Hershey, 1963; D_1 and D_2, distances through which the two DNAs sedimented; M_1 and M_2, molecular weight of marker DNA and the DNA in question). The DNA was further checked for single-strand breaks using an alkaline sucrose gradient (see Thomas and Abelson, 1966). Specific activity of the H^3-DNA was calculated as dpm/µg DNA using known amounts of H^3-thymine as standard.

CsCl density gradient centrifugation. For CsCl density gradients, solid CsCl (suprapur, Merck) was added to a H^3-DNA-containing solution to give a final density of 1.735 g/cm^3 (density was checked pycnometrically). 6.5 ml of this solution each were transferred to 12 ml polyallomer tubes and overlaid with silicon oil. The gradients were centrifuged for 60 hours in a W 60 rotor of the Christ Omega ultracentrifuge at 33 000 rpm and 22°C. After the tubes had been pierced at their bases, fractions of 3 drops each were collected under constant pressure of silicon oil and analyzed for UV-absorption or radioactivity.

Sucrose density gradient centrifugation. Sedimentation analysis was done in 4.5 ml of linear 5 to 20% (w:v) sucrose gradients layered on a shelf of 0.6 ml of 20% sucrose to which 1.33 g solid CsCl per ml had been added. For neutral gradients, sucrose was dissolved in 0.05 M NaCl, 0.001 M EDTA, 0.05 M Tris-HCl, pH 7.6, and for alkaline gradients in 0.3 N NaOH, 0.7 M NaCl, 0.001 M EDTA, 0.01 M Tris. 0.1 ml samples containing 0.2 µg E. coli donor H^3-DNA and 0.5 µg phage T_7 marker P^{32}-DNA in neutral buffer, or in 0.3 N NaOH, respectively, were layered on top of neutral or alkaline gradients, respectively. In the case of alkaline gradients, the DNA solution in 0.3 N NaOH had been incubated for 30 minutes at room temperature in order to achieve strand separation prior to centrifugation. The gradients were centrifuged for 2.5 hours in a SW 40 rotor of the Christ Omega ultracentrifuge at 38 000 rpm and 5°C. Fractions of 3 drops each were collected and analyzed for radioactivity.

Treatment of embryos with donor H^3-DNA. DNA was applied in two different ways. For the first experiments, embryos were cultivated in tubes containing 5 or 10 µg H^3-DNA, respectively, in 0.1 ml of PBS each. For all other experiments, however, a H^3-DNA solution diluted with PBS to to give a final concentration of 300 µg DNA/ml was injected into the yolk sac of embryos, using a Leitz micromanipulator. About 0.3 µl of either naked H^3-DNA, spermine-complexed H^3-DNA (0.1 µg spermine, Serva, per µg DNA, see Caspari and Nawa, 1965), or DNase-degraded H^3-DNA were injected. The diameter of the glass cannula used for injection was about 20 µm. The injection procedure was controlled under a stereomicroscope. In order to make the flow of the DNA solution into the embryo visible, 100 µg phenol red per ml had been added. After injection, each embryo was washed twice in penicillin- and streptomycin-containing PBS and cultivated for various periods of time.

Estimation of the relative molecular weight of injected DNA. To examine whether the molecular weight of the donor DNA is reduced by shearing forces when passing the injection cannula, H^3-DNA of the yolk-plasma fraction of embryos (see below) was analyzed immediately after injection. Samples containing about 15 µg of that H^3-DNA as well as samples containing 15 µg of H^3-DNA which had not passed the injection cannula were each subjected to CsCl density gradient centrifugation. The width of the H^3-DNA band formed in the respective gradient can be used to calculate the relative molecular weight of the H^3-DNA in both preparations (Meselson et al., 1957)

Determination of radioactive material in whole embryos and in the culture medium. 2 and 24 hours after cultivation in H^3-DNA-containing medium, samples of 4 embryos each were washed twice in PBS and homogenized in 0.3 ml of 0.25 M sucrose-PBS in a glass homogenizer. The homogenates were digested by incubation in 0.5 ml Digestin (Merck) plus 0.2 ml distilled water at 60°C for 1 hour and shaking the tubes vigorously every 15 minutes (see Schaumlöffel and Graul, 1967). The digested homogenates and 0.5 ml samples of the respective 4 combined culture media were then analyzed for absolute radioactivity. - After injection of H^3-DNA into embryos, samples of 1 embryo each and 0.5 ml samples of the respective medium were similarly examined, 2, 20, and 100 hours after injection.

Determination of acid-soluble and acid-insoluble radioactive material in nuclei and yolk-plasma fraction. After injection of H^3-DNA, spermine-complexed H^3-DNA, or DNase-degraded H^3-DNA, respectively, into the yolk sac of embryos, acid-soluble and acid-insoluble radioactive material in nuclei and yolk-plasma fraction of embryos were determined at various times after injection. At each time interval, samples including 4 embryos were taken. All steps were carried out at 0-4^oC. Embryos were gently homogenized in 0.3 ml of 0.25 M sucrose-PBS in a glass homogenizer and centrifuged for 20 minutes at 800 x g. The sediment, which contained cell nuclei, was washed once with 0.3 ml sucrose-PBS, resuspended in 0.5 ml sucrose-PBS and precipitated by adding 0.5 ml of 0.6 M trichloroacetic acid (TCA). The supernatant of the homogenate and the first wash, which represented the yolk-plasma fraction, were combined and precipitated with 0.6 ml of 0.6 M TCA. Both precipitates were centrifuged for 20 minutes at 2 000 x g. The supernatant (acid-soluble material) was saved for radioactive measurement, and the sediment (acid-insoluble material) was washed once with 0.4 M sodium acetate, ethanol and ether, each, and digested with Digestin as described above. Radioactivities in acid-soluble and acid-insoluble material of both nuclei and yolk-plasma fraction were measured as absolute activities.

DNase and RNase treatment. At 1, 40, 96, and 140 hours after the injection of spermine-complexed H^3-DNA, samples of 3 embryos each were homogenized in 0.4 ml 0.01 M Tris-HCl, 0.003 M $MgCl_2$, pH 7.2. After addition of 0.1 ml Tris-$MgCl_2$ containing 100 μg DNase (Calbiochem) or RNase (Calbiochem), respectively, the homogenates were incubated for 2 hours at 37^oC and then precipitated with 0.5 ml of 0.6 M TCA. Controls without these enzymes were run simultaneously. After centrifugation and one wash with 0.3 ml of 0.3 M TCA, sediment and supernatant were treated as described for the preceding experiments. For injection of embryos with DNase-degraded H^3-DNA, 150 μg H^3-DNA in 0.5 ml 0.003 M $MgCl_2$-PBS containing 30 μg DNAse were incubated for 2 hours at 37^oC.

Radioactivity measurement. For measurement of radioactivity in culture media, DNA solutions, TCA supernatants, and water-diluted fractions from CsCl and sucrose gradients, as well as in sediments treated with Digestin, samples were counted in Instagel (Packard), using a Philips PW 4510 or a Packard Tricarb 3380-544 liquid scintillation counter.

3. Results

Characterization of donor H^3-DNA. The isolation procedure yielded a high molecular weight H^3-DNA with less than 2% of RNA and protein impurities. The specific activity of this DNA was 8 x 10^4 dpm/μg DNA. Sedimentation of the H^3-DNA through a linear sucrose gradient (Fig. 1A) revealed a mean molecular weight of 120 x 10^6 daltons. Analysis in alkaline sucrose gradient (Fig. 1B) resulted in a single peak of H^3-radioactivity, indicating a homogenous population of molecules. The isolated donor DNA, therefore, contained no or an only insignificant number of single-strand breaks.

Application of donor DNA by cultivation of embryos in DNA-containing medium. The results of cultivating the embryos in H^3-DNA-containing PBS are presented in Table 1. Independent on incubation time and DNA concentration of the culture medium, only an amount of less than 3% of the added H^3-DNA radioactivity was found within the samples of 4 embryos. This low amount was not regarded as uptake of H^3-DNA.

Table 1. Radioactivity in embryo and culture medium after cultivating embryos in H^3-DNA-containing PBS for various periods of time

Time of cultivation	Amount of H^3-DNA added per embryo	Labelled material in % of the total H^3-DNA radioactivity added to the medium	
		embryo	culture medium
2 hours	5 µg	2.7	97.3
		2.8	97.2
24 hours		2.2	97.8
		2.1	97.9
2 hours	10 µg	1.8	98.2
		2.3	97.7
24 hours		2.8	97.2
		2.3	97.7

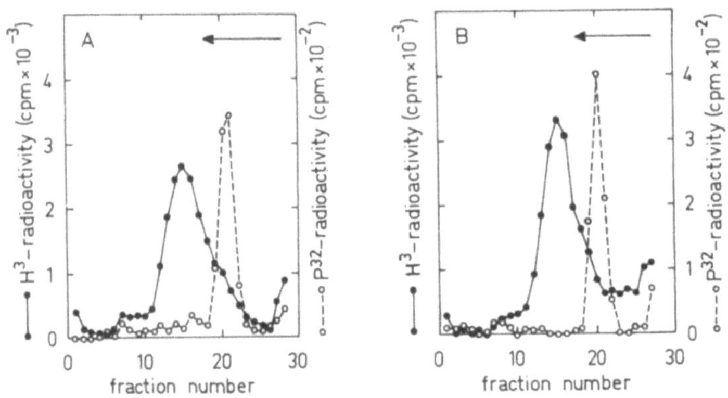

Fig. 1A and B. Sedimentation of 0.2 µg of H^3-labelled E. coli donor DNA and of 0.5 µg of P^{32}-labelled phage T_7 marker DNA in neutral (A) and alkaline (B) linear 5 to 20% sucrose gradient (arrow indicates the direction of sedimentation)

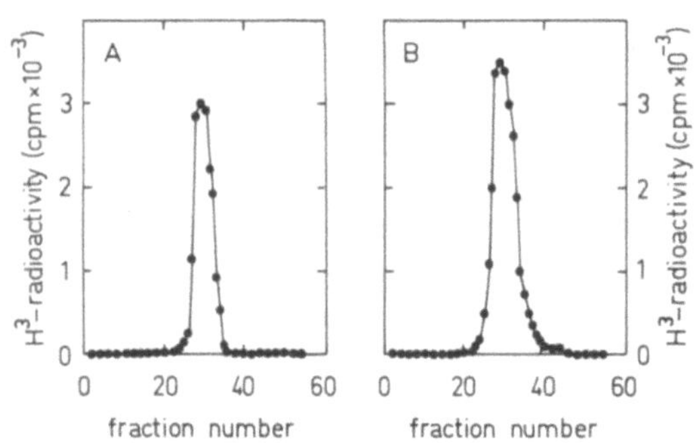

Fig. 2A and B. Analysis in CsCl density gradient of about 15 µg of donor H^3-DNA which had not passed (A) and which had passed (B) the injection cannula

<u>Application of donor DNA by injection of DNA into the yolk sac of embryos.</u> Since embryos did not take up donor H^3-DNA supplied by the culture medium, donor DNA had to be injected into the embryo. By passing the injection cannula, however, shearing forces may reduce the molecular weight of the DNA. Analysis by CsCl density gradient centrifugation of H^3-DNA which had not passed the injection cannula (Fig. 2A) and of H^3-DNA insolated from embryos immediately after injection (Fig. 2B) demonstrated that there was almost no increase in band width within the gradient when DNA had been pressed through the injection cannula. This indicates that the injection procedure caused no or, if at all, only a slight reduction in molecular weight of the donor DNA. - In the following experiments, injection was done exclusively into the yolk sac of embryos, since earlier investigations (Vielkind, 1971) had shown that H^3-DNA injected into regions other than the yolk sac was rapidly released into the culture medium. The results of measuring the radioactivity in embryo and corresponding culture medium at 2, 20, and 100 hours after injection of H^3-DNA into the yolk sac are presented in Table 2. In most cases all radioactive label was found within the embryo, and only in a few cases was some radioactivity found also in the culture medium.

<u>Distribution of radioactivity within the embryos after injection of naked H^3-DNA, spermine-complexed H^3-DNA and DNase-degraded H^3-DNA, respectively.</u> The fate of radioactive material injected into the yolk sac of embryos at stage 10 of development had been followed by comparing the distribution of radioactivity in acid-soluble and acid-insoluble material of both cell nuclei and yolk-plasma fraction of the embryos. This had been done over a period of about 150 hours during which the embryos developed from stage 10 to stage 17, i.e. the time during which the bulk of melanoblasts differentiate and migrate from the neural crest. Since no labelled material had been found to diffuse from the embryos into the culture medium (see Table 2), all values obtained had been calculated in per cent of the total injected H^3-DNA radioactivity. The results are demonstrated in Fig. 3A, B, and C, for naked H^3-DNA, spermine-complexed H^3-DNA, and DNase-degraded H^3-DNA, respectively.

<u>Fate of naked H^3-DNA.</u> As shown in Fig. 3A, only about 50% of the total radioactivity was found in the acid-insoluble material of the yolk-plasma fraction immediately after injection of naked H^3-DNA[2]. During the next 2 hours, the radioactivity in this material rose to 65% and decreased fairly rapidly during the following period, reaching a level of less than 5%, 40 hours after injection. The radioactivity in the acid-soluble material of the yolk-plasma fraction accumulated rapidly during the first few hours, reaching a maximum of 45%, 15 hours after injection. It then decreased to 20% at 40 hours and further ceased to about 2% towards the end of the experiment. During the first 10 hours after injection, there was an unexpected large amount of labelled acid-insoluble material in the nuclei[2], decreasing from about 50% immediately after injection to 15% after 10 hours. Then it rose rapidly to 70% at 40 hours and leveled off to about 95% towards the end of the experiment. Only between 20 and 40 hours after injection a significant amount, i.e. about 8%, of the radioactivity injected was found in the acid-soluble material of nuclei.

---

[2]The low percentage of radioactive acid-insoluble material within the yolk-plasma fraction and the high percentage of this material in the nuclei immediately after injection of donor H^3-DNA is not understood. A simple explanation is that immediately after injection the DNA was not yet equally distributed within the yolk so that "free DNA" could bind to nuclei during homogenization of the embryos. Note that spermine somewhat inhibited this effect.

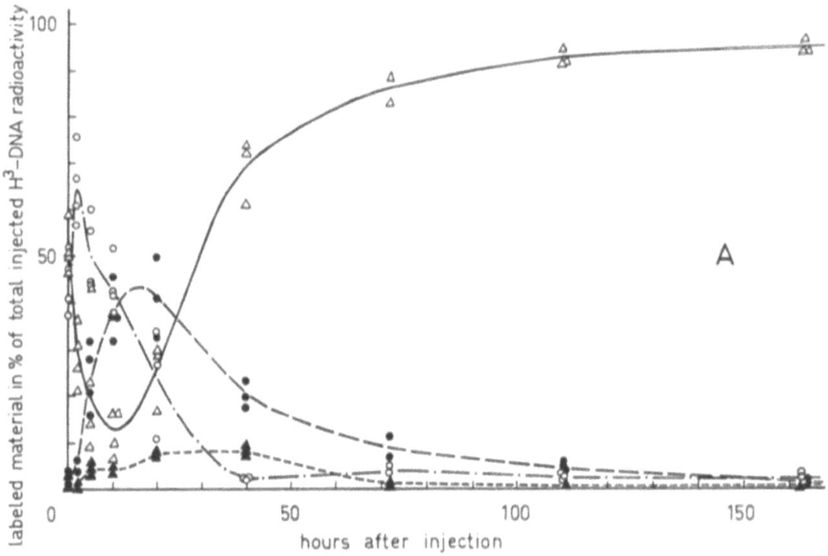

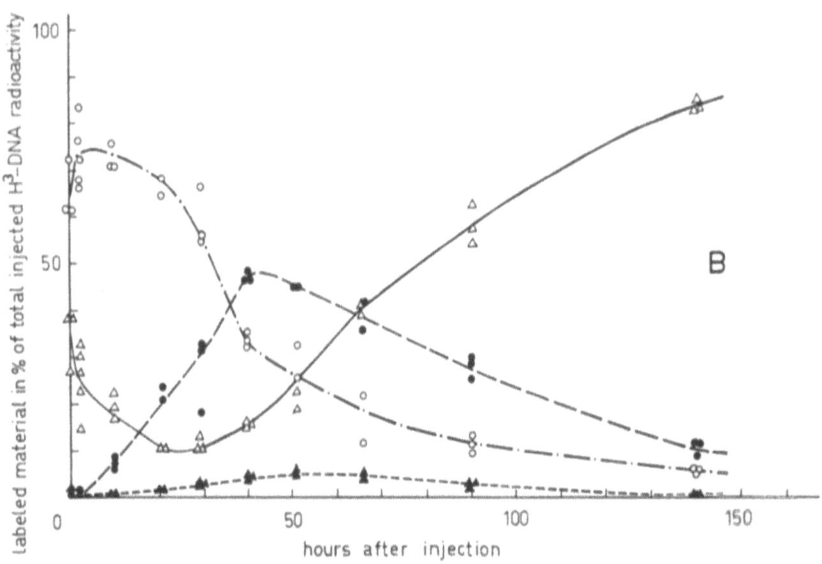

Fig. 3A-C. Distribution of radioactivity in acid-soluble and acid-insoluble material of nuclei and yolk-plasma fraction of embryos after injection of naked donor H^3-DNA (A), spermine-complexed donor H^3-DNA (B), and DNase-degraded donor H^3-DNA (C). o—.—o acid-insoluble material in the yolk-plasma fraction, ●— —● acid-soluble material in the yolk-plasma fraction, △——△ acid-insoluble material in the nuclei, ▲---▲ acid-soluble material in the nuclei

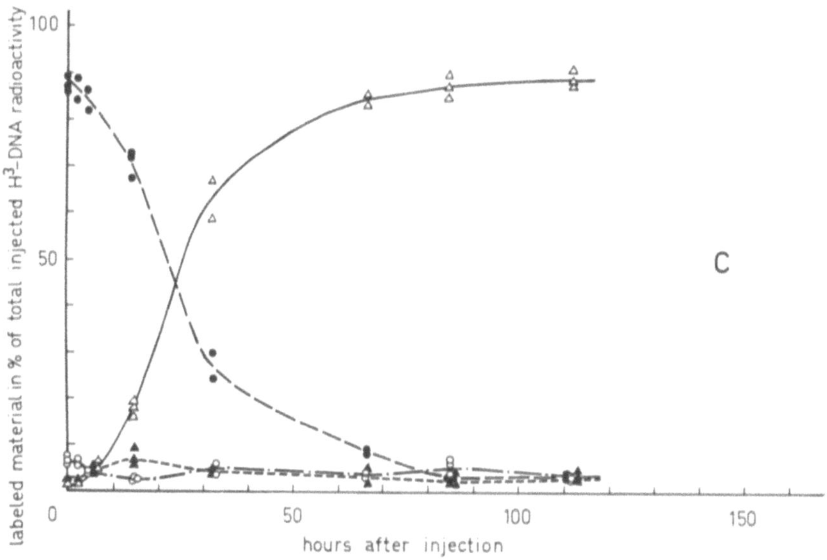

Fig. 3 C

Fate of spermine-complexed H^3-DNA. From Fig. 3B it becomes obvious that there was a much slower decrease in radioactive acid-insoluble material within the yolk-plasma fraction when spermine-complexed DNA had been injected instead of naked DNA. Immediately after injection, this radioactive material was about 65% and showed a wide maximum of about 70% within the first 10 hours. It then decreased to about 30% at 40 hours and declined slowly to reach a level of about 5% towards the end of the experiment. It should be noted that the same process that occurred within the first 40 hours after injection of naked H^3-DNA was now extended to more than 140 hours when spermine-complexed H^3-DNA had been injected. As a consequence of this, the increase in radioactive acid-soluble material in the yolk-plasma fraction was also retarded, reaching a maximum of about 45% not before 45 hours after injection. Thereafter, it declined steadily to about 10% towards the end of the experiment. The radioactive acid-insoluble material in the nuclei first decreased from 35% immediately after injection to about 10%, 25 hours later. It then increased slowly but continuously and reached about 85% at 140 hours after injection. During the entire experiment no significant amounts of radioactive acid-soluble material were found in the nuclei.

Fate of DNase-degraded H^3-DNA. The distribution of radioactivity after the injection of DNase-degraded H^3-DNA is presented in Fig. 3C. As expected, there were only insignificant amounts of radioactive acid-insoluble material in the yolk-plasma fraction during the entire experiment. During the first hours after injection, most of the radioactivity was present in the acid-soluble material of the yolk-plasma fraction. The label in this material decreased rapidly from about 90% immediately after injection to 30%, 30 hours later. Thereafter, the slope of the curve was similar to that obtained for the same material when naked H^3-DNA had been injected (see Fig. 3A). It decreased slowly and reached a level of less than 5% at 80 hours. Also very similar to the respective curve obtained after injection of naked H^3-DNA (see Fig. 3A) was the curve representing radioactive acid-insoluble material

Table 2. Radioactivity in embryo and culture medium after injection of H^3-DNA into the yolk sac of embryos and cultivation for various periods of time

Time after injection	Labelled material in % of the total H^3-DNA radioactivity injected into the embryo	
	embryo	culture medium
2 hours	98.0	2.0
	92.0	8.0
	98.5	1.5
	99.5	0.5
	99.0	1.0
20 hours	96.0	4.0
	98.5	1.5
	95.5	4.5
	92.0	8.0
	99.5	0.5
100 hours	97.0	3.0
	94.0	6.0
	85.0	15.0
	87.0	13.0
	96.0	4.0

of nuclei. The label in this material increased rapidly from almost 0% immediately after injection to 60%, 30 hours later. Thereafter, it rose more slowly and leveled off to 90% at 70 hours after injection. During the entire experiment, radioactive acid-soluble material of nuclei was less than 5%, except for the time between 10 and 20 hours after injection when it amounted to some 8%.

Sensitivity of radioactive acid-insoluble material of embryos against DNase. The following experiments were carried out in order to test whether the radioactivity, which had been found during the preceding experiments within the acid-insoluble material of nuclei and yolk-plasma fraction of H^3-DNA-treated embryos, was bound exclusively to DNA. Homogenates of whole embryos were incubated with DNase, 1, 40, 96, and 140 hours after injection of spermine-complexed H^3-DNA. The radioactivity in acid-soluble and acid-insoluble material of these homogenates was compared with that present in the respective material of RNase-treated homogenates and of control homogenates that had been incubated without enzyme. As shown in Table 3, all radioactivity of DNase-treated homogenates was found within the acid-soluble material independent on the time after injection. In RNase-treated as well as in control homogenates, however, different amounts of acid-soluble and acid-insoluble radioactive material were found at different times after injection. There was no difference between RNase-treated and control homogenates. The amount of labelled acid-soluble and acid-insoluble material of these homogenates exactly corresponded to that calculated from Fig. 3B as the sum of the respective material in nuclei and yolk-plasma fraction of embryos at a given time after injection.

4. Discussion

The present research is part of a transformation experiment with poeciliid fish in which certain genes shall be transferred by means of DNA from one organism, the platyfish, to another, the swordtail. As already discussed in the introduction, a transformation system of higher organisms, however, has many disadvantages as compared to bacterial transformation systems. These disadvantages had led to the unsatisfactory results of numerous investigations in this field (for review, see Ledoux, 1965), but they can be partially overcome by using a favourable system. The main advantage of the platyfish-swordtail system (see introduction) is the fact that, due to the genetic constitution of the recipient, the transformation of certain cells should result in an uncontrolled tumor-like proliferation of the transformants so that a transformation event should be easily detected. Since the cells to be transformed are undifferentiated melanoblasts, donor DNA has to be applied at an early stage of embryonic development. The aim of the present studies was to work out a suitable method for the application of donor DNA to fish embryos and, furthermore, to test whether the donor DNA is degraded within the recipient embryos and how this degradation can be reduced to a minimum.

It has been pointed out by several authors that donor DNA of a high grade of purification and of a minimal molecular weight of 5×10^5 daltons is necessary for successful transformation in bacterial systems (see Schaeffer, 1964). Furthermore, it has been found empirically that especially the uptake and integration of donor DNA by recipient cells of higher organisms requires a highly purified donor DNA of very high molecular weight (Ledoux, personal communication). These requirements were fulfilled in the present studies. The donor DNA we used was highly purified and had a mean molecular weight of 120×10^6 daltons. The fact that we used bacterial donor DNA does not reduce the significance of the present results, since it has been shown by various authors that there is no difference in DNA uptake, degradation, and integration by recipient cells when heterologous donor DNA is used instead of homologous one (for review, see Ledoux, 1965). Nevertheless, it should be taken into consideration to do experiments of such high biologic relevance also with homologous donor DNA, because there may be some structural differences between bacterial DNA and DNA of higher organisms which might influence the uptake and integration of the donor DNA.

Cultivation of embryos in H^3-DNA-containing medium revealed that the embryos did not take up DNA from the medium. This was thought to be due to the vitelline membrane around the embryo which might represent a barrier for external DNA. But recent results seem to show that also after the removal of this membrane, the embryos fail to take up DNA actively (unpublished data). This method of DNA application cannot be used, therefore, for transformation experiments with fish embryos. Thus, it seemed inevitable to inject the donor DNA into the embryo. As described elsewhere (Vielkind, 1971), the injection procedure did not affect the viability of the embryo; but the DNA ought to be injected into the yolk sac, because it was rapidly released into the culture medium when injected into other regions of the embryo. Up to this point, the method to inject the donor DNA into the embryo seemed to be practicable. However, the injection procedure requires the use of a small injection cannula, and it was feared that shearing forces might break the donor DNA into short pieces when forced through this cannula. Analysis by CsCl density gradient centrifugation demonstrated that the injection procedure caused almost no reduction in molecular weight of the DNA. Thus, the method to inject the donor DNA into the yolk sac of embryos is regarded to be a suitable mode of DNA application.

The most important question concerning the transformation problem is what happens to donor DNA when injected into the yolk sac of the

Table 3. Radioactivity in acid-soluble and acid-insoluble material of DNase- or RNase-treated homogenates or of control homogenates of embryos into which spermine-complexed H^3-DNA had been injected

	Labelled material in % of the total injected H^3-DNA radioactivity					
	DNase-treatment		RNase-treatment		Control	
Time after injection	acid-soluble material	acid-insoluble material	acid-soluble material	acid-insoluble material	acid-soluble material	acid-insoluble material
1 hour	98.6	1.4	2.2	97.8	1.6	98.4
	98.9	1.1	1.9	98.1	1.2	98.8
					1.8	98.2
40 hours	98.9	1.1	44.8	55.2	43.0	57.0
	98.7	1.3	41.4	58.6	44.0	56.0
			46.3	53.7	52.2	47.8
96 hours	95.0	5.0	27.4	72.6	31.0	69.0
	97.8	2.2	21.5	78.5	38.0	62.0
	97.9	2.1	28.2	71.8	36.5	63.5
140 hours	98.0	2.0	19.9	80.1	9.9	90.1
	98.2	1.8	16.8	83.2	19.4	80.6
	97.9	2.1			22.0	78.0

embryo. For successful transformation, this DNA should be transported from the yolk into the cell nuclei of the embryo where it may be integrated into the recipient genome and later may give rise to a new transformant phenotype. Within the yolk and during transport into the cells of the embryo, the donor DNA is expected to underlie the action of nucleases so that it will be degraded before having reached the nuclei of the target cells. Hence, before doing transformation experiments, it seemed worthwhile to prove what chances exist for high-molecular donor DNA to become integrated and how this donor DNA can be protected against nuclease activities. For this purpose, we followed the fate of naked H^3-DNA as well as H^3-DNA protected by spermine (see Caspari and Nawa, 1965) after injection into the yolk sac of embryos. The amount of radioactive acid-soluble and acid-insoluble material in both yolk-plasma fraction and nuclei were compared over a period which covered the embryonic development from stage 10 to stage 17, a period during which melanoblasts migrate from the neural crest and begin to differentiate (see Tavolga, 1949). Since all the radioactive acid-precipitable material that had been found to decrease at the beginning and to increase towards the end of this period had been shown to be DNase-sensitive (see Table 3), the label in this material represented H^3-radioactivity bound to DNA. The labelled acid-soluble material, however, that accumulated within the embryos soon after injection of donor H^3-DNA represented degradation products of this donor H^3-DNA. It should be emphasized at that time that determination of radioactivity in the acid-insoluble material does not allow to distinguish between the following possibilities: the radioactive acid-precipitable material in the nuclei might represent either pieces of donor H^3-DNA, recipient DNA that had incorporated degradation products of the donor H^3-DNA, or both. Experiments which allow such a distinction are now in progress.

By comparing the distribution of radioactive material after injection of naked H^3-DNA and spermine-complexed H^3-DNA into the yolk sac of embryos, it became obvious that naked donor DNA was much more rapidly degraded than donor DNA protected by spermine. As a consequence, the degradation products of naked donor DNA reached a maximum already 15 hours after injection (see Fig. 3A), whereas those of spermine-complexed donor DNA reached this maximum not before 45 hours after injection (see Fig. 3B). At the respective time, about 30% of the injected donor DNA was still present in a relatively high-molecular state. Thus, the degradation of donor DNA to about one third could be delayed by 30 hours when spermine had been used for its protection. Most important for a transformation experiment, however, was the fact that even 100 hours after injection into the yolk sac, about 10% of the original spermine-complexed donor DNA was still acid-precipitable within the yolk-plasma fraction, while the unprotected naked donor DNA completely disappeared from the yolk-plasma fraction already at 40 hours after injection. As demonstrated by the increased slope of the curve representing the DNA of the nuclei, the earlier availability of degradation products of naked donor DNA caused an accelerated incorporation of these labelled compounds into the recipient DNA (see Fig. 3A). Since the injection of naked donor DNA yielded almost the same curve for radioactivity in the DNA of nuclei as the injection of DNase-degraded donor DNA (compare Fig. 3A with Fig. 3C), it was concluded that there was a high DNase activity present within the yolk which made the degradation products of naked donor DNA rapidly available for recipient DNA synthesis.

From the present experiments, it is decided that in order to achieve genetic transformation in the platyfish-swordtail system, the donor DNA which has to be injected into the yolk sac of recipient embryos must be protected against nuclease activities. The finding that the donor DNA can be protected from degradation when complexed with spermine agrees with the observations of other authors. By using spermine-complexed donor DNA, Caspari and Nawa (1965) were able to demonstrate mutational effects in *Ephestia* which had been interpreted as genetic transformation. There seems, therefore, to be a good chance for spermine-protected high-molecular donor DNA to be integrated by the embryonic melanoblasts which shall be transformed. However, further studies will be necessary to find out how the uptake of donor DNA into the recipient cells can be enhanced, perhaps by substances which influence the properties of cell membranes. Besides this, experiments are now in progress to follow the fate of radio-labelled donor DNA of high density by CsCl density gradient centrifugation and hybridization techniques in order to demonstrate uptake and integration of high-molecular donor DNA by recipient cells.

5. Summary

For the purpose to do transformation experiments in poeciliid fish, a method for the application of donor DNA to fish embryos had been worked out. During cultivation of embryos in DNA-containing medium, there was no uptake of DNA into the embryos. The donor DNA had to be injected, therefore, into the embryo. The fate of donor H^3-DNA injected into the yolk sac of embryos had been followed by determination of radioactivity in acid-soluble and acid-insoluble material of yolk-plasma fraction and nuclei of the treated embryos. Within the first 40 hours after injection, there was a rapid and complete degradation of donor DNA when naked DNA had been used. As demonstrated by the increasing radioactivity in the nuclear DNA fraction of embryos, the labelled degradation products of the donor DNA were rapidly utilized for recipient DNA synthesis. The degradation of the donor DNA, however, was markedly delayed when the DNA had been complexed with spermine. Even at 100 hours after injection, the yolk-plasma fraction of embryos still contained about 10% of the injected donor DNA in an acid-precipitable, i.e. relatively high molecular weight state. Thus, there will be a good chance for spermine-complexed high-molecular donor DNA to reach the target cells when injected into the yolk sac of fish embryos.

References

Anders, A., Anders, F., Klinke, K. (1973): Regulation of gene expression in the Gordon-Kosswig melanoma system. II. The arrangement of chromatophore determining loci and regulating elements in the sex chromosomes of the xiphophorin fish, *Platypoecilus maculatus* and *Platypoecilus variatus*. This issue.

Anders, F. (1967): Tumor formation in platyfish-swordtail hybrids as a problem of gene regulation. Experientia 23, 1-10.

Avery, O.T., MacLeod, C.M., McCarty, M. (1944): Studies on the nature of the substance inducing transformation of pneumococcal types. Induction of transformation by a desoxyribonucleic acid fraction isolated from pneumococcus type III. J. Exp. Med. 79, 137-157.

Benoit, J., Leroy, P., Vendrely, C., Vendrely, R. (1957): Des mutations somatiques dirigées sont-elles possibles chez les Oiseaux? Compt. Rend. Acad. Sci. 244, 2320-2321.

Burgi, E., Hershey, A.D. (1963): Sedimentation rate as a measure of molecular weight of DNA. Biophys. J. 3, 309-321.

Burton, K. (1956): A study of the conditions and mechanisms of the diphenylamine reaction for the colorimetric estimation of deoxyribonucleic acid. Biochem. J. 62, 315-323.

Caspari, E.W., Nawa, S. (1965): A method to demonstrate transformation in Ephestia. Z. Naturforsch. 20b, 281-284.
Godson, G.N. (1967): A technique of rapid lysis for the preparation of Escherichia coli polyribosomes. In: L. Grossman and K. Moldave (eds.), Methods of Enzymology 12A, 503-516, Acad. Press, New York.
Hess, D. (1972): Transformation an höheren Organismen. Naturwissenschaften 59, 348-355.
Hotchkiss, R.D. (1952): The role of desoxyribonucleate in bacterial transformation. In: M. McElroy and B. Glass (eds.), Phosphorus Metabolism 2, 426-436. Johns Hopkins Press, Baltimore.
Ledoux, L. (1965): Uptake of DNA by living cells. Progr. Nucl. Acid Res. Mol. Biol. 4, 231-267.
Ledoux, L. (ed.)(1971): Informative Molecules in Biological Systems. Amsterdam, London: North-Holland Publishing Company.
Lowry, O.H., Rosebrough, N.J., Fair, A.L., Randall, R.J. (1951): Protein measurement with the Folin phenol reagent. J. Biol. Chem. 193, 265-275.
Marmur, J. (1961): A procedure for the isolation of DNA from micro-organisms. J. Mol. Biol. 3, 208-219.
Mejbaum, W. (1939): Über die Bestimmung kleiner Pentosemengen, insbesondere in Derivaten der Adenylsäure. Z. Physiol. Chem. 258, 117-120.
Meselson, M., Stahl, F.W., Vinograd, J. (1957): Equilibrium sedimentation of macromolecules in density gradients. Proc. Nat. Acad. Sci. 43, 581-588.
Raspé, G. (ed.)(1972): Workshop on Mechanisms and Prospects of Genetic Exchange. Advances in the Biosciences 8. Oxford, New York, Braunschweig: Pergamon Press, Vieweg.
Schaeffer, P. (1964): Transformation. In: I.C. Gunsalus and R.Y. Stanier (eds.), The Bacteria 5, 87-153. New York: Acad. Press.
Schaumlöffel, E., Graul, E.H. (1967): Vergleichende Untersuchungen mit neuen Szintillationschemikalien. Atompraxis 13, 260-262.
Schwab, M., Vielkind, J. (1973): A simple effective procedure for the isolation of high molecular weight DNA. Int. Res. Commun. System (73-3) 3-2-2.
Szybalska, E.H., Szybalski, W. (1962): Genetics of human cell lines. IV. DNA-mediated heritable transformation of a biochemical trait. Proc. Nat. Acad. Sci. 48, 2026-2034.
Tavolga, W.N. (1949): Embryonic development of the platyfish (*Platypoecilus*), the swordtail (*Xiphophorus*) and their hybrids. Bull. Americ. Mus. Nat. Hist. 94, 165-229.
Thomas, C.A., Jr., Abelson, J. (1966): The isolation and characterization of DNA from bacteriophage. In: G.L. Cantoni and D.R. Davies (eds.). Procedures in Nucleic Acid Research, 553-561. New York, London: Harper and Row.
Vielkind, J. (1971): Versuche zur genetischen Transformation bei lebendgebärenden Zahnkarpfen (Poeciliidae). Applikation heterologer DNA und ihr Schicksal im Rezipienten. Dissertation, Universität Giessen.

Regulation of the Pigment Cell Arrangements in Species and Interspecies Hybrids of *Xiphophorus* (Pisces, Poeciliidae) by Cellular Interactions on the Fish Body[1]

W. Lueken, E. R. Schmidt, and K. Lepper

1. Introduction

Pigment cell arrangements in species and interspecies hybrids of *Xiphophorus* can be correlated to gene constellations (references until 1971 see Wolf, 1972, surveys and introductions see Anders et al., 1972; Kallman and Atz, 1966; Zander, 1969). The realization of genotypes from zygotes to adult fish involves a lot of steps. Part of them, preferently those which occur in the skin are summed up here. At first, the whole process of gene manifestation was divided into many different steps in order to make them accessible to experiments. It was proved later, however, that the variety of the factors itself plays a substantial role in the interpretation for the realization of "colour genes" and thus for tumor-formation in *Xiphophorus*. Therefore, this survey about experimental results pursues two aims: demonstration of single processes which can be investigated, and demonstration of heterogeneity in the ways in which pigment cell arrangements are established.

Of course, the paper cannot claim to be complete, on the contrary, the description of results is subdivided into so many parts for the purpose to integrate, easily, further findings into the following scheme:

1. Existence and behaviour of pigment cell precursors
2. Inhibition of differentiation or growth
3. Destruction of pigment cells
4. "Infectious distribution" of pigment cells

Examples are presented, at first, for melanophores (page 141 ff.), as a complete survey is only available for this kind of pigment cells, which is due to the fact that they are best observable in all aspects. For the two other kinds, xanthophores (plus erythrophores)[2] and iridophores, parallels and contradictions, respectively, to the melanophores are given according to the same scheme (page 149 ff.). A further part (page 151 ff.) deals with genealogical relations of pigment cell kinds, as these are important for transference of knowledges from one kind of pigment cell to another one. Finally (page 152 ff.) some examples are given for the composition and the development of pigment cell arrangements, in order to demonstrate some principles in the activities of pigment cell populations. Concerning the conclusions (page 157 ff.), only a few aspects could be pointed out, as there is no space to discuss the results extensively.

Classification of pigment cells in three kinds and in various types among these corresponds for melanophores with that given by Becker-Carus (1965), the terminology for developmental stages being based on the 6th pigment cell conference in 1965 (Levene, McGovern, Mishima, Oettle, 1966). Designations for xantho- and erythrophores are taken from Öktay (1964), and Zander (1969), considering recent results of Henze (unpublished). Classification of iridophores has been undertaken

[1]This research was accomplished with the aid of Deutsche Forschungsgemeinschaft und Stiftung Volkswagenwerk.

[2]Summarized by several authors as "pterinophores".

by E.R. Schmidt (1973). Distinct types of pigment cells within one kind are only discriminated in the text, if it is indispensable, therefore they are not discussed in detail here.

2. Material and Methods

2.1. Fishes

Species and hybrids were taken from the breedings of the Institute[3], stocks and genes are described by Kallman and Atz (1966) and by Zander (1969). The species and genotypes used in various experiments are referred to in the description of results.

2.2. Methods

Those which have been taken from literature or have already been described are only listed: artificial insemination: Zander (1960); short-time cell culture: Lueken and Foerster (1969); staining of melanoblasts, according to Mishima (1960): Lueken and Kaeser (1972); impregnation with $AgNO_3$: Mishima (1960) - but treatment with gold chloride being omitted; autoradiography: common methods with products of Amersham-Buchler and stripping film AR 10.

For regeneration experiments animals have been narcotisized in MS 222 and tissues which should be brought to regeneration were removed. For control corresponding uninjured regions at the opposite body side of the specimen were used.

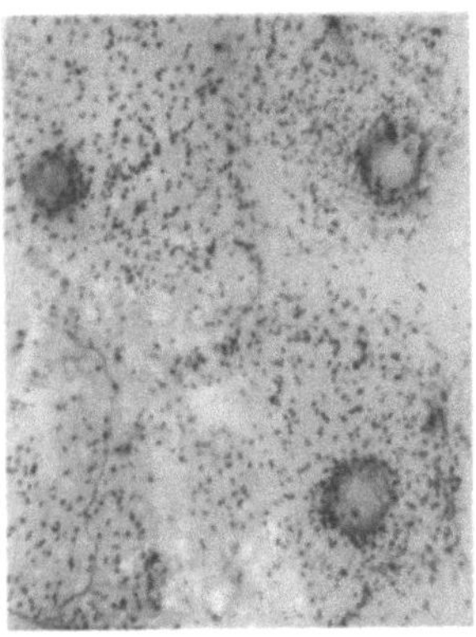

Fig. 1

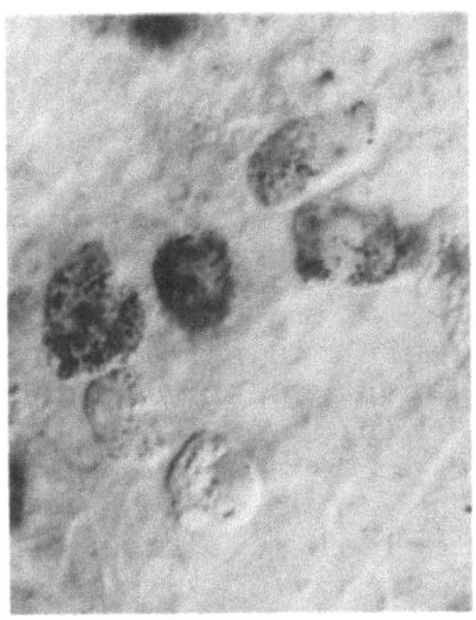

Fig. 2

Fig. 1. Example for density (appr. 300/mm^2) and arrangement of reservoir melanoblasts. Each black dot is a dopa-positive cell (cp. Figs. 2 and 4a). They are concentrated around the centers of scales. Scales from the belly of *X. helleri*, homozygous for "golden", 25:1

Fig. 2. Part of a dopa-incubated scale of a *X. helleri/X. maculatus* F_2-backcross-hybrid. All intermediate stages from dopa-negative (unstained) to totally impregnated melanoblasts are to be recognized. Interference contrast, 1 000:1

[3]We are greatly indebted to Miss Käte Klinke for providing us with the fish specimens.

3. - 6. Results

3. Melanophores

3.1. Existence and Behaviour of Melanophore Precursors (Melanoblasts)

3.1.1. Melanoblast-Densities

In many very different genotypes, high melanoblast concentrations occur in the skin region (Lueken and Kaeser, 1972), mostly in a density of about 150-200/mm^2, sometimes below, sometimes considerably above (Fig. 1). Their distribution all over the body is unequal. Correlations between melanoblast-numbers and melanophore distribution could not be found. The melanoblast densities may, however, fluctuate within one genotype according to the age of the animals. This was observed, for example, in young specimens of *X. maculatus*, where the density diminished during development from approximately 100/mm^2 (in animals of 10 mm length) to almost 0 (18 mm) and enlarged to 100/mm^2 in adult fishes. One can generally assume that the melanoblast reservoir in the skin is sufficient to be the basis for all degrees of melanophore spotting up to fast growing tumors. The reservoir of dopa positive melanoblasts is still enlarged through dopa negative ones, which correspond to the normal ones in seize, shape, and mobility (Lueken and Kaeser, 1972), and are thought to be less differentiated melanoblasts, which is confirmed by intermediate stages (Fig. 2).

3.1.2. Locomotion of Melanoblasts

In tissue preparations, especially from regenerating skin regions, migration speeds of 5 µm/min. were measured (in interference contrast microscopy), about 10 µm being the total length of the cell. The melanoblasts move like amoebae. Beside this active migration one can presume passive transportation of the cells through blood-vessels. This is concluded from the fact that melanoblasts always are found in vessels, occasionally penetrating a wall, which has also been observed by Greenberg and Kopac (1963). Another observation which points in the same direction is that melanoblasts on scales are often concentrated around a perforation through which a vessel crosses the scale (Fig. 1). Mobility and passive transportation explain why melanoblasts are, in general, present all over the fish body and reappear in regenerating tissues soon after wound-healing has onset.

3.1.3. "Phobic Reaction" of Melanoblasts

One of the authors was able to observe that melanoblasts that touched a differentiated melanocyte or melanophore stopped or even changed the direction of their migration. It is unknown, if this is caused by a real phobic reaction or if the differentiated cell is only a mechanical obstacle for a melanoblast. As further indication for a phobic reaction one can interprete the fact that melanoblasts in the skin are localized within other cell layers than differentiated melanophores are. Phobic reactions of melanoblasts could also be the reason for the well known halos around melanophore spots (Fig. 3). In these regions only few melanoblasts are found, perhaps they have been expelled, therefore no melanophores can be formed either.

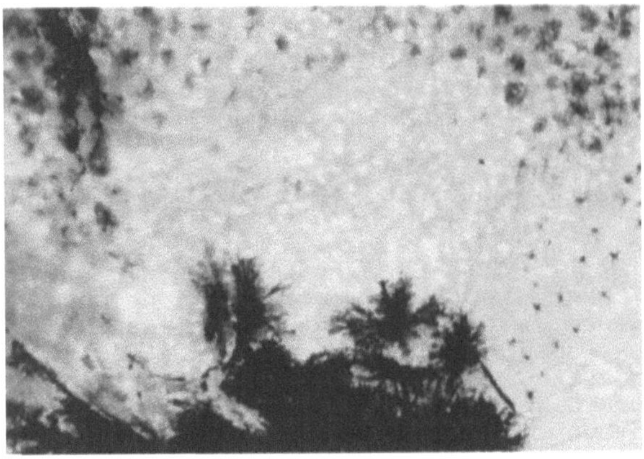

Fig. 3. Melanophore-free area at a macromelanophore spot. The distance from the margin of the spot to the next micromelanophores is appr. 1 mm. *X. helleri/X. maculatus* F_1-hybrid with the gene Sp, 30:1

3.2. Inhibition of Differentiation or of Growth

Inhibition-mechanisms at several steps of pigment cell development become evident through typical reactions which occur, when inhibition is abolished, and which correspond to an "induction" process. In other cases, such mechanisms can be concluded from the arrangements of pigment cells.

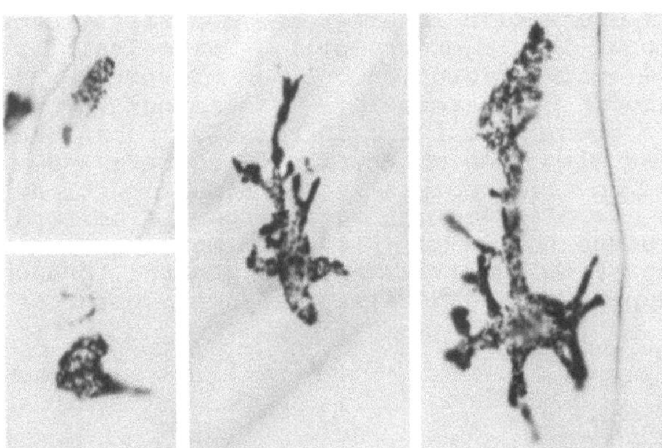

a b c d

Fig. 4a-d. Stages of melanocyte development
a. Dopa-stained melanoblast.
b. Young melanocyte, 24 hrs. after induction of differentiation by transplantation of a scale with melanoblasts.
c. Melanocyte 48 hrs. after transplantation.
d. Melanocyte 72 hrs. after transplantation.
In this experiment scales from *X. helleri* without the gene Db^2 were transplanted onto *X. helleri* possessing Db^2. The combinations of donor and recipient have, however, no recognizable influence on the result. All 560:1

3.2.1. "Blocked" Melanoblasts

If melanoblast containing tissues are transplanted or regeneration is induced by vulnerating a fish, melanoblasts start to differentiate to melanocytes. After transplantation, melanoblasts (Fig. 4a) may have developed to the stage of Fig. 4b within 24 hrs, after 48 hrs usually they look like the cell in Fig. 4c, and Fig. 4d shows a melanocyte 72 hrs after the transplantation. From then, melanocytes grow only very slowly. Principally, the same occurs in regenerating tissues after large injuries. However, as melanoblasts have to immigrate, first of all, into the new built skin tissues (or/and differentiate out of unstainable precursors) the delay from operation to appearance of the first melanocytes is extended to about 3 days. Thereafter a gradient in melanocyte differentiation from the margins of the wound-area into its center is established. These events are thought to be caused by a change of the physiological milieu in the tissues, which enables "blocked" melanoblasts to continue their development to melanocytes. It must be well remarked that quick differentiation and growth occur only until a cell type that is also the main constituent of fast-growing tumors. This is known from microscopical research (Vielkind, Vielkind and Götting, 1968) as well as from in vitro-experiments, in which such melanocytes emigrate from tumors.

3.2.2. Gradients in the Degree of Melanophore Differentiation

Melanophores are present in a fish in different stages of differentiation (Becker-Carus, 1965). Their distribution can be accorded to characteristic gradients in the degree of melanophore differentiation.

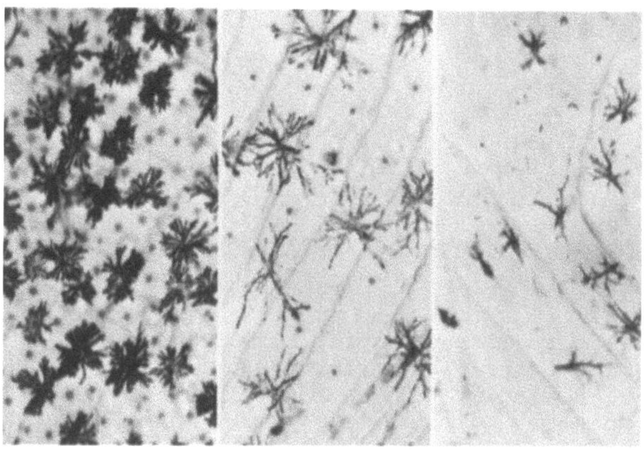

a b c

Fig. 5a-c. Melanophores in different degrees of differentiation in one fish (*X. helleri*), demonstrated on scales
a. From the dorsal region.
b. From the region of the middle-line.
c. Below the middle-line.
All 120:1

The most outstanding one determines the decline in melanophore differentiation from dorsal to ventral regions of a fish specimen. Examples are given from melanophores which are located on scales, since these can be thought to represent always the same substrate all over the fish

(Figs. 5a-c). Less differentiated melanophores are usually arranged in lower concentrations, in the ventral region of many fishes no melanophores are present. The steepness of this gradient varies according to the species. Also in longitudinal direction of the fish body a similar gradient is to be seen. Its maximum is not situated on or near the head, but more or less displaced. Thus a decline in the melanophore differentiation degree as well towards the head as towards the tail is expressed. These common gradients are overlapped in some species by another one, the differentiation optimum of which lays in the area of the middle-line. This is especially pronounced in *X. variatus*.

3.2.3. Growth Inhibition through Pigment Cells of the same Kind (Contact Inhibition and Attraction of Cell Branches)

Usually melanophores are situated side by side without covering each other (Fig. 6). Then it is remarkable that branches of neighbouring cells touch each other at their tops. The contact can be so close that the cells shade off into one another (where this cannot be observed the pigment granules have often moved from the cell periphery towards the

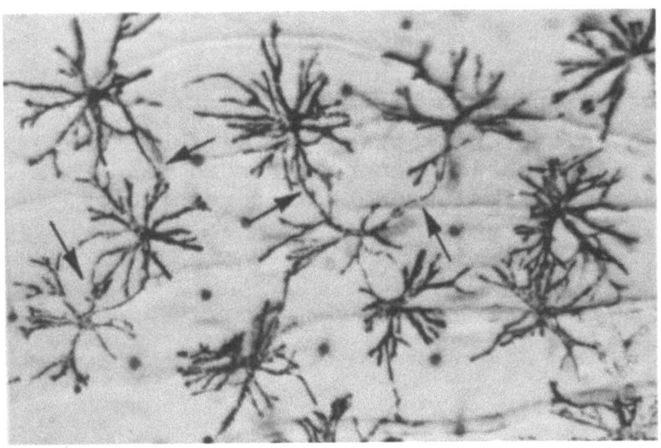

Fig. 6. Contact of melanophore branches (some marked by arrows) on a scale of *X. helleri*. 220:1

cell center). The observation leads to the assumption that the cells grow until their branches touch each other, whereafter, because of contact inhibition the cells stop growing. This results in the typical net work of melanophores in a completely pigmentated region of a fish's skin, and in one layer arrangement of pigment cells as well. Since pigment cells, when starting their development, are relatively far from each other, their branches cannot be expected to meet by chance. This improbability is demonstrated in Fig. 7.: in a melanophore net a "hole" is filled by a growing pigment cell. Its branches are clearly directed towards the tips of surrounding melanophore branches. This leads to the conclusion that some kind of attraction may exist. This assumption was still strengthened in experimental conditions (Fig. 8): melanocytes that differentiate in a melanophore-free area, grow, first of all, independently from each other. Later, however, branches from neighbouring melanocytes are directed towards each other. Even a ridge on the scale does not prevent expansion (which sometimes may occur - see below). It has to be considered that in the case of Fig. 7 a branch of the young pigment cell has to cross a distance which corresponds to approximately ten cells. These are, of course, not arranged in layers. The way in which directed growth is regulated cannot yet be understood.

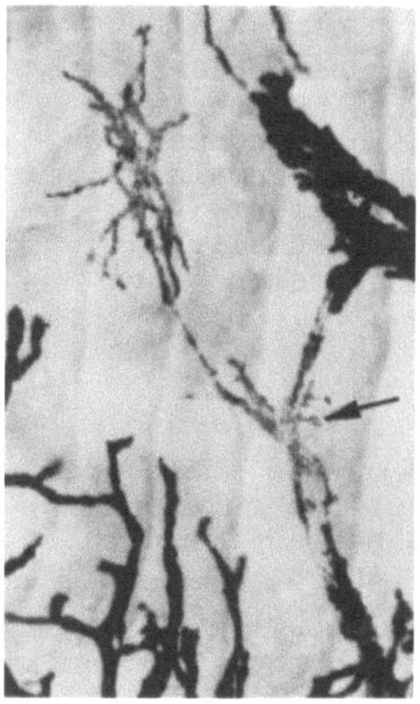

Fig. 7

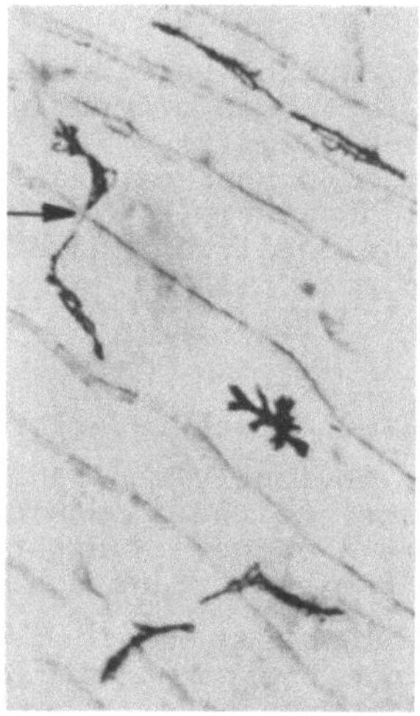

Fig. 8

Fig. 7. A growing melanocyte settles in a hole between "old" melanophores. Its branches are directed towards those of the melanophores. Scale of *X. maculatus*. 320:1

Fig. 8. Mutually "attracted" branches of melanocytes. Attraction is not even inhibited by a scale ridge (arrow). *X. maculatus*, differentiation of melanocytes induced by autotransplantation. 150:1

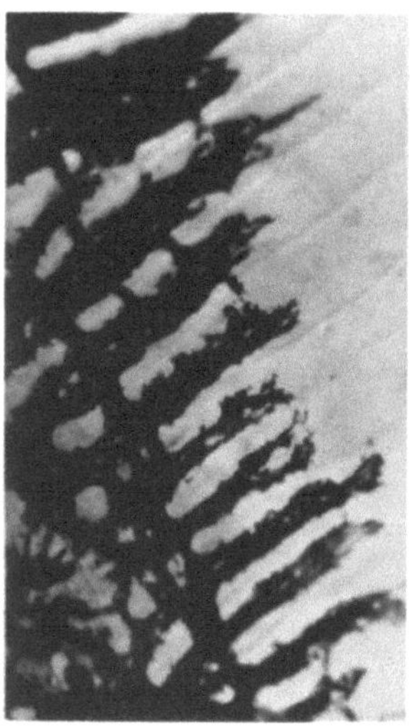

Fig. 9

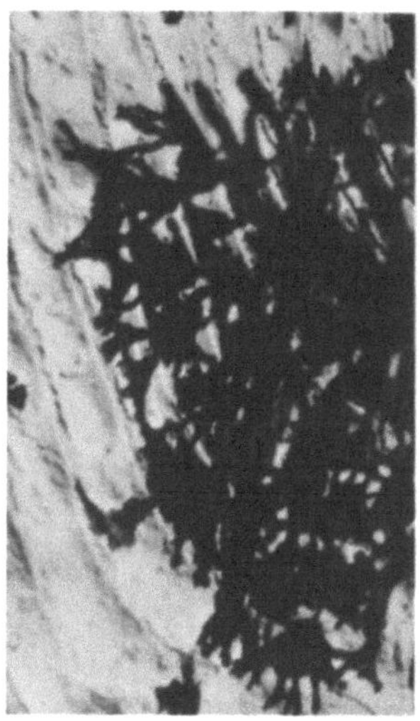

Fig. 10

Fig. 9. Expanding branches of macromelanophores at the margin of a wound area. *X. maculatus* with the gene N. 220:1

Fig. 10. Expansion of a macromelanophore following the principle of lowest resistance. If a branch has crossed a scale ridge in a narrow incision it expands in the "plain" until it hits another ridge or another expanded branch of the same cell. Fish as Fig. 3. 220:1

3.2.4. Inhibition of Growth by other Cells

From uninjured macromelanophores, which are situated immediately at the margin of a regenerating tissue, branches often migrate into the area that does not contain melanophores (Fig. 9). This is interpreted as follows: cells that have prevented further expansion of macromelanophores have been removed from their neighbourship by the vulneration. As soon as in the regenerating tissue conditions have been established which correspond with those of normal tissues, growth of melanophore branches is inhibited again.

3.2.5. Inhibition of Growth Caused by other Factors

One example is given in Fig. 10: obviously the ridges on the scale are obstacles (perhaps mechanical ones) to the expansion of the melanophore. When a narrow branch has overcome a ridge, it expands behind it in the "plain", until it is stopped at the next ridge again, and so on.

Also the fact that melanophores often expand along vessels, may be caused by mechanical effects, since the connection of cells may be less tight between the vessel and the surrounding tissue.

3.3. Destruction of Melanophores

3.3.1. Regenerating Tissue

Here, the effect is most obvious: when the physological milieu of the regenerating tissue has reached the stage which admits development of melanocytes, an extraordinary dense net of melanocytes is built up. That can be quantitatively demonstrated by calculating the mean distances of pigment cell centers. In *X. helleri*/*X. maculatus*-F_1-hybrids (possessing the gene Sp) the melanocytes were appr. 30 µm from each other. But melanophores, the final stages of these melanocytes, have diameters of appr. 120 µm. Therefore, there cannot be enough room for all melanocytes for further growth, unless they are arranged in several layers or move away. Indeed the regular density is established by destruction of cells. This is brought into action appr. ten days after the beginning of melanocyte differentiation. The reduction of

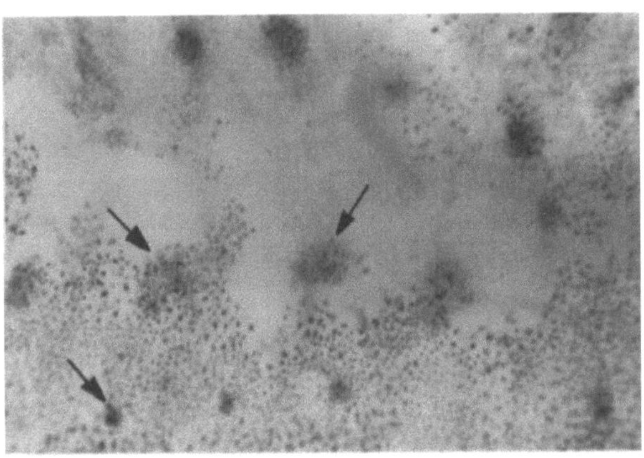

Fig. 11. Accumulations of macrophages (some are marked by arrows) after destruction of melanocytes. *X. variatus*, 15 days after removal of the skin for regeneration experiments. 15:1

cell numbers goes even too far - so that considerable holes are made in the melanocyte net, which are filled again through differentiation of a second generation of melanocytes. From then the pigment cell arrangement remains relatively constant. As has been described, this is reached by an oscillating process, typical for a lot of regulation mechanisms. It is not to be seen, in which way the necessary feed back is performed in this case. Either is it known what causes the destruction itself. It is indicated, beside the disappearance of the pigment-cells, by an immense number of macrophages carrying away the melanin granules of the dead cells. The macrophages usually migrate to the peripheral regions of a fish, where they can be found in extended groups (Fig. 11). They are often arranged in a distinct pattern. They are washed away from a living fish due to its movements. This is concluded from the fact that in fishes which are fixed in formalin the macrophage groups get easily lost by manipulation or slightest rinse.

3.3.2. Establishment of Natural Patterns

One example is given from *X. xiphidium* (Fig. 12). Here, in very young specimens, the region between the rays of the dorsal fin is populated homogenously with melanophores. An older fish, however, has a characteristic pattern in its dorsal fin with light zones. These are the

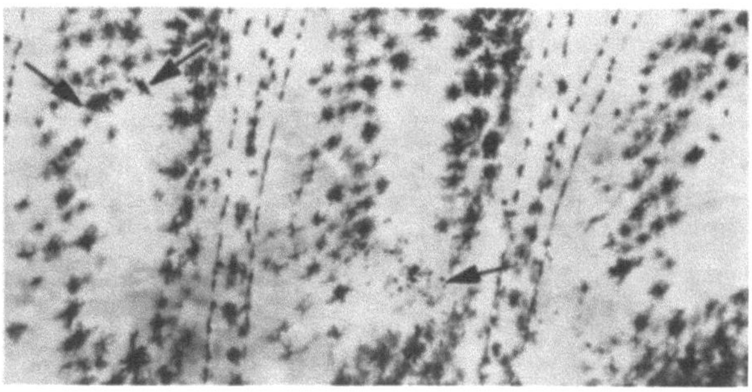

Fig. 12. Indication of melanophore destruction in the dorsal fin of a young *X. xiphidium*. Macrophage groups (marked by arrows) are to be recognized at the edges of the light zones. 15:1

results of melanophore destruction. This process starts well synchronized in all corresponding areas of a fin. Therefore, in some specimens, the typical accumulations of macrophages occur manifolded (arrows), in other specimens no one can be found.

3.3.3. Maintaining of Patterns

Rudiments of destroyed melanophores are to be found, more or less, at all times everywhere on the fish body. From this, one can presume that melanophores can always be victims of destruction. For the estimation of its quantitative effect two photographs of an identical region are given, which were taken in an interval of 2 weeks (Figs. 13a and b): 20-30% of the melanophores formerly present have totally disappeared (except of some melanin which is carried by macrophages). These photos by no means reproduce an extreme (compare Becker-Carus, 1965, for other fishes s. already Fukui, 1927).

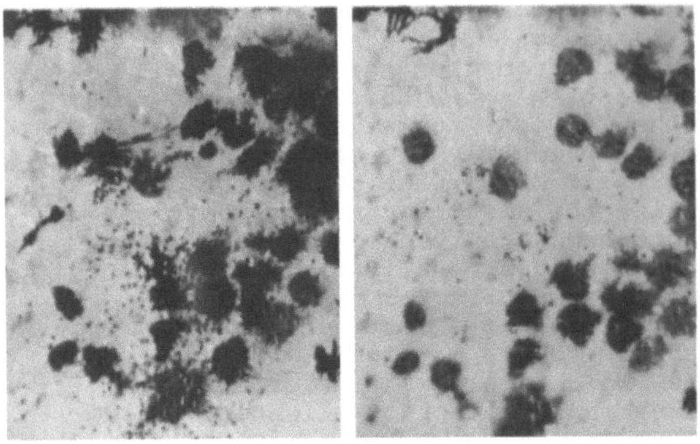

a b

Fig. 13a and b. Results of melanophore destruction in the skin within 2 weeks (cp. 13a and 13b). Fish as Fig. 3. 35:1

3.4. "Infectious Distribution" of Melanophores

3.4.1. Local Accumulation of Melanophores on Scales

Obvious examples are scales which are completely covered with macromelanophores (Fig. 14). In concentrations like these, the differentiation of each melanophore cannot be regarded as a single event. Instead,

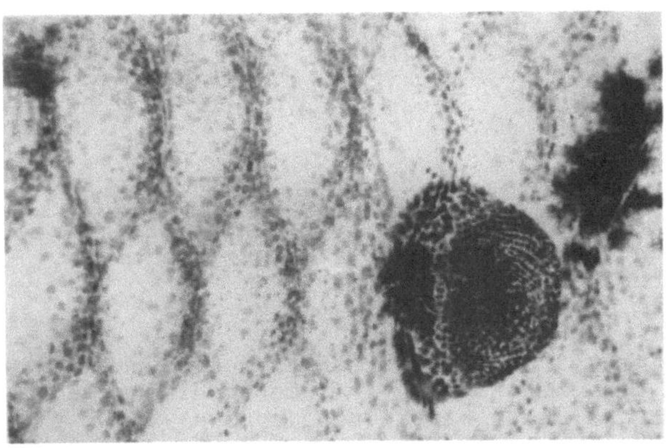

Fig. 14. Part of the skin of a *X. maculatus* (gene Sp) with a scale which is completely populated with relatively homogenous melanophores. Note that the melanophores outside the scale are from another type. 15:1

one favourable combination of melanophore promoting circumstances leads to a lot of melanophores (which may also differentiate to unequal sizes). Such observations are important, as they indicate that the number of melanophores must not be representative for the intensity of the number of melanophore determinating factors.

3.4.2. Melanophore Spots in a Pattern

When development of macromelanophores is conditioned by several gradients, pigment cells are expected to be arranged in a pattern, which is the result of interference of the gradients involved. This can be found in specimens of *X. helleri* which bear the gene Db^2 (Fig. 15): the black spots are arranged in rows, and their distances within the rows are

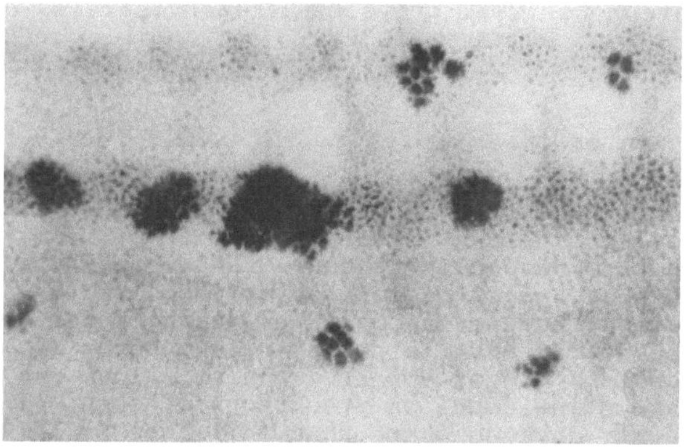

Fig. 15. Part of the skin of a *X. helleri*. Macromelanophore spots (caused by the gene Db^2) are arranged in a pattern which is connected with scale pocket edges. Spots mostly consist of many melanophores. 8:1

obviously regular. Since in a network of gradients, extreme constellations of melanophore promoting circumstances arise with little probability, many spots in the pattern should contain few and small melanophores, whereas spots with many large macromelanophores should appear rarely. This, however, is not the case: size and composition of the spots cannot be coordinated with any rule. From this observation it is clear that the number of macromelanophores in a spot is determined by other factors than the origin of the spot itself. It is assumed that the first melanophore(s) are responsible for a milieu in which further increase of melanophore number is favoured, either through protection against destruction or through abolishment of differentiation inhibition for melanoblasts.

4. Parallel Observations in Xanthophores (or Erythrophores) and Iridophores

Researches on xanthophores, erythrophores and iridophores are more difficult, because of technical reasons. For example, their cell margins are often hardly recognizable, so cell numbers cannot be established. Young cells do not yet contain coloured pigments. By use of interference contrast in microscopy, the cell expansion may be determined by means of its granulation. This, however, requires too much time. Especially iridophores are often so closely concentrated that they cannot be discerned. Up to now, no clearly distinguishable cells are proved to be precursors (but compare page 151 ff.). Even pigment cell destruction is hardly visible, since macrophages which could carry the pigments are scarcely perceptible because of low colour intensity.

4.1. Precursors of Xanthophores and Iridophores

In regenerating tissues melanophores are the first pigment cells that develop. They are followed by xanthophores, at last iridophores appear. From this it can be argued, that precursors for xanthophores and iridophores must exist somewhere. Since repetition of regeneration did not yet lead to a lack of these cells, the reservoir of precursors can be regarded as rather high, perhaps abundant, like that for melanophores. Nevertheless, they could not yet be identified. Perhaps, at least precursors of xanthophores and melanophores will prove to be identical, as is pointed out by several authors (cp. Bagnara, 1972).

4.2. Inhibition of Differentiation or of Growth

4.2.1. Blocked Xanthophore and Iridophore Precursors

When xanthophores and iridophores appear in a regenerate, this may be interpreted by the assumption that corresponding precursors have been stimulated for development, as has been pointed out for melanophores. Nevertheless, there is a basic difference: melanocytes or melanophores are always present when xanthophores differentiate, and melanophores and xanthophores are present when iridophores appear. The existing pigment-cells may have a suppressive influence on precursors of the "new" kinds. Then, these could only develop where this influence is low, i.e. in areas between melanophores where no pigment cells are existent. Actual distribution of pigment cells fits well to this interpretation. In some cases, however, pigment cells may have promoting effects on precursors of other cell kinds, as can be argued from associations of melanophores with iridophores (s. below) in some genotypes.

4.2.2. Gradients

Xanthophores. In most fishes there is a gradient in which xanthophores and erythrophores are larger and more differentiated in the dorsal regions than in the ventral ones, in addition, their number generally decreases in the same direction. Typical gradients are expressed in longitudinal stripes, the maxima of differentiation being in the middle of them. Concentrations of erythrophores are mostly established where scale pocket gradients for melanophores cross each other.

Iridophores. Two gradients are marked especially well: the number of iridophores increases from dorsal to ventral, conversely to melanophore numbers, and from rostral to caudal. The first gradient goes along with a change in the type of iridophores, since their sizes decrease from dorsal to ventral.

4.2.3. Contact Inhibition

Because cell margins of xantho- and erythrophores cannot be determined exactly, it cannot be ascertained that branches of separated cells touch each other or that they grow directedly. It is to observe clearly, however, that xantho- and erythrophores are arranged in general in one layer, which mostly is another one than that of melanophores of the same skin region. Even iridophores form layers, in which they may be densely packed side by side. Several layers may be arranged one upon another, they are separated from each other by non-pigment cells.

4.3. Destruction of Xanthophores and Iridophores

Macrophages containing yellow or red pigment granules are sometimes found. These indicate that xantho- and erythrophores have been destroyed. The quantitative effect of this process and its significance for regulation of patterns cannot be estimated at all. Macrophages carrying crystals similar to those of iridophores were observed, too.

4.4. "Infectious Distribution" of Xantho- or Erythrophores and Iridophores

4.4.1. Local Accumulations on Scales

This has only been observed for iridophores especially in fishes bearing a lot of macromelanophore spots. Such scales appear as bright shields (Fig. 16).

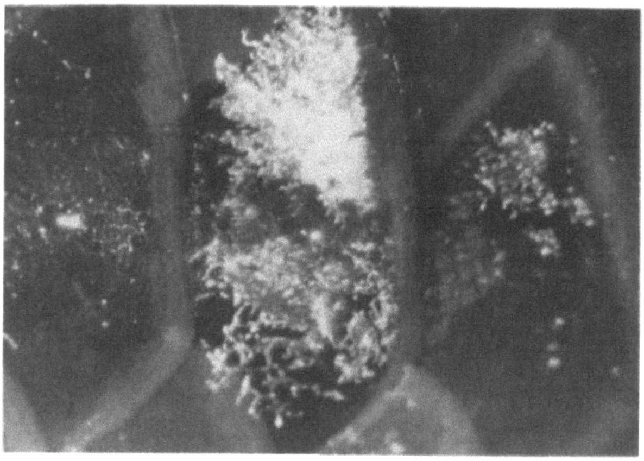

Fig. 16. Local concentration of iridophores on a scale of a *X. helleri*/*X. maculatus*-F_1-hybrid with the genes L and P. The background is a compact layer of macromelanophores. 15:1

5. Relationship of Pigment Cells

5.1. Xantho- or Erythrophores vs. Melanophores

5.1.1. Common Characters

Both kinds of cells correspond in shape and size, both contain pigment granules which are able to move, and they contain special compounds to which for example belong pteridins, tyrosinase.

5.1.2. Cells which Explain Characters of Both Cell Kinds

Occasionally xanthophores contain granules which correspond with melanosomes of melanophores exactly in colour and in size. The pigments behave in solvents as melanins do. Cells of this kind mostly occur isolated among melanophores, sometimes three or four of them lie together.

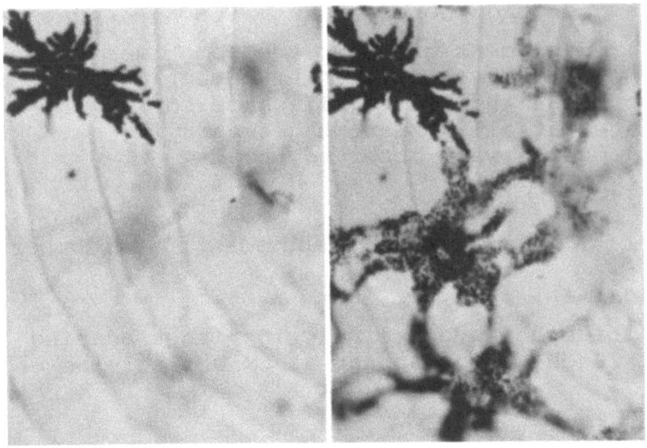

a b

Fig. 17a and b. Impregnation of xanthophores with $AgNO_3$.
a. Part of a scale (with one melanophore) before impregnation. Xanthophores (photographed with a blue filter) are faintly visible.
b. Same section after impregnation. Xanthophores are tightly filled with black granules, that are considered as premelanosomes, according to the specifity of the method. *X. xiphidium*, 320:1

On the other hand especially in regenerating tissues melanocytes can be found that contain a diffuse yellow pigment typical for young G-xanthophores. Impregnation of xanthophores with $AgNO_3$ gives still another hint for the relationship (Figs. 17a and 17b). As the method used leads only to impregnation of premelanosomes or premelanins, it is shown that, although the cells are undoubtedly xanthophores, they have melanophore specific characters. It should be marked that not all xantho- or erythrophores react in this way.

5.1.3. Experimental "Melanophorisation" of Xanthophores

Xanthophores in the vicinity of regenerating tissues are stained by dopa, with concentration of stain in granules and diffusely darkened plasma. The staining is significantly less effective in xanthophores which are not neighboured by regenerating tissues (cp. Figs. 18a and 18b). This is explained by the assumption that in regeneration areas the same circumstances which induce melanocyte differentiation also act on xanthophores.

5.2. Iridophores vs. Melanophores (page 156 ff.)

6. Special Principles in Pattern Formation

6.1. Heterogeneity of "Black Spots"

The term "black spot" comprises quite different collectives of melanophores, that cannot be compared with each other. Thus a spot may be composed of a few large macromelanophores or of melanophores very different in size or of many micromelanophores. Figs. 19a and 19b represent some examples. One specimen of fish can bear these different

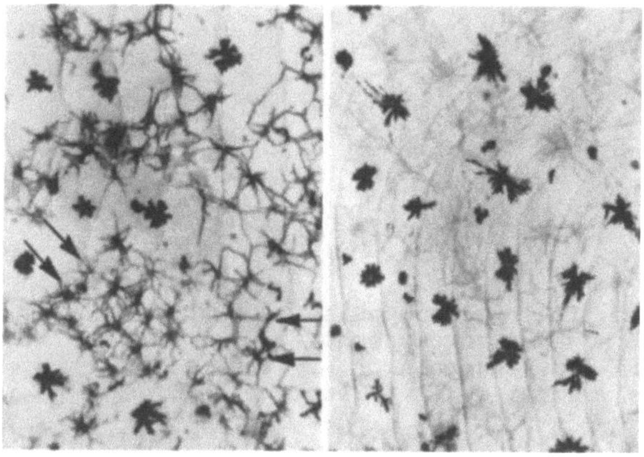

a b

Fig. 18a and b. Different effect of dopa-incubation on xanthophores in regenerating tissue, in which melanophore-differentiation is induced and on xanthophores of the opposite, uninjured body-part of the same fish (genotype as Fig. 3, 30 days after wounding).
a. Xanthophores from the wounded region have been strongly stained with dopa (some are marked by arrows, the compact cells are melanophores), whereas the control xanthophores are only slightly stained (b.). 120:1

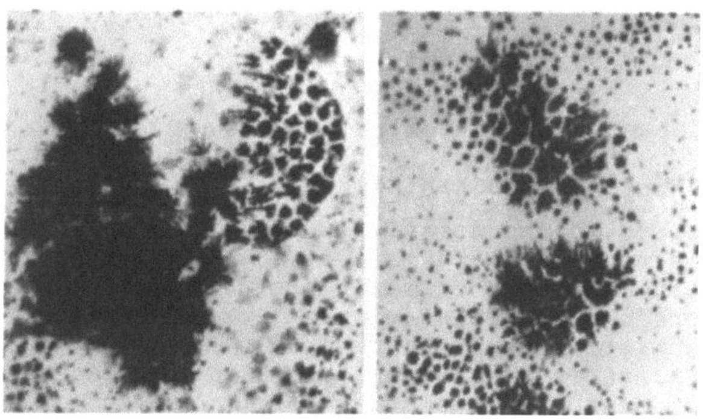

a b

Fig. 19a and b. Different compositions of "black spots".
a. Large macromelanophores beside the scale, small ones on it (skin of a *X. variatus*).
b. Spots consisting of large and small macromelanophores (skin of a *X. maculatus*) (cp. also Fig. 20). Both 25:1

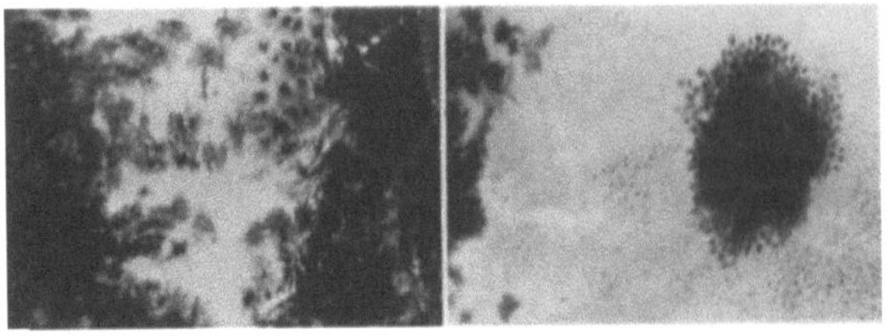

a b

Fig. 20a and b. "Black spots" of different composition on the same fish.
a. Part of the skin with typical huge macromelanophores.
b. Regenerated spot from the same body region appr. 50 days after wounding, consisting of relatively small cells. Genotype as Fig. 3. Both 10:1

arrangements simultaneously. The composition of a spot proved to be not determined by the genotype of the fish, as was shown in regeneration experiments: *X. helleri/X. maculatus* hybrids bearing the gene Sp have spots on their flanks according to Fig. 20a. After removal of the skin, spots develop again. They start, however, from a lot of melanocytes, that differentiate simultaneously, and which turn into relatively small melanophores; hence the new spots (Fig. 20b) are completely different from the original ones on the same place. Altogether it seems that, in a teleological formulation, a black spot seems to be necessary. The way in which it is developed depends on the local possibilities.

6.2. Parallel Reactions in Different Pigment Cell Kinds

6.2.1. Spots Consisting of Melanophores and Xanthophores

One characteristic example are spots in the dorsal fin of *X. helleri*: They are formed within an area of rather equally sized melanophores and xanthophores by substantial enlargement both of melanophores and of xanthophores. The latter ones produce a special pigment, drosopterine, when a certain degree of differentiation has been reached. Thus they appear red, and they are called xanthoerythrophores. Another impressive example is given by aggregates of erythrophores and melanophores in longitudinal stripes of *X. helleri* which are homozygote for the gene "golden" (s. Anders et al., 1972).

6.2.2. Parallel Increase of Melanophore and Xanthophore Numbers

Gene constellations which cause an important increase of macromelanophore numbers may effect xantho- or especially erythrophores in the same manner. This becomes drastically obvious in hybrids of *X. maculatus* and *X. helleri*, which possess the gene complex Sd and Dr of *X. maculatus*. In specimens of this genotype the macromelanophores expand in the dorsal area (dorsal fin, back, dorsal tail-stalk), whereas the red cells settle in the flanks - the two pigmented cell kinds substitute each other. It has to be regarded that in other genotypes other events are observed (page 156 ff.).

6.2.3. Iridophores and Melanophores in Heavily Blackened Fishes

In some cases, when very large numbers of macromelanophores are produced, iridophores may be increased very much, too. Then they may be located quite unusually on scales. This has been observed in *X. helleri/X. maculatus*-F_1-hybrids with the gene Sp, for example. Increase of iridophore numbers is not necessarily connected with high amounts of melanophores, thus *X. maculatus* with the gene Pseudo-Sp does not possess many iridophores (Anders, unpublished).

6.2.4. "Melaniridosomes"

They consist of melanophores and iridophores (Fig. 21) and have been named by Ballowitz (1913). Since there complexes mainly occur in body-regions, where iridophores are very rare or missing, we presume that restriction of iridophores to the immediate neighbourhood of a melanophore is the result of parallel reaction to local circumstances.

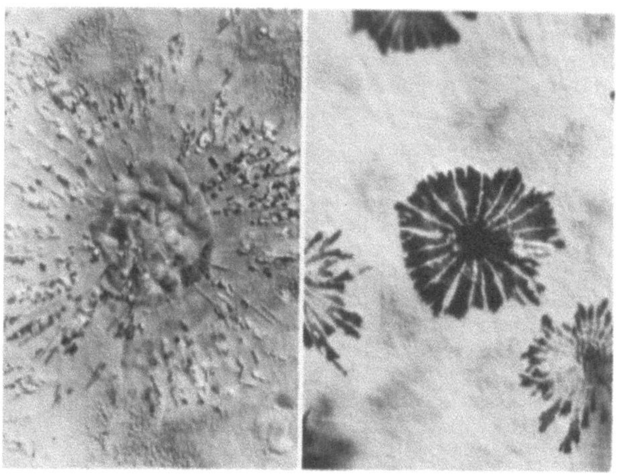

a b

Fig. 21a and b. Melaniridosomes.
a. From a *X. helleri* albino, the melanophores of which are atypical. Therefore the association of one of them with the iridophores is to be seen very clearly. 800:1.
b. A standard melaniridosome from *X. helleri*. The iridophore is hardly visible. Interference contrast, 320:1

6.3. Antagonistic Reactions of Different Pigment Cell Kinds ("Competition")

6.3.1. Iridophores and Xanthophores

It becomes obvious in so-called albinos, the melanophores of which do not produce melanin, so that other pigment cells are better to be seen than in normal animals. If a so-called "white albino" was taken as a standard for iridophore numbers, then in sister animals, possessing a gene which causes development of a lot of erythrophores (making the specimen red) the iridophore number is reduced to 1/3 - 1/4 of the standard. It cannot be decided yet if this is caused by competition for suitable sites in the skin or if, perhaps, a limited amount of common precursors is available for both cell kinds. It may even be that existing erythrophores are able to suppress differentiation of iridophores. Independent from causal analysis it is clearly stated that pigment cells of one kind exert a restrictive influence on other ones (s. Humphrey and Bagnara, 1967).

6.3.2. Alternative Increase of Melanophores or of Erythrophores

In contradiction of the example given on page 154 ff. *X. maculatus* hybrids with the gene-complex Sd-Dr either may have preponderantly macromelanophores or erythrophores, depending on the parent used for hybridization. Thus, in F_1-hybrids with *X. montezumae cortezi*, macromelanophores are increased, in F_1-hybrids with *X. couchianus*, erythrophores (taken from Zander, 1969).

7. Discussion

Most observations have been done on melanophores. Their relevance for other kinds of pigment cells depends on the relationships within them. Therefore, at first this will be discussed. Then some conclusions concerning spot and pattern formation in general will be pointed out.

Melanophores and xantho- or erythrophores possess many characters, which are common to both of them (page 151), but different from other cell types. So the main difference are the pigments which are stored in them. Perhaps observers tend to overestimate differences in colour (and pigments), thus these pigment cells are to be regarded as very similar. Consequently, many authors assume that "melanoblasts" represent the precursors for both of them (for example Ide and Hama, 1969; Alexander, 1970). Although iridophores and melanophores have fewer common points in their morphology, hints for a close relationship between them come from microscopical research. Thus, Arnott, Best and Nicol (1970) and Bagnara (1972) demonstrated that iridophores may contain melanosomes, and that melanophores may contain iridosomes. According to Bagnara this "bipolar" development takes place in tissues in which two gradient of differentiation meet. Other observations confirming close relationship among all three kinds of pigment cells come from electron microscopical work. Thus, it has been found that in one membrane-circled organelle, pigment bearing granules of different cell kinds can be developed (Arnott, Best and Nicol, 1970; Bagnara, 1972; see also Riley, 1972). Biochemical research also has pointed in the same direction: Ide and Hama (1969) showed that in goldfish in pterinosomes of typical xanthophores tyrosinase is active.

As for the problem of pigment cell pattern formation, these data seem to be sufficient for the assumption that regulation of the numbers of xantho- and erythrophores and of iridophores will be based on a similar complexity of mechanisms as can be demonstrated for melanophores. Thus, the total complexity is increased considerably.

Considering, first of all, an actual pigment cell population within the framework of a kybernetical model, one can state that it is the outcome of rates of pigment cell development and pigment cell destruction. The balance depends on the relation of both rates to each other. It may be positive or negative. This may vary gradually in a given fish specimen and it is obviously different according to the body regions of an individual. In nearly all cases it is not clear what operates as regulatory factor, sometimes pigment cell density itself seems to constitute a decisive part in it. Nevertheless, it is not yet possible to explain single observations in that scheme. It can only be stated that lack of pigment cell precursors usually is not involved in the regulation mechanism. That holds, at least, for melanophores (page 141 ff.), probably as well for other kinds, if observations (page 149 ff.) and interrelationships (s. above) are regarded.

As for general conclusions first the complexity of events must be regarded. Since so many variable effects play a role in establishing or maintaining a pigment cell arrangement, it is not exaggerated to state that every pigment cell population is based on a unique combination of parameters. The probabilities for combinations are, of course, affected by the genotypes. Thus, certain pigment cell arrangements will be found often or never in specimens. If genetic analysis has been done, prognosis is possible, but only in giving the potential spectrum of possibilities with estimated probabilities. In this respect, even parts of one fish's body are not equal, but have different probabilities for the appearance of certain pigment cell arrangements or different pigment cell arrangements, respectively. A fish appears subdivided in "localities" that partially influence each other or are partially independent. This interpretation leads to the conclusion that a whole specimen is too "large" to be the unit for research in pigmentation analysis, dynamics of cell populations going on heterogenously in various regions of the body. The specimen remains only as the unit as a "genotype" that enables the breeder to foresee which genotypes are to be expected among the descendants.

Two kinds of observations concerning the genetic backgrounds of pigmentation polymorphism in *Xiphophorus* still will be regarded.

As genetic constellations determine only a vast framework in which cells react according to their possibilities, there may result very similar or equal pigmentations that are based on quite different genotypes. Kallman has provided research material to this topic (1970) and has given further data by oral information. Therefore it is clear that one cannot always conclude with certainty from the phenotype of a fish to its genotype.

It is well known from many authors that many genes are involved in pigment cell pattern formation. Then it is quite understandable that small populations of any species that are isolated from each other (Kallman, oral information) must have different gene systems for pigment cell arrangements after a short time, due to genetic drift. If descendants are crossed, it is statistically improbable that the hybrids receive genotypes that make a pigment cell arrangement according to one of the parents possible. Instead, "unbalanced" genotypes are to be expected causing the formation of more or of less pigment cells than the parents had. If one regards, in these cases, the so-called "colour cell genes" one will see that their manifestation is enlarged or restricted, respectively.

The same regulation mechanisms that are applicable for numbers of pigment cells ought to be applicable also for other cells. Indeed, hybrids mostly have special body proportions, which may be due to variations of cell numbers and/or cell arrangement, and melanoma inducing gene constellations have been observed as affecting non-pigment cells (Lueken and Knoll, 1969). Nevertheless, the pigment cells have an outstanding significance: they are often multiplied so much that they are harmful to their carrier or even may kill it. In this aspect mainly melanophores or melanocytes are extraordinary. It cannot be decided what is the reason for it.

8. Summary

Many different steps are required to establish the pigment cell arrangements in species of *Xiphophorus* and interspecific hybrids. Those taking place in the peripheral region of a fish can be contributed to four classes: 1. Existence and behaviour of pigment cell precursors; 2. Inhibition of differentiation or growth; 3. Destruction of pigment cells; 4. Pigment cells inducing the development of more pigment cells. Most of the results presented to these points come from experiments with melanophores. Their precursors, melanoblasts, are abundant in most genotypes. Lack of melanoblasts seems to be only exceptionally a way to limit the amount of melanophores present in fish. The distribution of melanoblasts all over the body is accomplished by active, amoeba-like movements of the cells and by transportation through blood vessels. Melanoblasts show phobic reactions against melanophores. Their continued existence as melanoblasts depends on inhibition of differentiation which is caused in a so far unknown way by surrounding cells. Introduction of unbalanced intercellular conditions (e.g. through injury or transplantation) activates the melanoblasts. Then they differentiate into typical melanocytes within only a few hours (about 24). The growth of the melanophores may also be inhibited by their immediate environment to a degree which depends on the location of the pigment cells. This results in several partly overlapping gradients from well differentiated to less differentiated pigment cells: from the back to the belly, from the head to the tail, from the center of a stripe to its edges, etc. Greatly expanded melanophores show contact inhibition at the tips of their branches. On the other hand, however, growth of the tips is directed toward each other in some unknown way. Another mode of inhibition can be caused by mechanical obstacles such as ridges on scales. Destruction of melanophores occurs when too many have differentiated in regenerating tissues; also, when a pattern with melanophore-free areas originates from a previous one with more melanophores; moreover, it is a regular process in the maintenance of any pattern of melanophores - a given arrangement is the result from both differentiation and destruction. The distribution of melanophores on scales or in spots can be "infectious", therefore induction of melanophore development by existing melanophores is postulated. - Parallel observations on other pigment cells, xanthophores (plus erythrophores) and iridophores, are impossible in most cases because they are hardly visible as separated cells. Genealogical relations between the different kinds of pigment cells are discussed taking into consideration the observation of cells with mixed characteristics, the experimental induction of melanophore characters in xanthophores, and the existing literature. All three main kinds of pigment cells seem to be rather closely related. So conclusions can be drawn to a certain extent from the behaviour of melanophores to that of other pigment cells. Some examples are given for different modes of cooperation between pigment cells. It is pointed out that every existing pigment cell arrangement has its own history of development. Therefore a whole fish specimen mostly cannot be regarded as the unit for the realization of gene constellations responsible for pigmentation. Another consequence is that it is often impossible to conclude from the result, the phenotype, on the steps of the development which have taken place during the fish's life span. As identical pigment cell arrangements can be formed in different ways during development, they can be correlated with different genotypes.

References

Alexander, N.J. (1970): Differentiation of the melanophore, iridophore and xanthophore from a common stem cell. J. Investig. Dermatol. 54, 82.

Anders, F. (1967): Tumour formation in platyfish-swordtail hybrids as a problem of gene regulation. Experientia 23, 1-10.

Anders, F., Klinke, K., Vielkind, U. (1972): Genregulation und Differenzierung im Melanomsystem der Zahnkärpflinge. Biol. in uns. Zeit 2, 35-45.

Arnott, H.J., Best, A.C.G., Nicol, A.C. (1970): Occurrence of melanosomes and cristal sacs within the same cell in the tapetum lucidum of the stingaree. J. Cell Biol. 46, 426-427.

Bagnara, J.T. (1972): Interrelationship of melanophores, iridophores and xanthophores. In: Riley, V.: Pigmentation: Its genesis and biologic control. New York: Meredith Corporation, 171-180.

Ballowitz, E. (1913): Die chromatischen Organe, Melaniridosomen, in der Haut der Barsche (*Perca* und *Acerina*). Z. wiss. Zool. 110, 1-35.

Becker(-Carus), C. (1965): Untersuchungen zur Phänogenese von Melanophorenmustern bei Zahnkarpfen. Z. wiss. Zool. 172, 37-103.

Fukui, K. (1927): On the color pattern produced by various agents in the goldfish. Folia Anat. Japon. 5, 257-302.

Greenberg, S., Kopac, M.J. (1963): Studies of gene action and melanogenic enzyme activity in melanomatous fish. Ann. N.Y. Acad. Sci. 100, 887-923.

Humphrey, R.R., Bagnara, J.T. (1967): A color variant in the Mexican axolotl. J. Heredity 58, 251-256.

Ide, H., Hama, T. (1969): Tyrosinase activity of the melanin and pteridin granules in goldfish chromatophores. Biochim. Biophys. Acta 192, 200-204.

Kallman, K.D. (1970): Different genetic basis of identical pigment pattern in two populations of platyfish, *Xiphophorus maculatus*. Copeia 1970, 472-487.

Kallman, K.D., Atz, J.W. (1966): Gene and chromosome homology in fishes of the genus *Xiphophorus*. Zoologica 51, 107-135.

Levene, A., McGovern, V.J., Mishima, Y., Oettle, A.G. (1966): Terminology of vertebrate melanin-containing cells, their precursors, and related cells: a report of the Nomenclature Committee of the Sixth International Pigment Cell Conference. In: Della Porta, G., Mühlbock, O.: Structure and control of the melanocyte. Berlin-Heidelberg-New York: Springer, 1-5.

Lueken, W., Foerster, W. (1969): Chromosomenuntersuchungen bei Fischen mit einer vereinfachten Zellkulturtechnik. Zool. Anz. 183, 169-176.

Lueken, W., Kaeser, U. (1972): The role of melanoblasts in melanophore pattern polymorphism of *Xiphophorus* (Pisces, Poeciliidae). Experientia 28, 1340-1341.

Lueken, W., Knoll, C. (1969): Über den Einfluß tumorinduzierender Genkonstellationen auf verschiedene Zelltypen. Untersuchungen an lebendgebärenden Zahnkarpfen (Poeciliidae). Verhdlg. Dtsch. Zool. Ges. Innsbruck 1968, 238-244.

Mishima, Y. (1960): New technique for comprehensive demonstration of melanin, premelanin, and tyrosinase sites. J. Investig. Dermatol. 34, 355-360.

Öktay, M. (1964): Über genbedingte rote Farbmuster bei *Xiphophorus maculatus*. Ein Beispiel für Nicht-Autonomie der Merkmalsbildung. Mitt. Hamb. Zool. Mus. Inst. Kosswig-Festschrift, 133-157.

Riley, V. (1972): Pigmentation: its genesis and biological control. New York: Meredith Corporation.

Schmidt, E.R. (1973): Über die Rolle der Iridophoren bei der Realisierung des Farbzellenmusters von *Xiphophorus* (Pisces, Poeciliidae). Diplomarbeit Universität Gießen.

Vielkind, U., Vielkind, J., Götting, K.-H. (1968): Zur Feinstruktur von Fischtumoren. Naturwiss. 55, 349.

Wolf, B. (1972): Die Farbmuster lebendgebärender Zahnkarpfen der Gattung *Xiphophorus*. Diplomarbeit Universität Gießen.

Zander, C.D. (1960): Künstliche Befruchtung bei lebendgebärenden Zahnkarpfen. Zool. Anz. 166, 81-87.

Zander, C.D. (1969): Über die Entstehung und Veränderung von Farbmustern in der Gattung *Xiphophorus* (Pisces). Mitt. Hamb. Zool. Mus. Inst. 66, 241-271.

Experimentally Induced Sexual Maturity in Artificially Reared Male Eels *(Anguilla anguilla)*

Ch. Meske

In the experimental plants of the fish research station in Ahrensburg aquarium experiments have been conducted over many years on food fish, in particular on carp and eels, under the following conditions:

1. All experimental animals are reared exclusively under controlled laboratory conditions in the station's indoor aquaria and are at no point kept out of doors.

2. A continuous stream of water flowing through the experimental aquaria removes all growth-inhibiting excretions. This renders possible an unusually high population density, i.e. up to 1 kg live fish per 3 litres of aquarium water.

3. The water is maintained at a constant temperature all the year round, in the majority of cases at 23°C.

4. The fish are fed throughout the entire year, preferably, and in most cases, with standardized dried food.

As a result of these constant conditions of keeping and feeding the fish show a steady increase in weight over the entire year, far above that observed under outdoor conditions.

The aims of the aquarium experiments were the following:

1. Investigations on the influence of abiotic environmental factors on development.

2. Investigations on the nutritional physiology of food fish with particular respect to nutritional requirements.

3. Investigations on the reproductive biology of fish under the constant environmental conditions of the aquarium, all seasonal variations being thus excluded.

4. Experimental and applied genetics of food fish.

5. Hydrobiological investigations and the development of new methods of keeping fish under continuous circulation. Microbiological and chemical investigations were carried out on a closed circulation with biologically cleaned water, as developed in this station (v. Sengbusch, Meske and Szablewski, 1965; Meske, 1973).

So far, carp (*Cyprinus carpio*), european catfish (*Silurus glanis*) and eels (*Anguilla anguilla*) have been used in our experiments on warm-water rearing of fish. Experiments with some other fish species are in progress.

Fig. 1 shows the weight increase of carp in the experimental aquaria under the conditions listed above. It is seen that a weight of 2,5 kg can be attained after only one year, a weight usually reached by carp reared in ponds for commercial purposes in Central Europe after an average of 4-5 years. After barely 2 years a weight of 5 kg is attainable by carp reared according to the intensive warm-water method.

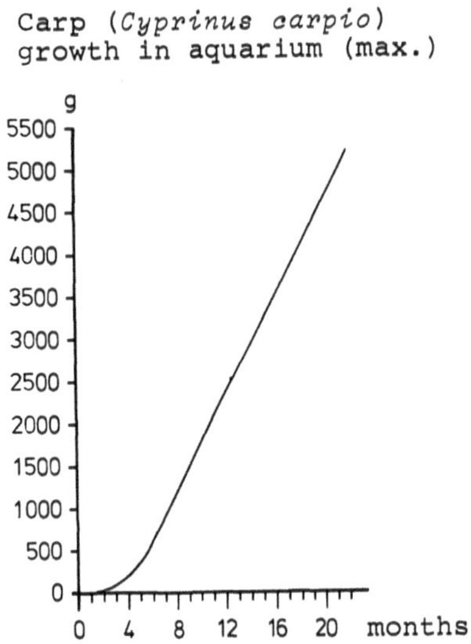

Fig. 1

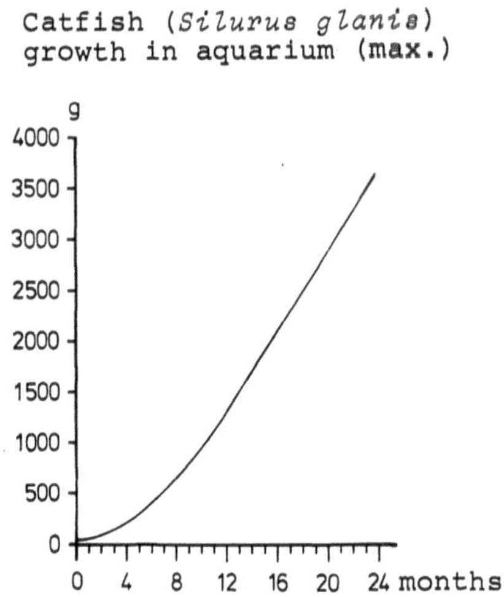

Fig. 2

Fig. 1. Growth curve of carp (*Cyprinus carpio*) in warm water

Fig. 2. Growth curve of catfish (*Silurus glanis*) in warm water

Fig. 2 shows the weight increase observed in european catfish. An animal that weighed 20 g when placed in the aquarium weighed 3,745 kg after 2 years

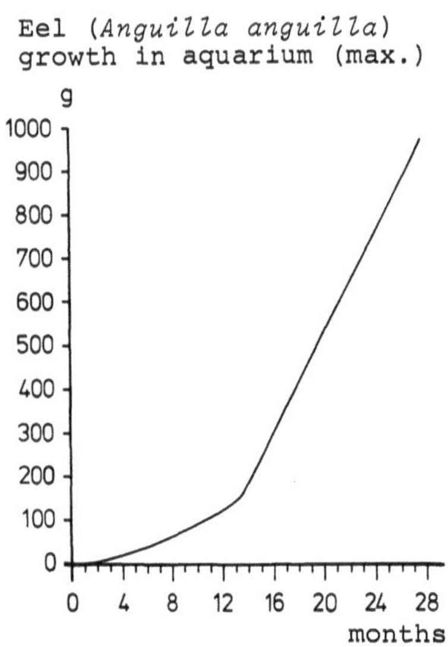

Fig. 3. Growth curve of eel (*Anguilla anguilla*) in warm water

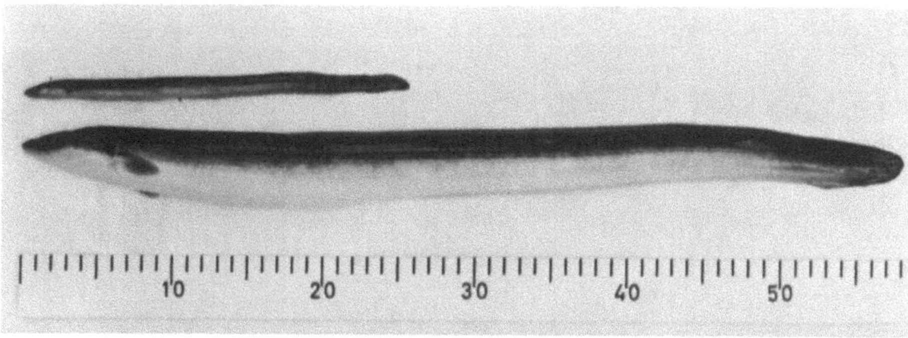

Fig. 4. Increase in weight of eel in one year (above: 24 g, below: 445 g)

Fig. 3 demonstrates the weight increase of eels placed in the aquarium as glass eels with a weight of approximately 0,4 g. After 28 months in warm water weights of almost 1 kg were attained. According to Ehrenbaum (in Meyer-Waarden, 1965), eels in outdoor waters attain a weight of at most 3 g after 2-2 1/2 years. To quote the same author: 'well developed females weigh a maximum of 266 g after 9-9 1/2 years'. Fig. 4, too, illustrates the rapid weight development under warm-water conditions. The weight of an eel of 24 g (above) is compared with one of 445 g (below), values separated by a growth period of one year (24 g: 2.10.68, 445 g: 9.10.69). Such a large increase in weight, however, is unusual. In fact, in each new group of glass eels placed in the warm-water aquaria the individuals showed a strong tendency to differ from one another in weight. It later appeared that there is no correlation between weight increase and sexual maturation.

The keeping and feeding of eels naturally presented different problems for those encountered with carp. The migratory instinct of freshly caught glass eels had to be combatted initially with the aid of specially constructed tightly-closed aquaria. Pellets of dried food proved to be unsuitable food and minced fresh fish was substituted. The fresh food was pushed through a feeding tube constructed for this purpose and removed as soon as the fish had received sufficient food (Fig. 5)(Meske, 1969).

Besides bringing about a constant and greatly accelerated increase in weight as compared with outdoor specimens the warm-water method results in an acceleration of sexual maturity. Female carp are ready to spawn after one year in the warm water aquaria as compared with 5 years in the open. Male carp can produce fertile sperms in an age of only 3 months whereas this state is achieved only after three years in the open. By means of intramuscular injections of fish hypophysis the release of sexual products can be induced and fertilization can then be brought about in vitro. This presents new openings in experimental and applied genetics of food fish thanks to the almost unlimited combinations possible for each of the parent animals (Meske et al. 1968; Meske, 1971, 1973). The carp hatching from eggs fertilized in this way were also reared exclusively in aquaria, the entire life cycle thus being completed indoors.

Our knowledge concerning gonadal development in the eel is extremely incomplete. Only in a few cases has experimentally induced sexual maturity been described. Boëtius and Boëtius (1967) forced the maturation of male eel gonads by hormone treatment and reported 'spawning'. The spermatozoa were not, however, described.

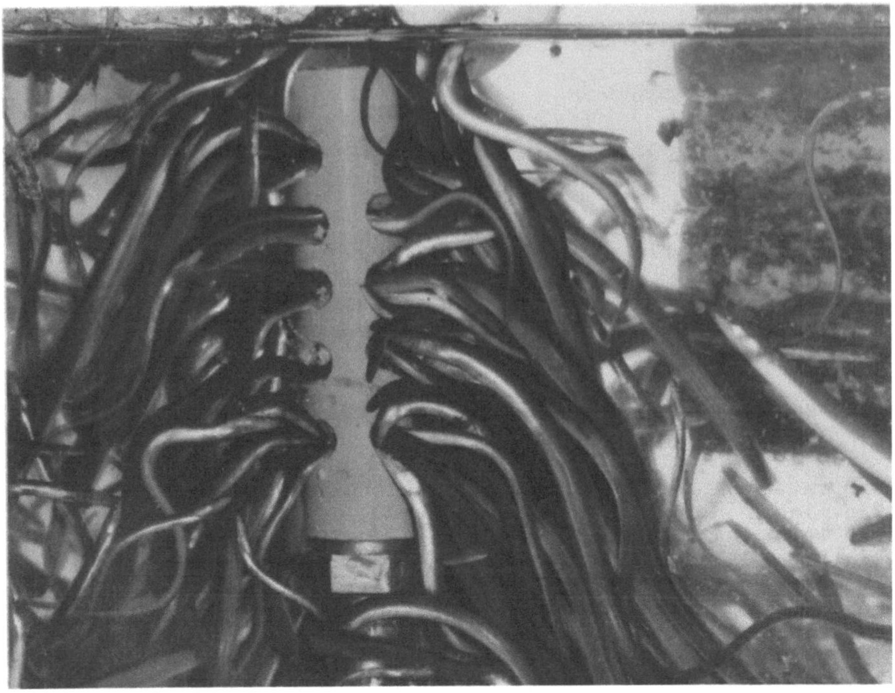

Fig. 5. Feeding of eels with ground fish meat through a feeding tube

The maturation of female gonads was reported by Fontaine et al. (1964) and in several cases they succeeded in obtaining eggs. Japanese authors reported the treatment of one female eel in 1970 up to egg deposition (Nose, 1971).

All of the cases mentioned involved silver eels that had been caught in the vicinity of the coast and already showed considerable sexual maturation. Age and origin were thus unknown. In all cases the animals were subsequently kept in sea water aquaria and treated with various hormone preparations.

In Ahrensburg nine male and nine female eels were chosen for treatment aimed at inducing sexual maturity in animals raised in the aquarium. The nine male eels had all been placed in the warm water system as glass eels in April, 1968. The average weight of these animals at the commencement of the experiment was 114 g. Some of the nine females used in the experiment were glass eels of the 1968 population and some were of unknown origin. Their average weight at the beginning of the experiment was 598 g.

Both male and female eels were chosen from the available material on account of somewhat enlarged eyes and a so-called silver eel colouring, considered to be indicators of the beginning of sexual maturation. In accordance with values so far known for male and female eels our distinction between the sexes was based upon the fact that male animals are thought not to achieve a length of more than 40-50 cm.

The 18 animals chosen were treated as follows:

I. After having been reared entirely in freshwater aquaria all experimental fish were transferred in December, 1971 to saltwater aquaria with a constant flow of salt water of a salinity of 35°/oo. The water temperature was 22°C. At this point the fish had, to a large extent, stopped feeding. Food offered from time to time was only rarely accepted.

II. Commencing with April 8, 1972 the 18 experimental animals were treated with the following organ and hormone preparations:

1. Solcosplen (SOLCO GmbH, Wyhlen). Solcosplen is a protein-free extract of fresh calves' spleen, the positive effect of which on the development of the gonads of rodents has been reported by Grigoriadis, Goslar and Jaeger (1968) and Goslar, Grigoriadis and Jaeger (1969). According to these authors the spleen extract had a considerable stimulatory effect, for example, on infant guinea pig testes: the Leydig ceels were greatly enlarged, the majority of the tubuli opened and lactate dehydrogenase and 3β-ol-steroid-dehydrogenase activity was elevated. The authors suggested that the active substance in the spleen extract has a stimulating effect upon the gonadotropin production of the hypophysis.

2. Synahorin (Teikoku Zoki Pharmaceutical Co., Tokyo). Synahorin is a mixture of gonadotropic hormone of the anterior lobe of the hypophysis and of the placenta of warm-blooded animals, in the ratio of 1:9.

3. Cyren B (Bayer, Leverkusen). Cyren B is a synthetic oestrogen, diethylstilboestroldipropionate.

The animals were treated according to the following scheme: three groups, each consisting of three male and three female eels, were formed and treated as follows: Group A: per animal 0,2 ml spleen extract at intervals of two weeks. Group B: per animal 0,2 ml spleen extract plus 50 RU Synahorin at intervals of two weeks. Group C: per animal 50 RU Synahorin + 0,25 mg Cyren B at intervals of two weeks. All substances were injected intramuscularly.

Fig. 6 shows the plan of treatment together with the results obtained. It can be seen that no sexual products were obtained from the animals treated with Solcosplen alone, but that a combination of Solcosplen and Synahorin (treatment B) brought positive results in all male eels used in the experiment, that is to say, they attained sexual maturity. Treatment C (Synahorin + Cyren B) had no effect in either sex.

	treatment A Solcosplen	treatment B Solcosplen +Synahorin	treatment C Synahorin +Cyren B
♂	−	+	−
♂	−	+	−
♂	−	+	−
♀	−	−	−
♀	−	−	−
♀	−	−	−

Fig. 6. Schedule followed in order to bring about sexual maturity in the eel. (Details in text)

Fig. 7. Untreated immature male eel, weight 72 g, body length 34 cm

The sexual maturity so far induced in the three male eels of Group B took the external form of greatly enlarged eyes. The so-called 'deep-sea eyes' already reported on several occasions (e.g. Schmidt in Meyer-Waarden, 1965) were observed here for the first time in eels reared in aquaria from the glass eel stage. Fig. 7 shows a 1 1/2 year

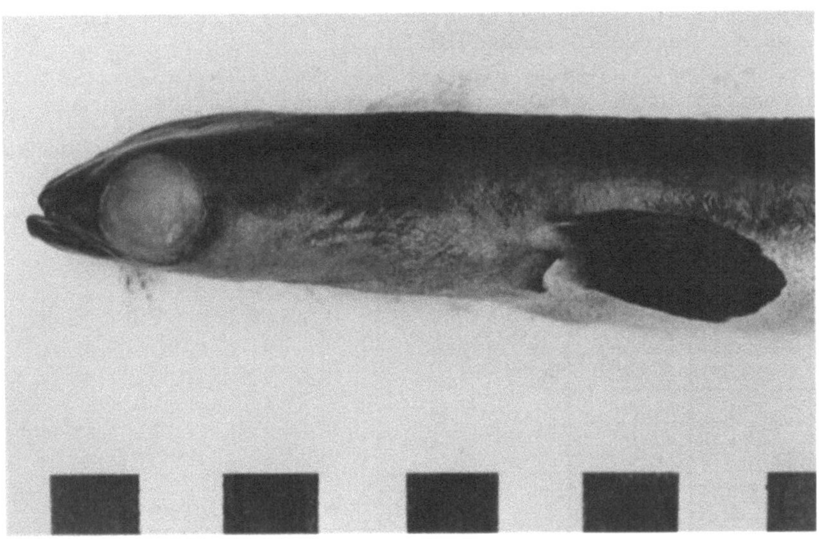

Fig. 8. Treated, sexually mature eel, weight 80 g, body length 44,5 cm

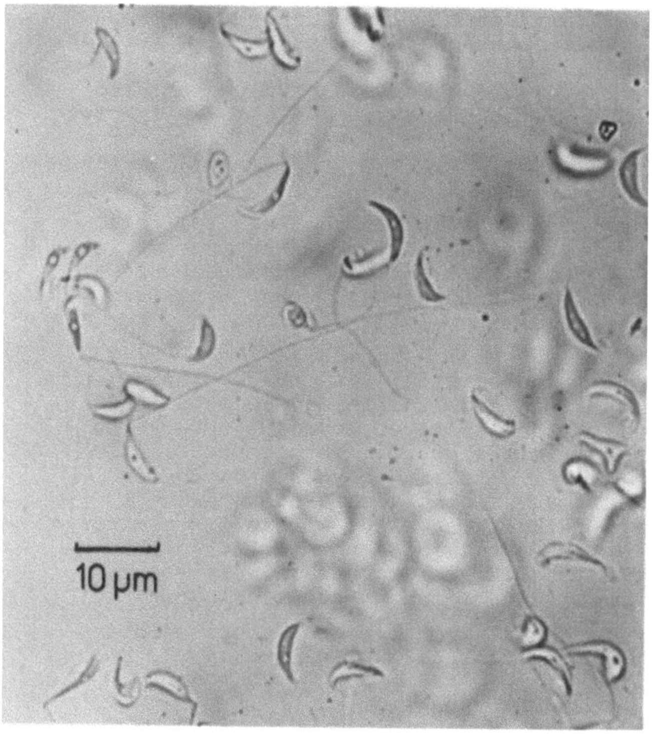

Fig. 9. Free eel sperms in unfixed preparation

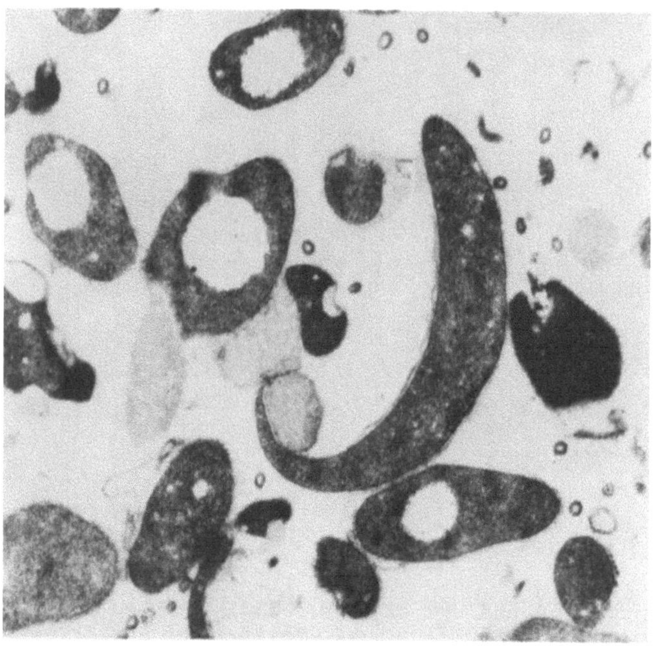

Fig. 10. Electron micrograph of eel sperms. (Fixation: gluturaldehyde-osmium)

old male, weight 72 g, length 34 cm, untreated, sexually immature. Eye diameter 5 mm, relatively small thoracic fins (the scale divisions given are in cm). Fig. 8 shows a male eel treated according to the plan in Fig. 6 with spleen extract and Synahorin, up to the emission of sperms. It weighs 80 g and has a body length of 44,5 cm. The eyes of this individual have the unusually large diameter of 12,5 mm, the cornea exhibitinga slightly opalescent turbidity[1]. The thoracic fins are also very much enlarged.

Animals exhibiting extensive external changes of this nature, following four injections, exuded spermatozoa on 3.7.72 in response to light massage of the underside of the body. The animals concerned were all of the males of Group B, treated with spleen extract and Synahorin (Meske and Cellarius, 1972). No sperms could be obtained from animals of Groups A and C.

The spermatozoa are slightly bent and pointed, oval shape, their length, excluding flagellum is ca. 9μm, The flagellum is ca. 30μm in length, and its rapid beating brings about the agitated movement observable in unfixed preparations (Fig. 9). The electron micrograph (Fig. 10) shows sections of eel spermatozoa at different levels. In the one sperm that is cut longitudinally over most of its length the cell membrane is recognizable, surrounding the nucleus within its nuclear membrane. A large mitochondrion is easily distinguishable just in front of the base of the flagellum. The numerous small circles scattered over the picture are cross sections of sperm tails[2].

The mature testes of an eel that had died just before sperm ejection are shown in Fig. 11. The dissected testicular line in Fig. 12 shows

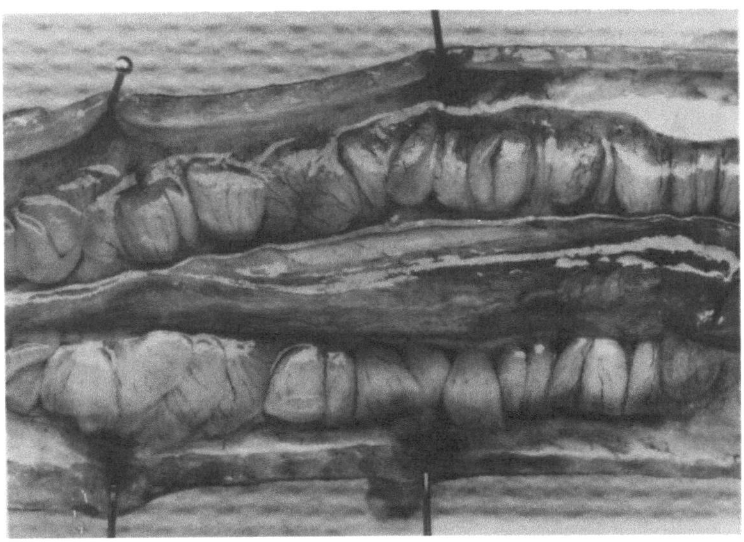

Fig. 11. View of opened, mature male eel *in situ*

[1]For a more detailed description of the eye of *Anguilla anguilla*, its enlargement and possible biological significance, see Stramke (1972).

[2]My thanks are due to Prof.Dr.med. Schirren, Universitäts-Haut- und Poliklinik, Hamburg-Eppendorf, for the electron micrograph (Fig. 10).

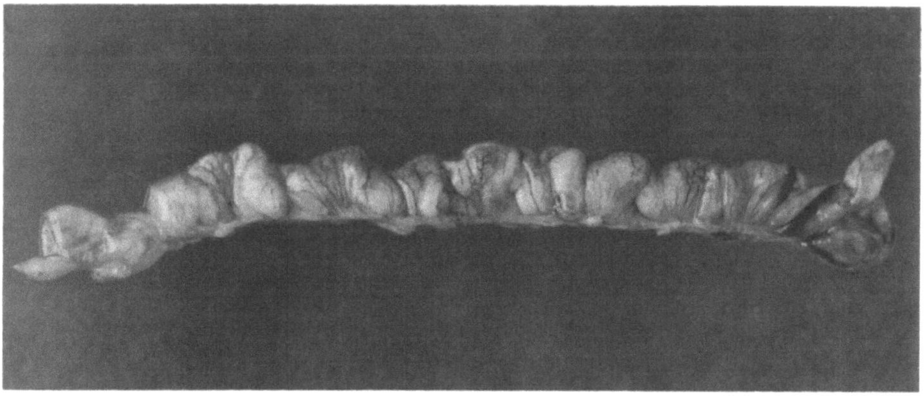

Fig. 12. Dissected, mature testicular lobule of male eel

clearly the lobular arrangement of this organ (so-called Syrski organ), and its attachment over its entire length to the vas deferens. The combined weight of the two testicular lines amounted to 3,56% of the body weight in this case.

The experimental routine so far adhered to (see Fig. 6) has revealed the combination of substances capable of eliciting spawning in male eels in the aquarium. It is, however, unsatisfactory in that the individual components of the substances used could not be tested. To investigate the influence of the environment on the maturation of eels further experiments were carried out on other individuals that had been kept exclusively in fresh water from the glass eel stage onwards and treated with Solcosplen and Synahorin at the age of 3 or 4 years. They were given 0,2 ml spleen extract per 100 g body weight and 50 RU Synahorin per animal intramuscularly. After only three such injections actively mobile spermatozoa could be obtained by massage. This could be repeated several times on one and the same individual, and from one particular animal motile sperms were obtained from July to October, 1972, six times.

The latter experiments show clearly that adaptation of the organism to salt water is not essential for the development of the gonads. It was also observed that the vitality of the male animals was not exhausted subsequent to spawning. By far the larger part of the successfully treated eels is still alive, several months after the first release of sperms.

Although the treatment of female eels with the various substances mentioned has so far not led to the release of eggs, several of the females so treated showed similar external changes to those seen in successfully treated male animals (particularly greatly enlarged eyes). A histological preparation from one of the almost mature females that died during the experiment revealed eggs in an advanced stage of development. The weight of the animals lagged far behind another series of eels of the same age. The heavier fish, in contrast to the sexually more mature animals that were subsequently treated as above, showed no enlargement of the eyes and did not stop feeding. It is hoped that a series of experiments designed to explain these observations will bring useful results.

Summary

Carp (*Cyprinus carpio*), european catfish (*Silurus glanis*) and eels (*Anguilla anguilla*) kept all the year round in warm water aquaria exhibit a constant and greatly accelerated increase in weight. Under such conditions carp and eels attain sexual maturity abnormally early. For the first time mature, spawning male eels were reared from the glass eel stage with the help of intramuscular injections of spleen extract and gonadotrophic hormone. Sexually mature eels are characterized by greatly enlarged eyes. Free, highly motile sperms are described. The testes of a sexually mature eel were seen upon dissection to be completely filled with sperms. Females just short of maturity exhibit less increase in weight than other females. Maintenance in seawater is not essential for attainment of sexual maturity.

References

Boëtius, I., Boëtius, J. (1967): Studies in the European Eel, *Anguilla anguilla* (L.). Experimental induction of the male sexual cycle, its relation to temperature and other factors. Meddelelser fra Danmarks Fiskeri- og Havundersøgelser 4, H. 11, 339-405.

Fontaine, M., Bertrand, E., Lopez, E., Callamand, O. (1964): Sur la maturation des organes génitaux de l'Anguille femelle (*Anguilla anguilla* L.) et l'émission spontanée des oeufs en aquarium. C.R. Acad. Sci. (Paris) 259, 2907-2910.

Goslar, H.G., Grigoriadis, P., Jaeger, K.H. (1969): Die Wirkung eines Milzdialysates auf das Enzymmuster des infantilen Meerschweinchen- und Rattenhodens am normalen und hypophysektomierten Tier. Arzneimittel-Forschung (Drug. Res.) 19, 1249-1253.

Grigoriadis, P., Goslar, H.G., Jaeger, K.H. (1968): Vergleichende enzymhistochemische Untersuchungen zum Verhalten des infantilen Meerschweinchenhodens nach Milzextrakt- und Gonadotropinapplikation. Verh. Anatom. Ges. 63, 611-617, Vers., Leipzig.

Meske, Ch. (1969): Aufzucht von Aalbrut in Aquarien. Arch. Fischereiwiss. 20, H. 1, 26-32.

Meske, Ch. (1971): Warmwasser Fischzucht - Neue Verfahren der Aquakultur. Naturwiss. 58, H. 6, 312-318.

Meske, Ch. (1973): Aquakultur von Warmwasser-Nutzfischen. Stuttgart: Eugen Ulmer.

Meske, Ch., Cellarius, O. (1972): Laboraufzucht von Aalen bis zur Geschlechtsreife. Naturwiss. 59, H. 10, 471-472.

Meske, Ch., Woynarovich, E., Kausch, H., Lühr, B., Szablewski, W. (1968): Hypophysierung von Aquarienkarpfen und künstliche Laicherbrütung als Methode zur Züchtung neuer Karpfenrassen. Theoretical and Applied Genetics 38, H. 1/2, 47-51.

Meyer-Waarden, P.F. (1965): Die wundersame Lebensgeschichte des Aales. In: Keune: Der Aal. Hamburg: Hans A. Keune Verlag.

Nose, T. (1971): Spawning of eel in a small aquarium. Riv. It. Piscip. Ittiop. 6, H. 2, 25-26.

Sengbusch, R. v., Meske, Ch., Szablewski, W. (1965): Beschleunigtes Wachstum von Karpfen in Aquarien mit Hilfe biologischer Wasserklärung. Experientia 21, 614.

Stramke, D. (1972): Veränderungen am Auge des europäischen Aales (*Anguilla anguilla*, L.) während der Gelb- und Blankaalphase. Arch. Fischereiwiss. 23, H. 2, 101-117.

5. Ethology

Change of Aggressive Readiness in Post-Irradiation Generations of the Convict Cichlid Fish, *Cichlasoma nigrofasciatum*

S. Holzberg

1. Introduction

During evolution, changes in behavior patterns are thought to precede those of morphological traits (Wickler, 1961). Furthermore, it is generally accepted that most species-specific behavior patterns are inborn. Accordingly, mutations concerning behavior patterns attain special evolutionary significance because they produce new candidates for evolution.

The aim of this investigation has been to prove whether mutations of behavioral traits really emerge earlier than those of morphological characters following mutagenic treatment.

2. Methods

Convict cichlids (*Cichlasoma nigrofasciatum*) of one brood were divided in two equal parts of immature sibs. One group was irradiated with 2 x 500 R, 24 h apart (300 kVp; 10 mA with 0.5 mm copper and 1.0 mm aluminium filtration; dose rate about 230 R/min), the other served as control. Histological examinations at the time of treatment showed undifferentiated gonads with oogonial and spermatogonial stem cells. After reaching sexual maturity, each unirradiated control female was mated with an unirradiated control male, and each irradiated female was mated with an irradiated male. Therefore, the combined genetically effective radiation dose of F_1 animals was 2 000 R. The corresponding F_1 controls, however, did not contain replicas of irradiated chromosomes. The aggressive readiness of the adult F_1 males was tested under two experimental conditions:

1. After 5 weeks of sexual isolation in a community tank, irradiated and control males were individually placed together with a group of young conspecifics. The test fish could attack the young fish at will. Because the young fish escape from the bites of the adult male, a ritualised fight never occurs. After 48 h habituation the adult males bites delivered to the young were recorded over a period of 15 min.

2. In a second experiment the aggression was tested in the same way, with the difference that mated couples were now observed from the day of spawning to the first day of free-swimming of the larvae, i.e. over a period of 9 days. In this experiment, however, the aggression was tested three times a day, every session lasting 15 min. (cf. Holzberg and Schröder, 1972).

Isolated F_2 males were tested in the same way as described in experiment 1. for the isolated F_1 males.

3. Results

Aggressive readiness of F_1 males

Experiment 1. The analysis of variance confirmed the overall impression of a reduced number of bites in the irradiated series as compared with the controls (F = 20.136; P 0.005; n = 10 for each series).

In comparing the mean values of the two series, i.e. 13.7 $\pm$ 11.3 for the irradiated line and 48.5 $\pm$ 20.5 for the controls, the controls were found to be 3.5 times as aggressive as the cousins derived from irradiated parents.

Experiment 2. The mean values of bites per 900 sec. with the standard error of the mean were 18.734 $\pm$ 3.456 for the irradiated group and 39.958 $\pm$ 3.898 for the controls, revealing a 2.132 fold greater aggressivity for the controls as compared with that of the irradiated series. Thus, the aggressive readiness of mated males of the irradiated series was reduced significantly but to a lower degree than that of non-mated males of the same series.

Preliminary results on the aggressiveness of F_2 males

The aggressive readiness was found to be higher in the control group than in the irradiated series. However, these differences have not yet been verified statistically, perhaps of the small sample size of the animals investigated up to now. As to the distribution of aggressiveness between the two series, the F_2 males derived from irradiated grandparents were found to show a shift towards less aggressive individuals (Fig. 1).

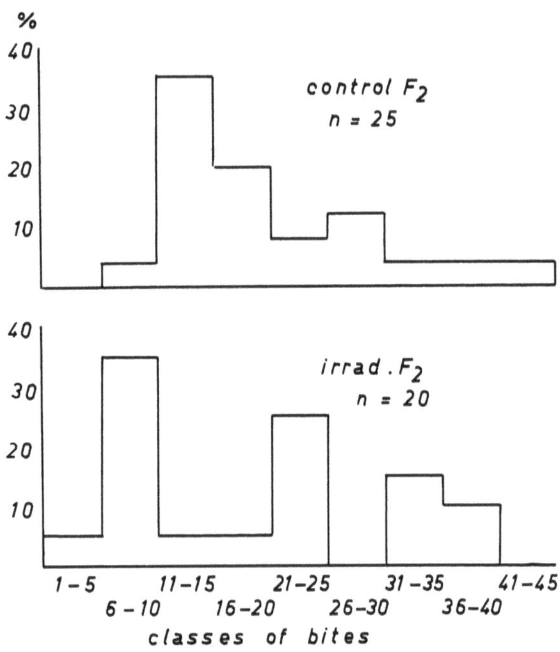

Fig. 1. The distribution of classes of bites expressed as percentage of irradiated and control F_2 males

Table 1. Comparison of the mean values and ranges of bites of F_1 mated males with their isolated individual F_2 sons

Mated F_1 males (mean values and range)		Non-mated F_2 sons (mean values and range)											$\frac{F_1}{F_2}$
		1	2	3	4	5	6	7	8	9	10	11	
B 11, 31.7 13-60	control group	9.0 5-17	13.5 9-17	13.8 11-20	14.0 18-21	14.5 9-21	19.8 17-23	23.5 15-28	48.8 16-87				1.6
B 3, 39.9 21-68	control group	11.5 7-19	12.5 8-17	15.5 8-32	17.0 12-22	32.8 10-57	34.0 18-45	34.5 12-61	37.0 23-51	41.0 23-51			2.4
B 12, 46.5 16-126	control group	13.5 9-19	14.2 10-20	14.8 8-21	17.4 8-23	19.2 13-29	21.2 14-30	25.5 13-35	30.0 19-45				1.5
B 10, 6.0 1-35	irrad. group	3.2 2-4	6.2 3-10	7.5 2-11	20.8 15-24	30.0 15-53	30.8 11-51						0.6
B 9, 13.4 1-31	irrad. group	5.2 4-8	6.5 2-14	10.0 6-15	17.2 8-29	20.8 5-31	22.0 4-37	23.8 13-39	31.0 12-76	40.8 29-61	41.0 24-64	20.8 15-26	0.4
C 11, 10.7 4-26	irrad. group	7.0 3-15	9.5 7-15	14.5 8-20									1.0

Considering the ratio of individual values of F_1 males to those of the F_2 sons, it was only possible to compare mated F_1 males with non-mated F_2 sons. Spacial limitations prevented the isolation of the non-mated F_1 males following testing of aggressiveness as described in experiment 1. However, because the difference of aggressiveness between mated and non-mated F_1 males within the same series was not statistically significant it seems justified to compare non-mated F_2 males with their mated F_1 fathers (cf. Holzberg and Schröder, 1972, Table IV). Nevertheless, the aggressive motivation may be different between mated and non-mated males.

As shown in Table 1, the ratio between the mean values of aggressiveness of F_1 mated males and isolated F_2 sons is greater than unity in the control and equal or less than unity in the irradiated group.

Although the variability of F_2 males belonging to one brood was greater in the irradiated than in the control series, no difference was found for the individual ranges (Table 1).

4. Discussion

The question whether the described changes are due to mutational events or to other causes arises.

The preliminary analysis of F_2 data favors hereditary origin of the observed effects. However, our present stage of knowledge does not permit to decide whether true point mutations or rather chromosomal rearrangements are involved. In the latter case differences in male gonadotropic hormone levels among animals of the two series could be responsible for the different aggressive readiness. This corresponds with the general view that chromosomal mutations are affecting the regulation system known to be influenced by hormones.

In the case of point mutations, however, quantitative traits such as aggressive behavior might be due rather to polygenes than to oligogenes. If so, the step-wise accumulation of behavioral micromutations would lead to a gradual alteration of the quantitative distribution of behavior. This corresponds to the view of Manning (1967) who assumes that behavioral mutations occur first by changing (lowering or enhancing) the threshold of innate releasing mechanisms against the corresponding releasers. These ideas are supported through the observation that species-specific behavior patterns as a rule differ quantitatively rather than qualitatively when closely related species are considered.

References

Holzberg, S., Schröder, J.H. (1972): Behavioral mutagenesis in the convict cichlid fish, *Cichlasoma nigrofasciatum* Guenther. I. The reduction of male aggressiveness in the first post-irradiation generation. Mutation Res. 16, 289-296.

Manning, A. (1967): Genes and the evolution of insect behavior. In: Behavior-Genetic Analysis (ed. J. Hirsch), New York: McGraw-Hill.

Wickler, W. (1961): Ökologie und Stammesgeschichte von Verhaltensweisen. Fortschr. Zool. 13, 303-365.

Attraction and Sexual Cycle of Poeciliids

J. Parzefall

1. Introduction

The existence of a cave-dwelling population of the live-bearing tooth-carp *Poecilia sphenops* provoked the question of what type of preadaptation the ancestors of this population might have had for cave life. Previous observations show that *P. sphenops*, contrary to the closely related *P. velifera* and *P. latipinna*, does not possess an optically effective display (Parzefall, 1969). This would suggest that the communication between the males and the females - by which the male is the actively seeking partner - must be achieved through chemical releasers. Through experiments with blinded males, Zeiske (1968, 1971) was able to show that such chemical releasers, which are probably perceived by tasting, are secreted by the females. Morphological studies, which show a larger number of taste buds in both sexes as well as a pad-like swelling of the genital region of the females in the cave-dwelling forms of *P. sphenops* (Parzefall, 1970), support this hypothesis and suggest the female genital region as the site of the secretion. This would also supply the explanation for a typical behaviour pattern in which the male nips at the genital region of the female (nipping).

This paper considers whether the female genital region actually is the site of the secretion and whether the secretion occurs constantly or according to a definite cycle.

2. Preference of the Conspecific Female

Before considering the main problem, it is necessary to test the assumption that the behaviour of the intact animal in darkness is the same as in the light and that the conspecific female is preferred. *Xiphophorus helleri*, a poeciliid with an optically effective display (backing)(Franck, 1964), and which, according to Zeiske (1968), probably cannot propagate in darkness, was used for a comparison. This species lives along with the surface form of *P. sphenops* in the stream before the cave (Gordon and Rosen, 1962).

Material and Methods

Using alternate choice experiments, the sexual reaction of *P. sphenops* males from the stream before the cave was tested. In one case the males were given the choice between conspecific males and females, and in the other case between conspecific females and *X. helleri guentheri* females from British Honduras. The duration of the test was always 30 minutes. The observations were made under infrared radiation using an Infrared-apparatus (AEG, Type IDF 34). The transmission limit of the infrared filter used was 760 nm. Based on the spectral sensitivity of the eyes of vertebrates, it can be assumed that an optical orientation would be impossible (Wolf, 1925). Preliminary experiments have borne out this assumption.

I wish to thank Miss M. Hänel for the execution of the drawings and Miss I. Breckwoldt for technical assistance. To the DFG I would like to express my gratitude for financial help (Pa 148/3).

Results

It was found that the sexual behaviour of *P. sphenops* in the dark is exactly the same as by daylight. In the case of *X. helleri*, no sexual behaviour was observed. However, the animals took up dry feed immediately as it was offered. In alternate choice experiments with *P. sphenops* it was clear that the conspecific female was preferred to the *X. helleri* female as well as to the conspecific male (Fig. 1).

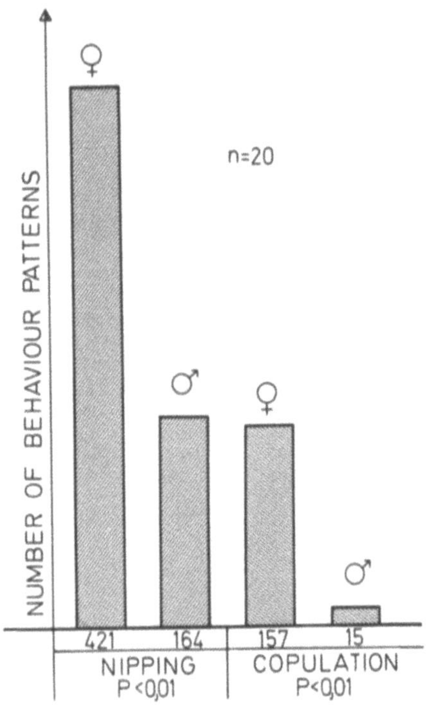

Fig. 1. *Poecilia sphenops*. Alternate choice experiment: sexual behaviour patterns of a ♂ as directed towards a conspecific unattractive ♀ or conspecific ♂. n = 20 tests/30 Min

The differences are significant (matched-pairs signed-ranks test from Wilcoxon). The females used in the test were not in a specific phase of their sexual cycle. With that, it was shown - in agreement with the findings of Zeiske (1968) using blinded animals - that *P. sphenops* males always recognize conspecific females, even when an optical orientation is not possible. Furthermore, they react sexually, a behaviour which does not take place in the case of *X. helleri*.

3. Attraction Cycle of the Female

Methods

In order to determine whether there is a specific cycle in the production of the chemical releasers in question, a "group-test" was made in which 4 marked females of the surface form of *P. sphenops* together with one male were observed by daylight over a period of several months. The sexual reactions, following, nipping, and copulation or copulation attempts, were recorded for each of the females. The females were marked according to a method described by Reinboth (1954) using different-colored glass beads. In addition, alternate choice experiments were run using females in different phases of their sexual cycles. In these experiments *X. helleri* was also tested.

Results

A rather even distribution of all of these elements was observed for all of the females up to the day of bearing. From this day on, there is an increase in sexual behaviour which is concentrated on the female which has borne young. The female in question avoids these overtures at first, however, the male persists, following her constantly and darkening in color. Fig. 2 shows how the following decreases from

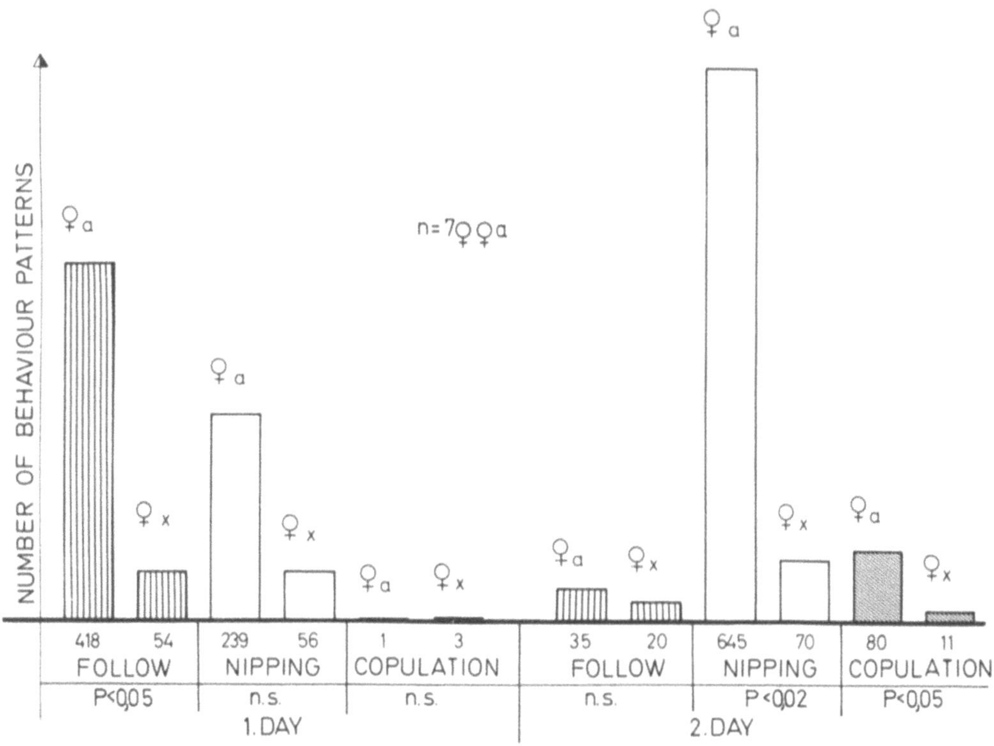

Fig. 2. *Poecilia sphenops*. Group test: sexual behaviour patterns of a ♂ as directed towards a ♀a (directly after bearing) or one of three ♀♀x (before bearing).

the first day to the second. On the second day the values for nipping and copulation movements increase due to the fact that the female no longer flees. On the whole, the preference for females directly after they have borne young is quite clear. The differences are significant. These experiments were also run using virginal females and - to prevent insemination - a male in which the gonopodium had been amputated. These females, too, although they produced no young, proved to have the same peaks of attractiveness corresponding to the dates when the fertilized females bore young. These results, which were obtained using few females in a long-term experiment, were confirmed by means of an alternate-choice experiment in which the sexual reactions of a male able to choose between an attractive female (directly after bearing) and an unattractive female (directly before bearing) were observed.

These alternate-choice experiments were also run with *X. helleri*, although the experiments in the darkness had shown this species to be so dependent on optical releasers that it is apparently incapable of carrying out sexual behaviour in the dark. However, males of this species also showed a distinct preference for the female directly after the bearing of young. In this case the increased frequency of nipping, which occurred with *P. sphenops*, was replaced by an increase in the frequency of backing. A *P. sphenops* male which was used in this experiment as a control hardly reacted at all to the *X. helleri* female and showed no significant preference for the attractive female (Fig. 3). Following preliminary experiments with *P. (=Limia) melano-*

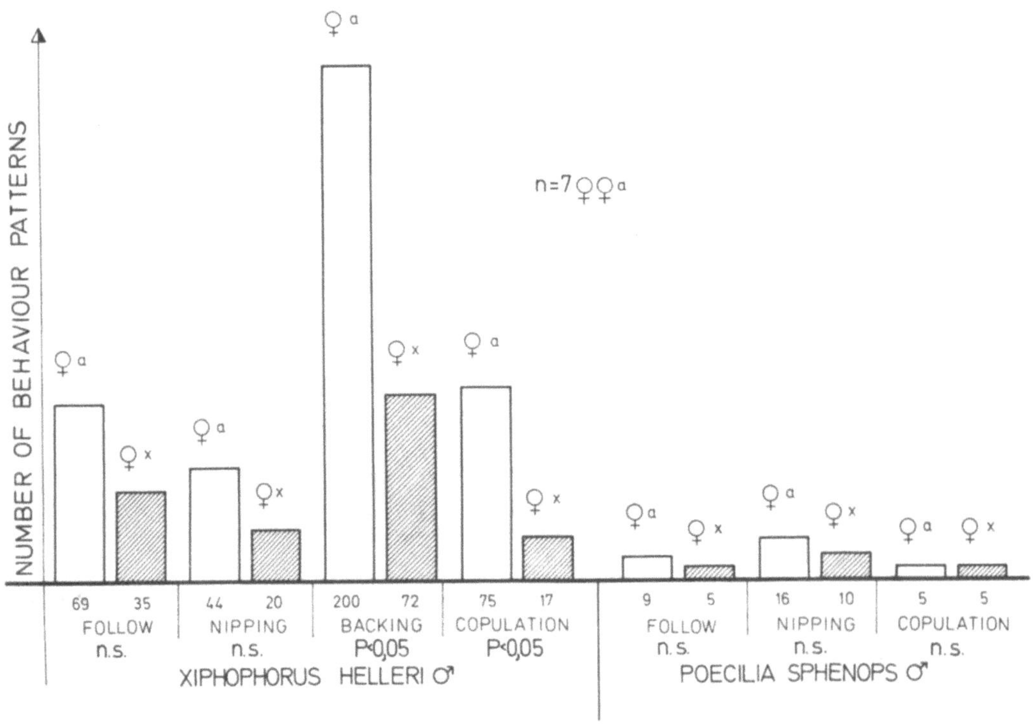

Fig. 3. *Xiphophorus helleri, Poecilia sphenops.* Alternate choice experiment: sexual behaviour patterns of the two ♂♂ as directed towards a *X. helleri* ♀a or *X. helleri* ♀x

gaster, it can be expected that males of this species, too, show a preference for females directly after they have produced a brood.

Morphologically, aside from being thinner due to the production of a brood, the attractive *P. sphenops* female has a distinct dark spot in the genital region. These signals, however, apparently do not determine the attractiveness of the female: in the "group-test", when the attractive female is removed from the aquarium and then immediately presented again behind a pane of glass or in a form-fitting nylon bag, she is no longer recognized as attractive. The male constantly tests the available females through nipping contact in the genital region and swims, searching, throughout the aquarium. After a while the darker color of the male pales, and he begins to give up his search. As soon as an attractive female (the one which was previously removed, or another attractive female) is offered to the male so that she is directly available to him, he darkens in color again within seconds after the first nipping contact. His sexual reactions are then exclusively directed at this female.

4. Secretion Site of the Presumable Releasers

Since the nipping is exclusively directed at the genital region of the female, the female genital organs played a central role in the investigation of this question.

Methods

Fertilized *P. sphenops* females on different numbers of days after the day of bearing, as well as virginal females, were killed and fixed after their attractiveness or unattractiveness had been determined through direct observation. The body was then cut sagitally and stained (Pasini or Azan stain).

Results

It was found that on all attractive females, including the virginal ones, the genital papilla is evaginated and opened at the top. In the case of fertilized females, this observation is in agreement with those for *X. helleri* (Peters and Mäder, 1964) and *P. reticulata* (Weishaupt, 1925) where it was seen that shortly before bearing the oviduct opens at the end, probably due to the influence of hormones. Several days after bearing it closes again. A similar mechanism appears to occur with the virginal females in connection with the degeneration of the mature unfertilized generation of eggs and the maturation of the next egg generation; accordingly it could be assumed that the production of the "attractive substance" takes place in the ovary. However, it is not yet possible to say where in the ovary this production occurs.

5. Discussion

On the basis of the above results, it is evident that even the surface form of *P. sphenops* is capable of identifying the conspecific female, when optical orientation is not possible, at any time. In association with nipping and copulation movements, which can be observed in the dark as well, a propagation in the dark would be assured.

The experiments of Zeiske (1968) show it to be quite probable that this function is principally due to a tasteable substance which is given off by the female, and that the main information is conveyed by direct nipping contact.

In this respect, *P. sphenops* differs from other fish, by which such a recognition or activation substance is apparently perceived olfactorily and, to some extent, over a larger distance. Amouriq (1964, 1965) and Gandolfi (1969) observed that *P. reticulata* females give off a substance, later identified as estrogen (Amouriq, 1967), which precipitates an increase in activity in the males when aquarium water is used in which females have been. Males of the characid *Astyanax mexicanus* are activated by females during spawning due to a substance which is also perceived olfactorily (Wilkens, 1972). Tavolga (1956) made similar observations for *Bathygobius soporator*.

These phenomena are in agreement with the second finding here: *P. sphenops* males are most attracted to the females in the swarm which have just produced a brood (or have resorbed one egg generation) and will produce a new egg generation in a few days. In this respect a successful fertilization is also more probable since the male does not have to pierce the normally sunken genital papilla.

These considerations would point to a possible parallelism between attractiveness and different egg maturation durations (Turner, 1937). Hereby, superfoetation would be of particular interest. The observations of such an attractiveness in *X. helleri* and *P. melanogaster* would indicate that this phenomenon occurs with many - if not all -, poeciliids in connection with a substance which is effective only for the species in question (Fig. 3).

Such a substance could also, for reasons mentioned above, have an important function for a swarm in the light. However, the observations of *X. helleri* show that this mechanism is not sufficiently effective to insure a successful propagation in the dark. For this, additional sexual behaviour patterns which can also be carried out in the dark are necessary. Only in this interrelation, as with *P. sphenops*, can the substance play a principal part in making the survival of a species in the dark possible.

Therefore, future investigations designed to find the location where this substance is formed should concentrate particularly on species which do not possess an optically effective display.

6. Summary

The males of *Poecilia sphenops* recognize conspecific females during the whole sexual cyclus even in darkness. The females are particularly attractive for the males after they have produced a brood. In this period the genital papilla of the female is evaginated and opened at its top. Similar observations in *Xiphophorus helleri* and *Poecilia melanogaster* indicate that this chemical attraction seems to be a common phenomenon in Poeciliids. In *P. sphenops* it has become very important for reproduction under cave conditions.

References

Amouriq, L. (1964): L'activité et le phénomène social chez *Lebistes reticulatus* (Poeciliidae, Cyprinodontiformes), C.R. Acad. Sc. Paris 259, 2701-2702.

Amouriq, L. (1965): Origine de la substance dynamogéne émise par *Lebistes reticulatus* femelle (Poisson, Poeciliidae, Cyprinodontiformes), C.R. Acad. Sc. Paris 260, 2334-2335.

Amouriq, L. (1967): Sensibilité de *Lebistes reticulatus* malê à la substance dynamogêne émise par les femelles de Poeciliidae et de Gasterosteidae. Rev. Comp. Animal 4, 83-86.

Franck, D. (1964): Vergleichende Verhaltensstudien an lebendgebärenden Zahnkarpfen der Gattung *Xiphophorus*. Zool. Ib. Physiol. 71, 117-170.
Gandolfi, G. (1969): A chemical sex attractant in the guppy *Poecilia reticulata* Peters (Pisces, Poeciliidae). Monitore zool. ital. 3, 89-98.
Gordon, M.S., Rosen, D.E. (1962): A cavernicolous form of the poeciliid fish *Poecilia sphenops* from Tabasco, Mexico. Copeia 2, 360-368.
Parzefall, J. (1969): Zur vergleichenden Ethologie verschiedener *Mollienesia*-Arten einschließlich einer Höhlenform von *Mollienesia sphenops*. Behaviour 33, 1-37.
Parzefall, J. (1970): Morphologische Untersuchungen an einer Höhlenform von *Mollienesia sphenops* (Pisces, Poeciliidae). Z. Morph. Tiere 62, 211-244.
Peters, G., Mäder, B. (1964): Morphologische Veränderungen der Gonadenausführgänge sich fortpflanzender Schwertträgerweibchen (*Xiphophorus helleri* Heckel). Zool. Anz. 173, 243-257.
Reinboth, R. (1954): Eine Methode zur Markierung von Aquarienfischen. Zool. Anz. 153, 190-194.
Tavolga, W. (1956): Visual, chemical and sound stimuli as cues in sex discriminatory behaviour of the Gobiid fish *Bathygobius soporator*. Zoologica (N.Y.) 41, 49-65.
Turner, C.L. (1937): Reproductive cycles and superfetation in Poeciliid fishes. Biol. Bull. Woods Hde 72, 145.
Weishaupt, E. (1925): Die Ontogenie der Genitalorgane von *Girardinus reticulatus*. Z. wiss. Zool. 126, 571-611.
Wilkens, H. (1972): Über Präadaptationen für das Höhlenleben, untersucht am Laiverhalten ober- und unterirdischer Population des *Astyanax mexicanus* (Pisces). Zool. Anz. 188, 1-11.
Wolf (1925): Das Farbunterscheidungsvermögen der Elritze. Z. vergl. Physiol. 3, 279-329.
Zeiske, E. (1968): Praedispositionen bei *Mollienesia sphenops* (Pisces, Poeciliidae) für einen Übergang zum Leben in subterranen Gewässern. Z. vergl. Physiol. 58, 190-222.
Zeiske, E. (1970): Ethologische Mechanismen als Voraussetzung für einen Übergang zum Höhlenleben. Untersuchungen an Kaspar-Hauser-Männchen von *Poecilia sphenops* (Pisces, Poeciliidae). Forma et functio 4, 387-393.

6. Evolution

Genetic Problems in the Regressive Evolution of Cavernicolous Fish

N. Peters and G. Peters[1]

1. Introduction

When speaking of evolutionary processes, we mean the change in organisms with the progression of geological time. These changes are stable, in that they are based in the genetic material of the organisms. In order to judge the evolutionary processes, we must first differentiate between inherited changes and those changes brought about by the environment, these latter being mere modifications. This distinction is essential for the regressive characteristics of cave-dwellers. The following example will help to illustrate this point: if the normal-eyed river fish *Astyanax mexicanus* is raised in total darkness in a laboratory, the eyes remain definitely smaller than those of siblings raised under daylight conditions. The retina shows a marked change in proportion: the total number of receptor cells, as well as the number per unit area, is obviously smaller (see also Franck-Krahé, 1962). However, when the troglobiont derivative of the river fish *Astyanax mexicanus* (also known as *Anoptichthys*) is raised under epigean conditions, no normal eye is produced. What does occur is a slight enlargement of the cyst-like eye rudiment, which is found deep in the orbit. This enlargement is accompanied by an improved structure of the cyst. The remains of the layered retina can often be found in such rudiments (Wilkens, 1971).

We see therefore, that the less distinctive regressive traits of the cave-dwellers can definitely be dependent on the environmental situation. The conspicuous changes of the troglobionts, however, are most likely genetically determined.

The numerous cave-living derivatives of the characid *Astyanax mexicanus* are among the most intensely studied cavernicoles. At present there are more than 20 known cave populations to be found in Central America. The various populations show marked differences in the amount of reduction present. In some cases the cave populations almost completely resemble the river form; in the majority, however, these cavernicoles are blind and lacking in pigmentation. In one of the geographically isolated caves, the Micos Cave, there lives a blind population whose pigmentation is still present; in the Chica Cave, a hybrid population exists with members ranging from blind to normal-eyed and from pale to well pigmented, respectively. The river fish obviously had the chance to enter this cave a second time, where it met the long established troglobiont and readily mated. In still other caves normal-eyed and pigmented fish co-exist with true troglobionts without the occurrence of cross-breeding; the reason for this is unknown.

It can be said that, on the one hand, troglophile forms are derived from the river fish *Astyanax mexicanus*, the regressive traits of which

---

[1]Supported by the Deutsche Forschungsgemeinschaft.

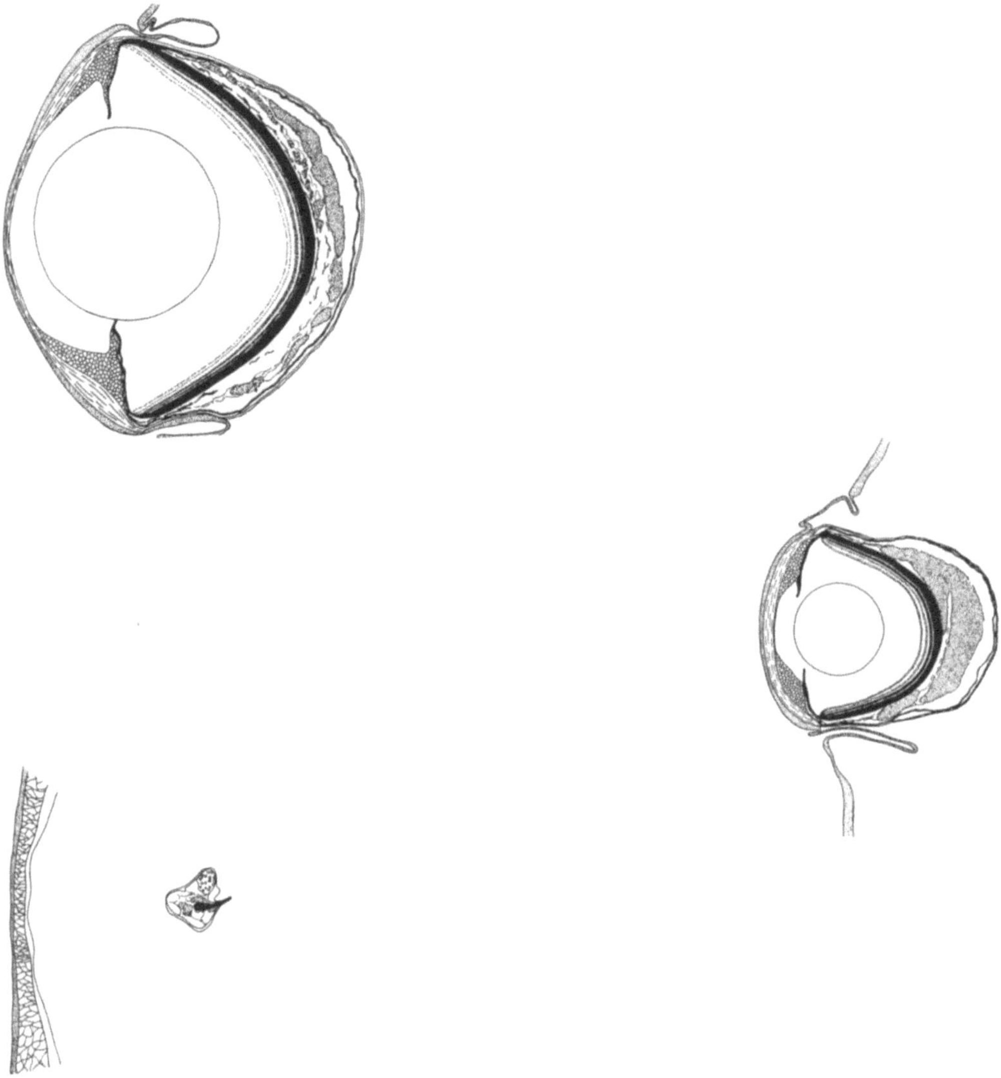

Fig. 1a-c. Microscopic anatomy of the eyes of *Astyanax* and "*Anoptichthys*" (Pachonfish) (P), their F_1-generation (F_1) and a selection of F_2-individuals (F_2). In total 30 parental specimens, 10 F_1, and 70 F_2 specimens were sectioned. (X 10)

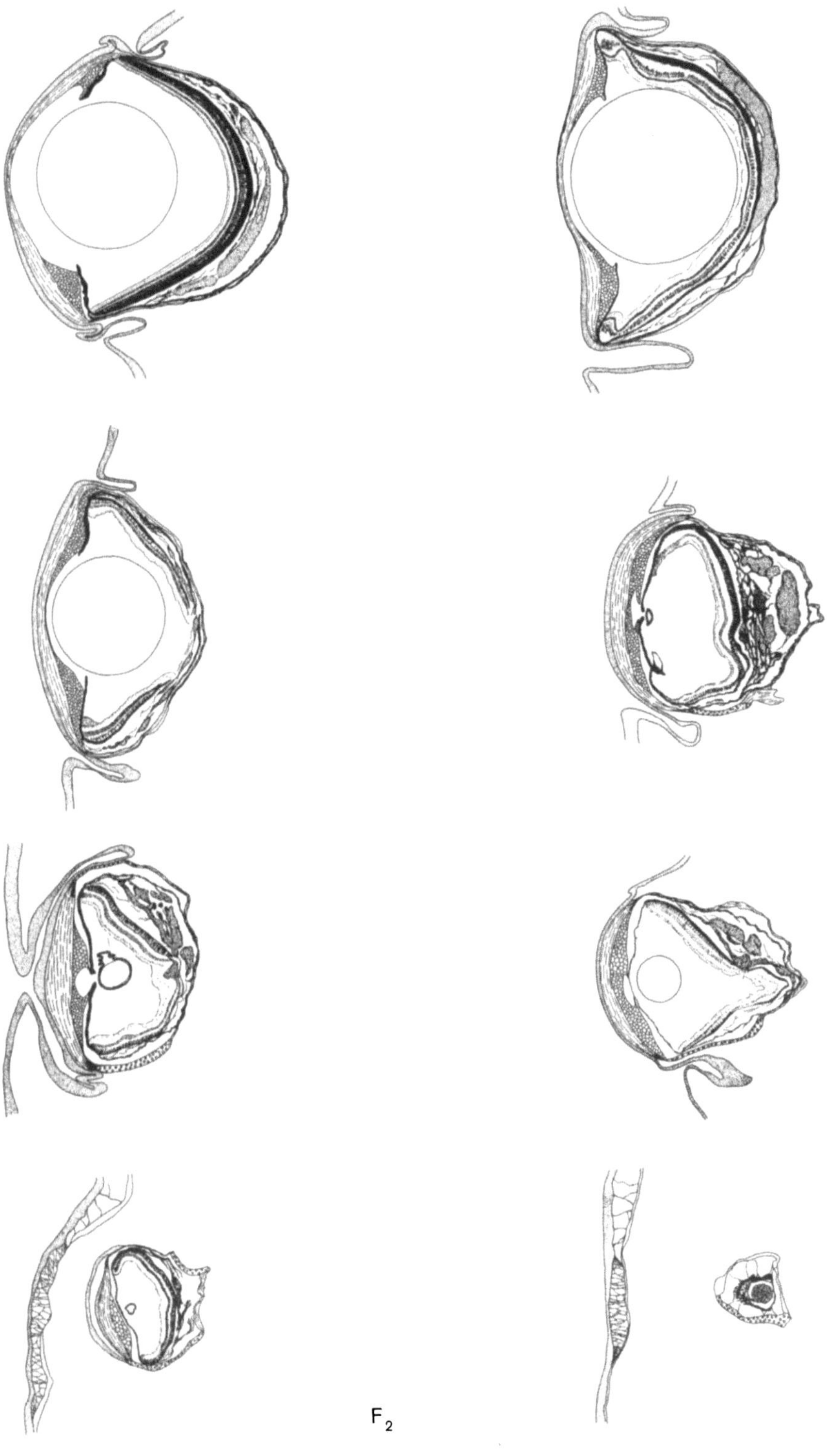

F_2

are essentially determined by the presence of darkness; their genetic material, however, is scarcely different from that of their river progenitors. On the other hand, troglobionts have also developed, that retain their regressive characteristics even when raised under daylight conditions. Both these forms are merely the extremes in a process of degeneration to which all cave-living animals are subjected. Between these extremes lie the various transitional forms.

2. Cross-Breeding Experiments

Four different troglobiont forms, one troglophile, one somewhat intermediate and one hybrid population have at present been experimentally crossed with both the river form and with each other in order to analyse their inherited qualities. These crosses - initiated by Kosswig - were started in 1956 in New York by Shadoglu and later carried on in great diversity by a team of the Zoological Department of Hamburg University. In all crosses no sterility barriers were encountered, so that all of these populations must be considered as members of the same species, *Astyanax mexicanus*. In order to describe the genetic principles of degenerative evolution in cavernicolous animals, we would like to limit the discussion to this test species, and at first to concentrate on only one of the several regressive traits, the eye.

Astyanax x *"Anoptichthys"*

When the horizontal diameter of the eye is taken as a measure of eye development, or rather eye reduction, there exists a linear relationship of approximately 6:1 between the eye size of the river fish and that of the troglobiont *"Anoptichthys"*. As stated above, the strongly reduced eye of the cave fish is also structurally degenerated (Fig. 1,P). It essentially consists of merely a rather bony eye capsule, in which only remnants of a retina, pigment epithelium and choroid are found. This rudimentary eye lies deep under the body surface.

In crosses between river and cave fish, the inheritance pattern basically follows Mendel's Laws (Fig. 1). The F_1 generation consists of a nearly uniform, intermediate hybrid, whose eyes are markedly smaller than those of the epigean form, but lying practically in the normal position, with rather normal structure (Fig. 1,F_1). Accordingly, the F_1-hybrid retains vision and shows the school behaviour of the normal-eyed parent.

The F_2 generation consists of various phenotypes (Fig. 1,F_2 and 2). The variability of the F_2 generation ranges from large-eyed individuals to completely blind ones, with an F_1-like phenotype being the most frequent. The frequency of the F_2 eye size corresponds with a normal distribution (Fig. 2).

Beyond this we find that in the F_2 generation there is a distinct relationship between the size of the eye and its structural development (Fig. 1,F_2). The large eyes closely resemble those of the river form in their structure, the small eyes those of the troglobiont. With diminishing eye size, first of all the receptor cells undergo degenerative changes: a shortening and swelling of the outer segments and a reduction in number per unit area. Next follows a destruction of the lens; the pupil closes. With a total loss of the receptor cells, the degeneration of the second and third neurons of the retina follows. Finally, the optic cup collapses as in the case of *"Anoptichthys"*. Parallel to this briefly described process, the eye rudiment sinks farther into the orbit and becomes totally encapsuled by a very tough sclera.

From these results we reach the following genetic conclusions: since the F_1 bastard is uniform, the parents must be rather homozygous for those genes responsible for the eye. The splitting into various F_2 phenotypes indicates that the genetic differences between the river and cave forms are of a polygenic nature, i.e. at numerous gene loci the normal alleles of the epigean form are substituted in the troglobiont by less active or inactive alleles. The responsible genes work in the sense of an additive polymeric system: the positive alleles are additive in their effect, so that their number essentially determines the size of the eye (Kosswig, 1960, 1963, 1965)[2]. Likewise, the structure of the eye is principally determined by the number of, and less by which of these additive genes are combined in the individual. This can be attributed to the complexity of the developmental-physiological processes in this organ. The primary embryonic structures induce and determine the size and shape of later developing anlagen. For example, the size of the optic cup and its spacial and temporal contact with the overlying epidermis determines the mode and extent of lens construction. Moreover, these induced secondary structures can in turn react upon the inducer. So, for example, the later growth of the optic cup is essentially dependent upon the size of the lens (Rotmann, 1939, 1942).

However, only the essentials of the degenerative process are stipulated in the way described. When the eye of the F_1 generation is compared to those of the F_2 generation of the same size, the latter are much more variable in their structure. Furthermore, as seen in Fig. 1,F_2, the eyes which are relatively normal in size (compared to the river fish) often show an overenlarged lens, or a partially deteriorated retina, whereas the remaining parts are highly differentiated. Massive warpings and dorsoventral or proximal indentations in the optic cup are not uncommon even in the larger eyes. Both eyes of the individual are usually influenced to the same extent. All this might appear as evidence that not only the number, but also which of the genes are recombined with each other, at least co-determines the particulars of the phenotype.

This interpretation of the cited crosses is supported by more extensive experiments (Wilkens, 1970), which at this point will not be discussed in detail. To summarize, it should once more be stated that, genetically speaking, the transformation of cavernicoles occurs in short discrete steps, whereby the troglophile differs from the epigean in only a few gene loci and the troglobiont in numerous, these two forms being the extremes. There is a possibility to determine the number of genes involved, when out of thousands of F_2 individuals those are counted, which have the eye size of either of the P forms. Using such a method one finds only seven gene pairs in which the troglobiont differs from the epigean form (Wilkens, personal communication). In that this is not a case of ideal additive polymery, this number is probably too low. That means, the true facts are hidden by dominant-recessive relationships, gene linkage and other gene interactions.

---

[2]Because the F_1 generation is not exactly intermediate, this deviates from the principle of additive polymery. The average eye size of the F_1 hybrid is not only nearer to that of its epigean parent, but also still larger than that of the F_2 generation (Fig. 2). The former can be interpreted as (partial) dominance, the latter as heterosis.

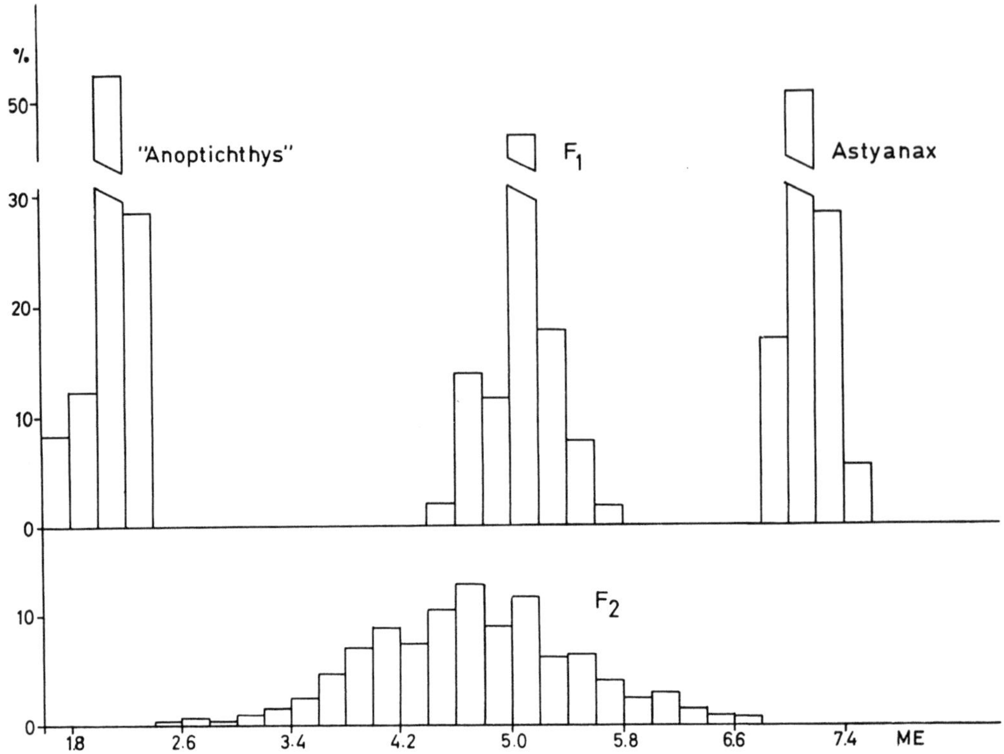

Fig. 2. Frequency distribution of the eye size of *Astyanax*, *"Anoptichthys"* (Sabinosfish) and their resulting hybrid generations. ME = units of measurement. (From Wilkens, 1970)

"Anoptichthys" x *"Anoptichthys"*

The eye remnants of the already examined extreme troglobionts are nearly identical in both size and structure. In crosses between various geographically isolated troglobionts and the river form, the same results were achieved. Accordingly, a considerable genetical identity is to be expected. In crosses between two troglobionts of different origin, the F_1 hybrid was both blind and lacking in pigment (Shadoglu, 1957). However, under closer observation, the eye rudiments of the hybrids were on the average somewhat larger than those of either parent and also improved structurally (Wilkens, 1971). These round-oval rudiments almost always contained a layered retina and a few even retained a degenerated lens capsule behind the closed pupil (Fig. 3), the latter being uncommon in the parents.

Accordingly, the crossed troglobionts are, indeed, morphologically (nearly) identical, but show a slight genetical difference; otherwise the results of the crosses would be incomprehensible. For this there are several interpretations: either the troglobionts differ in some gene loci in iso-alleles (see Avise and Selander, 1972), whereby heterosis is achieved as a consequence of over-dominance; or both

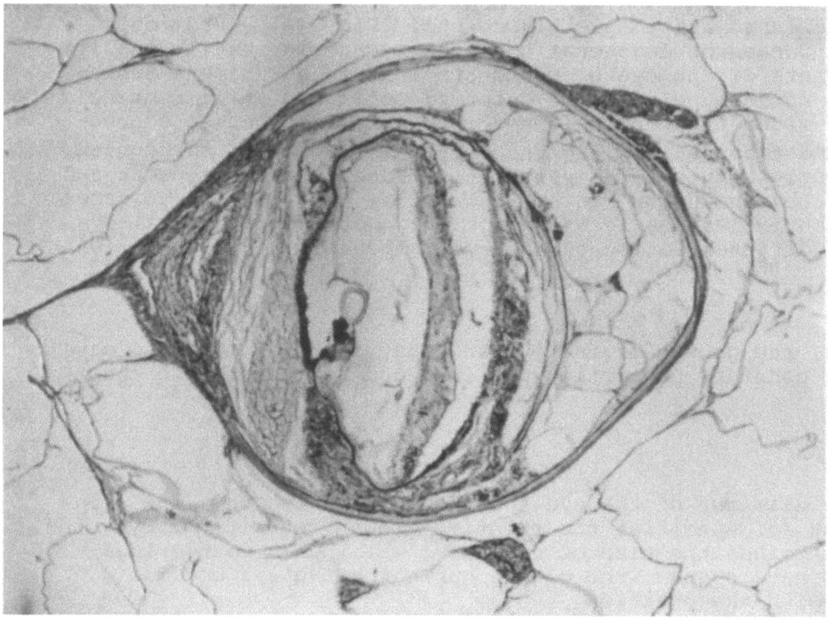

Fig. 3. Cross-section of a large eye rudiment from the F_1-generation of *"Anoptichthys"* (Pachonfish) x *"Anoptichthys"* (Sabinosfish). Paraffin 5 μ, Azan stained. (X 35). Slide courtesy of Dr. H. Wilkens, Hamburg

populations are in a few loci differently heterozygous or even differently homozygous. By (partial) dominance of the normal alleles, or in the case of complimentary gene activity, an enlargement and improvement of the eye may well be imaginable. These interpretations are ranked simply according to their degree of probability.

3. Ontogeny of the Eye

The results of the above mentioned crosses naturally refer to the characteristics of full-grown, mature fish, i.e. the phenotype of the adult. A comparison of the ontogenic eye development of the river fish, the troglobiont, and their F_1 hybrid discloses further information about the type and operation of the eye genes.

Morphogenesis

During the development of the embryo, the eye of *"Anoptichthys"* is at the very beginning normally formed (Fig. 4a). However, the primordial optic cup and lens - in comparison to the river fish - are somewhat smaller, and the time of their formation is slightly delayed (Cahn, 1958). The outer layer of the optic cup differentiates normally into the pigment epithelium; the inner part forms the layered retina. The rim of the optic cup develops into the iris (Fig. 4b and c). The lens anlage, however, undergoes practically no further differentiation: neither lens fibers nor lens capsule are created. Instead, the lens begins to degenerate from the center outwards (Fig. 4d). Outside this lens-relict the pupil disappears as the iris grows together. The anterior eye chamber becomes filled with the hypertrophied ligamentum annulare. In half-grown fish, the visual cells of the retina, in addition to the lens, are completely deleted (Fig. 4e). From this point on, the process of degeneration becomes chaotic, until finally, only some remnants of the eye elements are left in a rudiment sunken deep under the surface. In part a deformed optic cup still remains (Fig. 4f), but often the cup collapses into a ball of nerve, connective and pigment tissues (Fig. 4g and 1,P). This cyst is surrounded by a tough, mostly bony sclera. From this process of development and degeneration, the conclusion can be drawn, that the loss of eye genes in the cave fish influences less the formation of the primordial eye than the later differentiation of the eye components.

Growth

After hatching, the growth of the eye and body in both river and cave fish follows a constant proportion and can be described by the equation

$$y = b \cdot x^a,$$

where y is the diameter of the eye and x the total length. The allometric constant (a) specifies the relative growth rate of the eye; the constant b determines the size of y, when $x = 1$. In a double logarithmic co-ordinate system, the growth curve for the eye becomes a straight line with the equation

$$\log y = \log b + a \cdot \log x,$$

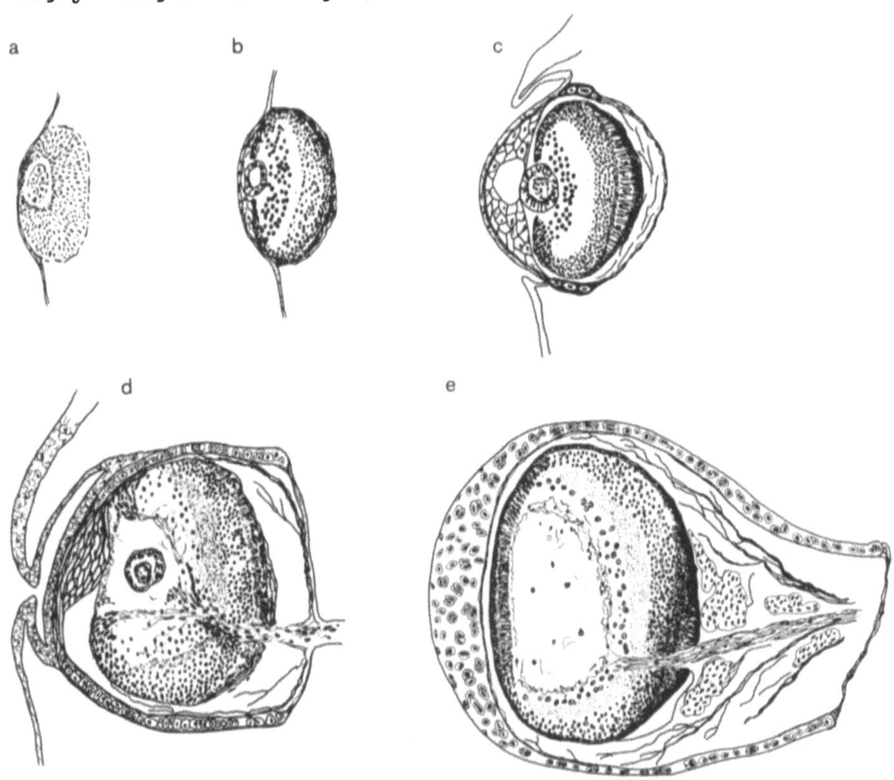

Fig. 4a–e

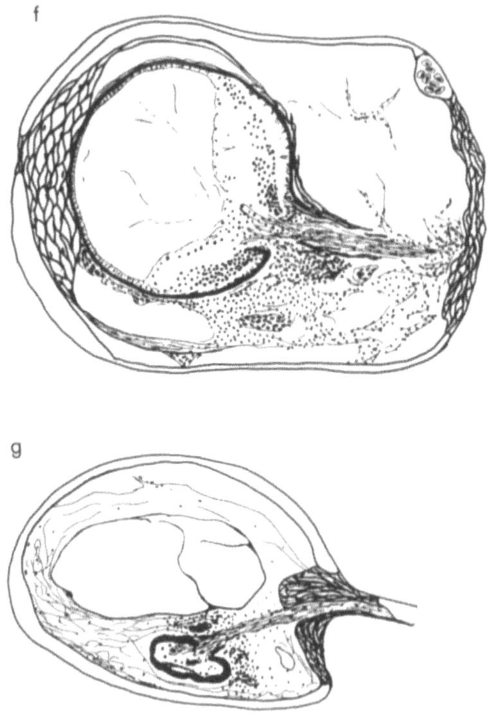

Fig. 4a-g. Eye ontogenesis of *"Anoptichthys"* (Sabinosfish). a) freshly hatched larva (3,8 mm), b) larva (3 days, 4,5 mm), c) fry (10 days, 6,7 mm), d) fry (30 days, 18 mm), e) half-grown fish (4 months, 37 mm), f) and g) adult fish (15 months, 64 and 54 mm resp.) (X 55)

in which a is the slope of the line. During the early intensive growth and morphogenetic process, the line of allometry for the river fish forms an angle to the abscissa of more than 45°, i.e. $a = 1.03$ (Fig. 5). This means that the eye grows as fast as or even a little faster than the entire body. The eye growth of the troglobiont, on the other hand, is noticeably slower, thus showing a negative allometry with $a \cong 0,6$; the F_1 hybrid lies somewhat intermediate with an a of 0,85. The conclusion can thus be drawn, that the allometric constant is approximately proportional to the number of genes involved in eye development. The number of genes present determines, therefore, the intensity of eye growth. The fewer active genes an individual has, the more inefficient the growth of the eye becomes. Again it is clear that the eye anlage is less dependent on the number of genes involved.

The relationship between the number of genes participating to form a structure and the intensity of this structure's growth has already been mentioned by V. Bertalanffy (1951): "Bei zahlreichen entwicklungsphysiologischen Erscheinungen gilt nun für die Wirksamkeit der Gene das Prinzip der abgestimmten Reaktionsgeschwindigkeiten, daß die Quantität der Gene am Anfang der Entwicklung die Geschwindigkeit der Reaktionen bestimmt, die nebeneinander herlaufen und damit die betreffende Struktur erzeugen. Wenn nämlich das relative Wachstum eines Teiles y gegenüber einem anderen x dadurch zustande kommt, daß die Wachstumsgeschwindigkeit von y während des ganzen Verlaufes um ein konstant bleibendes Vielfaches größer (oder kleiner) ist als jene von x, so kann das nichts anderes bedeuten, als daß die entsprechenden Gene Y, X gleichfalls aufeinander abgestimmt sind, d.h. in einem bestimmten quantitativen Verhältnis zu einander stehen, so daß sie mit verschiedenen Geschwindigkeiten nebeneinanderlaufende Reaktionsketten bedingen." Moreover, there exists, as stated above, a relationship between size and the extent of structural development of the resulting organ. This means that mutations, which cause either loss or destruction of a feature, will manifest themselves as growth inhibition on previous phases at the site of their later structural deficiency. With a decrease in size at one site, the eye as a whole adapts itself to such a reduction (Peters and Peters, 1968).

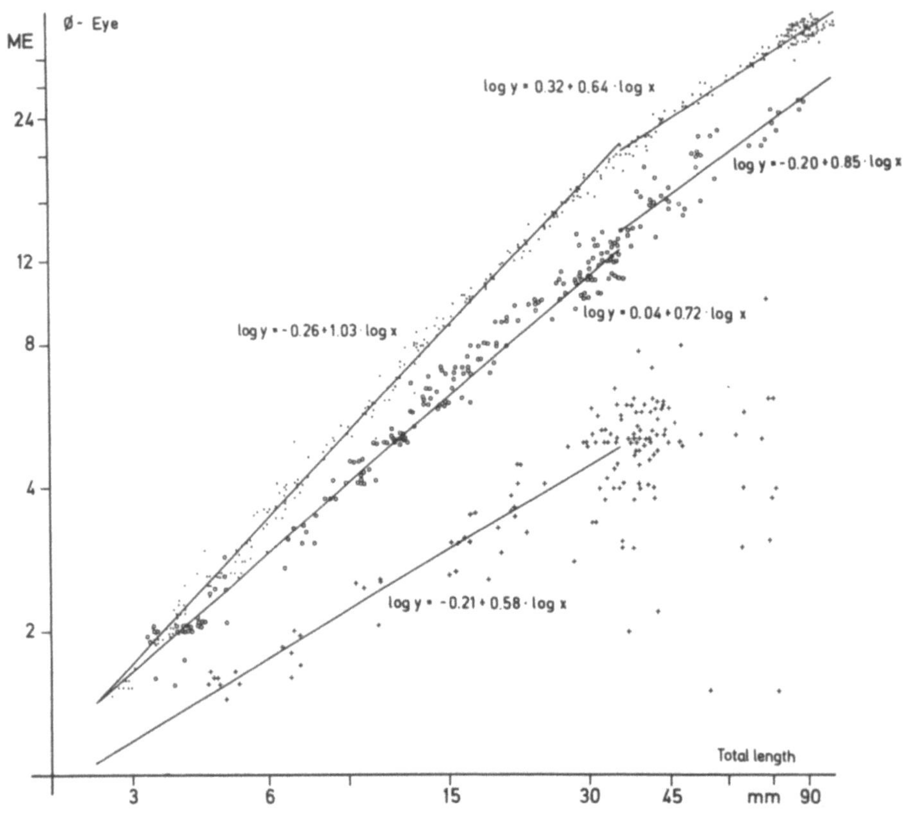

Fig. 5. Growth allometry of the eyes of *Astyanax* (•), *"Anoptichthys"* (Sabinosfish) (+) and their F_1-hybrid (o). The individual values were calculated from three offspring raised under the same conditions. The measurements were made on fixed material made transparent with benzylbenzoate and refer to the maximum horizontal diameter of the left optic cup without sclera. The individual regressions are highly significant ($r > 0{,}9$; $p \ll 0{,}001$). Besides, the slope of all the lines (a) differ significantly from each other ($p < 0{,}05$ to $p \ll 0{,}001$). For *"Anoptichthys"* there is no significant correlation between eye and body size after the break in allometry ($r = -0{,}125$; $p \gg 0{,}05$). (We would like to thank Prof. Dr. M. Gillbricht, Hamburg, for the computation of the statistical parameters)

As we can see from Fig. 5 at a total length of 3-4 cm in all three groups there occurs a break in the allometry. The allometric constant rather abruptly lowers its value. This results most likely from the oncoming of sexual maturity, i.e. this is due to a change in the physiology and not to genetical differences between the groups. In cave fish, this negative break in allometry means largely a stop in the growth of the eye or even the structural collapse of the, until now, relatively well organized eye rudiments. Often both eye rudiments of an individual are affected to a different extent, so that sometimes an extreme asymmetry results in size and structure (Peters and Peters, 1966; Wilkens, 1970).

4. The Phylogenetic Process of Degeneration

We are now aware of some of the genetic principles of the process of reduction; we know the genetic constitution of the epigean fish *Astyanax* and its troglobiont derivatives. The question now remains how the troglobiont derived from the river form. It is generally known - actually since Darwin's time - that biologically useless structures are extremely variable. This also pertains to the regressive characteristics of the cavernicolous animals as especially Kosswig (1937, 1960) and Vandel (1965) have proved with various examples. As we now know, this variability is mostly genetically controlled; it depends on the fact that for a series of gene loci differentially active and, above all, inactive alleles are present in the population (Kosswig and Kosswig, 1940; Peters and Peters, 1968). How does this concentration of negative alleles come about in a cave population?

A regressive development of the eye of a cavernicole can only occur when the gene flow between the cave population and the epigean population is hampered, or by geographic isolation completely stopped. Thereby the cave population becomes a founder population, that lies on the periphery of the species area and is mostly or totally separated from the entirety of the species. This founder population, because of its relatively small number of individuals, possesses only a fraction of the genetic variability of the entire species, including the negative alleles that are thinly distributed in every natural population. The very few negative alleles that were brought in with the founders, however, undergo a relative increase in their frequency simply by separation from the river population. While mutations, which interfere with adaptive structures, are again reduced in frequency or even eliminated by natural selection in the sequence of generations (genetic restoration, Mayr, 1967), those negative alleles of the eye-genotype can be easily recombined and thus have a good chance of accumulating to a 100% homozygous population.

This marks the beginning of the regressive evolution, but we are still far from the true troglobiont. The appearance of still more genes with degenerative effects in an isolated population can only occur by mutation. Mutations, being random in nature, generally provoke disorder and thus lead to a non-functioning of the genetic material. Only one or a few nucleotid sequences in the DNA-code lead to constructive information, whereas the possibility of creating incomplete or missing genetic information is excessive (Kosswig and Peters, 1967). This mutation pressure towards genetic disorder causes, as there is no opposing natural selection, a rapid accumulation of deleterious mutations. Thus, finally the phenotype of the troglobiont is established.

It is an often discussed problem, whether structures, after losing their function, are merely no longer under the influence of the stabilizing and improving effect of natural selection, or whether new selective forces promote a rapid reduction of all traits without biological significance. In any theoretical discussion, one must bear in mind that an organism, after a long time of stable environmental conditions, reaches a compromise between the various competing selective forces (Mayr, 1967). At this time the population is only subject to the stabilizing selection, which hinders deviations from the norm. If, however, by a sudden change in the environment, certain structures become practically functionless, an imbalance results under the different selective forces. The competing structures will experience a better development at the cost of the now meaningless traits, i.e. there occurs an indirect negative selection for the regressive trait.

However, this competition between the various structures involves a very complicated force pattern, in which the loss of one or a few partial forces causes only a relatively slight shift in the equilibrium. Because the non-regressive structures compete not only with the regressive traits but also with each other, their actual mobility returns rapidly into a new force equilibrium. In this way it becomes understandable that at present we have no real criterion for natural selection promoting the process of degeneration. Moreover, the excessive variability of the regressive traits is most likely evidence that natural selection does not essentially intervene in the reductional process: for strong selection should shorten genetic variability; mutations, however, increase it.

The troglobiont, according to definition, is completely blind. One cannot help but notice, however, that cave fish always possess eye remnants, at least these remnants appear during the course of ontogeny. In this way functionless rudiments are maintained over a long period of evolutionary time. From this we may conclude that for the development of these rudiments such genes are responsible which also participate in the development of non-regressive features (Kosswig, 1960). As especially the early eye formation in cave fish appears nearly normal, we can also assume, that the primordial eye anlage is indispensible for the normal organization of the head region (see Dobzhansky, 1954, 1956 on the persistence of gill slits in amniota). Naturally, this type of pleiotropic genes cannot be altered in a degenerative sense.

As soon as all unnecessary eye genes are inactivated by mutation, there will remain a rest genotype for the eye, which, by selection, is kept nearly homozygous. This gene configuration should be largely identical for all true troglobiont *"Anoptichthys"* forms, in that, although they developed independently from one another, they stem from the same ancestors. When hybridization of two troglobionts reveals far-reaching identity and only minor genetic differences, then both should (nearly) have reached the final stage of regression.

We have tried to demonstrate a few fundamental genetical experiments and conclusions on the problem of regressive evolution as seen in the example of the eyes of *Astyanax mexicanus* and its cave derivatives. The eye is not, however, the only varied trait of cavernicolous animals. In the *"Anoptichthys"* forms the skin pigmentation (guanophore and melanophore patterns) is also markedly reduced (Wilkens, 1970), in fact, two of the known cave forms are albinos. Furthermore, the midbrain is both in size and in structure greatly regressive (Franck-Krahé, 1962; Pfeiffer, 1967; Schmatolla, 1972). Whereas the river fish *Astyanax* shows a distinct fright reaction, *"Anoptichthys"* has lost the capacity to perceive these stimuli (Pfeiffer, 1966). In contrast to the aforementioned traits, the sense of taste is obviously improved in the cave form (Schemmel, 1967)[3].

---

[3]The labyrinth and nose show no significant changes (Schemmel, 1967; Popper, 1970).

In cross-breeding experiments all of these traits follow Mendel's Law, being inherited independently from the state of eye degeneration and from one another, so that all possible combinations of traits can be found. Only the degree of midbrain development is directly correlated to that of the eye.

In each case the same or similar genetic principles, as seen in the process of eye reduction (principle of additive polymery), are valid. Phylogenetically, indeed, these degenerative processes occur simultaneously, but they are genetically independent of each other, and also independent from the likewise simultaneously occurring constructive processes of evolution. The latter also differ from the degenerative processes in that the affected traits show no special genetic variability (Peters, Peters, Parzefall and Wilkens, 1973).

5. Summary

The regressive traits of the facultative cave dwellers, the troglobionts, are predominantly the direct result of living in darkness; in their genetic constitution they are scarcely different from their epigean ancestors. This is not the case of the strongly adapted troglobiont, which, even when bred under daylight conditions, retains most of its regressive characteristics, i.e. blindness and pigment reduction. Between these two extremes all possible transitionary stages may occur.

Occasionally it is possible to cross a troglobiont with its epigean, normal-eyed and pigmented relative. The F_1 and F_2 generations enable us to conclude that the differences in heredity between river and cave fish are of a polygenic (mostly additive polymeric) nature for the individual traits. In other words, at a number of loci the normal alleles are substituted by less active or even inactive alleles. The crosses demonstrate Mendelian inheritance for all regressive features, these being independent of each other.

The accumulation of "degenerative" alleles in the troglobiont occurs for the following main reasons: 1. If out of a large epigean population only a few fish become isolated in a cave (creating a founder population), this means a relative increase in the frequency of the few "degenerative" alleles brought in with the founders. 2. Newly occurring mutations that have a degenerative effect on now functionless structures are no longer subject to the limiting forces of selection, as can be concluded from the unusually high degree of genetic variability of the regressive features.

Even in extreme troglobionts remnants of the degenerated characteristics are always maintained. By hybridization experiments the troglobionts were found to be practically homozygous for their rudiments. During early ontogenesis these structures are formed quite normally, but then are increasingly retarded in their growth, which finally results in their structural break-down. The early anlagen of rudimentary organs are probably indispensible for the undisturbed development of the whole organism.

References

Avise, J.C., Selander, R.K. (1972): Evolutionary genetics of cave-dwelling fishes of the genus *Astyanax*. Evolution 26,1, 1-19.

Bertalanffy, L.v. (1951): Theoretische Biologie. Bern: A. Francke Verlag.

Cahn, Ph.H. (1958): Comparative optic development in *Astyanax mexicanus* and in two of its blind cave derivatives. Bull. Am. Mus. Nat. Hist. 115, 70-112.
Dobzhansky, Th. (1954): Evolution as a creative process. Caryologia, 435-449.
Dobzhansky, Th. (1956): What is an adaptive trait? Amer. Nat. 90, 337-347.
Franck-Krahé, C. (1962): Mexikanische Höhlencharaciniden im Vergleich zu ihren oberirdischen Vorfahren. Staatsexamanesarbeit Universität Hamburg.
Kosswig, C. (1937): Betrachtungen und Experimente über die Entstehung von Höhlentiermerkmalen. Züchter 9, 91-101.
Kosswig, C. (1960): Zur Phylogenese sogenannter Anpassungsmerkmale bei Höhlentieren. Int. Rev. ges. Hydrobiol. 45, 493-512.
Kosswig, C. (1963): Genetische Analyse konstruktiver und degenerativer Evolutionsprozesse. Z. zool. Syst. Evolutionsforsch. 1, 205-239.
Kosswig, C. (1963): Génétique et évolution régressive. Rev. Quest. Sci. 26, 227-257.
Kosswig, C., Kosswig, L. (1940): Die Variabilität bei *Asellus aquaticus* unter besonderer Berücksichtigung isolierter unter- und oberirdischer Populationen. Rev. Fac. Sci. Univ. Istanbul B, 5, 1-55.
Kosswig, C., Peters, N. (1967): Die Evolution der Höhlentiere. Bild der Wiss. 10, 829-835.
Mayr, E. (1967): Artbegriff und Evolution. Hamburg-Berlin: Paul Parey Verlag.
Peters, N., Peters, G. (1966): Das Auge zweier Höhlenformen von *Astyanax mexicanus* (Philippi)(Characinidae, Pisces). Roux'Arch. Entw. mech. 157, 393-414.
Peters, N., Peters, G. (1968): Zur genetischen Interpretation morphologischer Gesetzmäßigkeiten der degenerativen Evolution. Z. Morph. Tiere 62, 211-244.
Peters, N., Peters, G., Parzefall, J., Wilkens, H. (1973): Über degenerative und konstruktive Merkmale bei einer phylogenetisch jungen Höhlenform von *Poecilia sphenops* (Pisces, Poeciliidae). Int. Rev. ges. Hydrobiol. (im Druck).
Pfeiffer, W. (1966): Über die Vererbung der Schreckreaktion bei *Astyanax* (Characidae, Pisces). Z. Vererbungsl. 98, 97-105.
Pfeiffer, W. (1967): Die Korrelation von Augengröße und Mittelhirngröße bei Hybriden aus *Astyanax* x *Anoptichthys* (Characidae, Pisces). Roux'Arch. Entw. Mech. 159, 365-378.
Popper, A.N. (1970): Auditory capacities of the Mexican blind cave fish (*Astyanax jordani*) and its eyed ancestor (*Astyanax mexicanus*). Anim. Behav. 18, 552-562.
Rotmann, E. (1939): Der Anteil von Induktor und reagierendem Gewebe an der Entwicklung der Amphibienlinse. Roux'Arch. Entw. Mech. 139, 1-49.
Rotmann, E. (1942): Über den Auslösungscharakter des Induktionsreizes bei der Linsenentwicklung. Biol. Zbl. 62, 154-170.
Schemmel, C. (1967): Vergleichende Untersuchungen an den Hautsinnesorganen ober- und unterirdisch lebender *Astyanax*-Formen. Z. Morph. Tiere 61, 255-316.
Schmatolla, E. (1972): Dependence of tectal neuron differentiation on optic innervation in teleost fish. J. Embryol. exp. Morph. 27, 555-576.
Shadoglu, P. (1956): A preliminary report on the genetics of the Mexican cave Characins. Copeia 113-114.
Shadoglu, P. (1957): Mendelian inheritance in the hybrids between the Mexican blind cave fishes and their overground ancestor. Verh. dtsch. zool. Ges. Graz 432-439.
Vandel, A. (1964): Biospéleologie. Paris: Gauthier-Villars.

Wilkens, H. (1970): Der Bau des Auges cavernicoler Sippen von *Astyanax fasciatus* (Characidae, Pisces). Roux'Arch. Entw. Mech. 166, 54-75.
Wilkens, H. (1970): Beiträge zur Degeneration des Auges bei Cavernicolen, Genzahl und Manifestationsart. Z. Zool. System. Evolutionsforsch. 8, 1-47.
Wilkens, H. (1970): Beiträge zur Degeneration des Melaninpigments bei cavernicolen Sippen des *Astyanax mexicanus* (Filippi)(Characidae, Pisces). Z. Zool. System. Evolutionsforsch. 8, 173-199.
Wilkens, H. (1971): Genetic interpretation of regressive evolutionary processes: studies on hybrid eyes of two *Astyanax* cave populations (Characidae, Pisces). Evolution 25, 530-544.

Phylogenetic Age and Degree of Reduction of Cave Animals

H. Wilkens

Until the present day, no unanimous answer has been given to the question whether degenerative evolutionary processes occur with or without the influence of selection. On the one hand, people make no distinction between degenerative and constructive evolution. They argue that degenerative mutations manifested in biologically useless organs take on selectionistically positive value by pleiotropy. They show for example material compensation and are thus favoured by selection (Barr, 1968; Mitchell, 1969).

On the other hand, there is the opinion that degenerative mutations though being negative concerning a biologically useless organ are neutral concerning the whole organism. Thus, by remaining outside the influences of selection, they are not eliminated. This fact provokes the typical increased variability of degenerating structures (Vandel, 1964). Solely on account of the higher number of degenerative mutations in comparison to constructive ones functionless organs must rudimentate by mutational pressure (Kosswig, 1940, 1948, 1960, 1963; Peters and Peters, 1968; Wilkens, 1971).

Besides an increased variability, the rapidity of evolution of degenerative structures also seems to be higher than that of constructive ones. This is because all mutations are manifesting. Normally the larger number, namely all degenerative ones, are eliminated. It is paced down only when, at the end of a process of structural reduction, organs become homozygous for genes which are not allowed to mutate on account of pleiotropic functions, and until these functions have been taken over by others. Only at this moment are these genes, too, free for further degeneration, or their functionless DNA-rudiments free for new experiments of evolution.

It has to be pointed out, though, that the process of reduction of a certain structure is usually coupled with the simultaneous development of a compensatory one. In these cases there are no differences in speed. But in other cases we find examples for the uncoupling of both processes: eyes and pigment of troglobites lose their biological function abruptly when these animals are isolated in caves. It is certainly no mere chance that cave forms being primarily described as distinct species when studied more thoroughly turn out to be only ecological races of epigean forms although there are extreme phenotypic differences.

Mitchell (1969) tried to couple the problem of degenerative evolutionary rates with the theory of material compensation, as mentioned above. According to this, eyes and pigment of cave animals dwelling in caves with a scant food supply are supposed to be reduced more quickly than in those dwelling in caves where food is abundant, because in this case the alleged material compensating degenerative mutations are particularly favoured by selection.

This paper attempts to show that under the premise of structural reduction occurring without the influence of selection, the rapidity of evolution becomes directly dependent on the mutation rate. At least in corresponding organs of closely related species - and possibly in more distant ones, too - structural degeneration should occur at similar and thus comparable paces.

When considering the evolution of cave animals, a country like the peninsula of Yucatan deserves special attention. It is, especially in its northern parts almost completely carsic and there is hardly any surface fresh water (Pearse, 1936). This can only be found in an immense subterranean cave system. Characteristic of the fauna of Yucatan therefore is a great number of troglobites, mainly of crustacean and teleostean origin. Whereas on account of the lack of surface fresh water in the interior Yucatan most fresh water fishes live in the coastal regions, four of them have penetrated into the caves of the inner parts of the peninsula (Hubbs, 1938). They are *Typhliasina pearsei* (Ophidiidae), *Furmastix infernalis* (Synbranchidae), *Rhamdia guatemalensis* (Pimelodidae), and *Cichlasoma urophthalmus* (Cichlidae). In the following, *C. urophthalmus* will be excluded from consideration. It has only been found in cenotes - places which are under the influence of daylight - and it is not clear whether this fish has reached them by human agency.

The first true cave fish to be described here is *T. pearsei*. It is of marine origin and very closely related to the Cuban blind-fish *Stygicola dentatus* and *Lucifuga subterranea* (Eigenmann, 1909). Its eyes are extremely reduced and lie covered by the epidermis deep in the orbital cavity. The optic nerve and the retina except for the pigmentary epithelium are completely reduced. The latter consists of a layer of cubic cells which encloses the melanin pigment containing vitreous body. A double-layered ventral fold of the pigmentary epithelium is interpreted as the residue of the processus falciformis. The chorioid consists of a vessel and portions of melanin pigment. The whole is surrounded by a mostly cartilagineous sclera (for further details see Wilkens, 1973).

Along with *T. pearsei* another fish may be found in the same localities. It is *F. infernalis* which contrary to *T. pearsei* is a cave fish whose ancestral form, *Synbranchus marmoratus*, is distributed all over the Middle and South American continent. Studies have revealed that the distinction between the hypogean and epigean synbranchide populations, either on the generic or on the species level, is not justified (Parzefall and Wilkens, 1972). There are almost no differences except for the reduction of eyes and pigment in the cave form.

The eyes of the subterranean population of "*F. infernalis*" are also sunk beneath the skin, though they are less reduced than those of *T. pearsei*: they even contain a lens rudiment and the retina still consists of ganglionic, inner plexiform, and inner nuclear layers. The pigmentary epithelium is a layer of cubic cells. Vitreous body, anterior eye chamber, and pupillar opening have not yet vanished, though they are diminished (Parzefall and Wilkens, 1972).

Rhamdia guatemalensis is a fish which is widely distributed in the fresh water of Middle America. On the Yucatan peninsula it is differentiated only subspecifically. Contrary to *T. pearsei* and "*F. infernalis*" its eyes and pigment are not reduced - a fact which is rather astonishing, since it is a troglophilic form and we find other true troglobites among the Pimelodidae. Only two of the great number of populations of *R. guatemalensis* on the Yucatan peninsula are described as being slightly reduced concerning eye size and pigmentation (Hubbs, 1938).

When comparing the three fish species described here one observes different degrees of reduction: whereas the eyes of *T. pearsei* are among the most reduced ones to be found in any cave fish, those of

"F. infernalis" are far less diminished as concerns size and degree of differentiation. They have an intermediate stage as compared with *T. pearsei* and *R. guatemalensis*, which is almost unaffected. This sequence is valid for the intensity of melanin pigmentation, too:

T. pearsei is completely pale and no melanophores can be found, though this form is not albinotic as demonstrated by the eyes. (Only one specimen turned dark, when being preserved (Hubbs, 1938).) In *"F. infernalis"* the pigmentation is only slightly reduced. It has fewer melanophores than the epigean population of *S. marmoratus*. As the melanin content is still unaffected, the fish is coloured grey. *R. guatemalensis* has the greatest amount of pigmentation.

These different degrees of reduction are not caused by different evolutionary rates. They are - as will be demonstrated - caused by different phylogenetic ages as they are caused by the Yucatan geological history.

Having been covered by the sea in previous geological periods, an area of the Yucatan peninsula became dry in the Tertiary Age. This dry area expanded slowly, and during the periods of glaciation which took place in the Pleistocene Age more of the peninsula was free of the sea than is the case today; but in the warmer intervals all these dry regions, possibly including those where the caves with which we are concerned here are situated were submerged again. The present coast line of the Yucatan is the result of post-Pleistocene elevation (Weyl, 1964, 1970a, 1970b; Wilhelm and Ewing, 1972).

It is this geological development which shows that the troglobites of marine derivation must be phylogenetically older than those whose ancestors lived in fresh water. This assumption is furthermore supported by studies of a cave-dwelling Yucatan shrimp, *Creaseria morleyi* (Palaemonidae), which is also derived from a marine ancestor: it is also completely pale and the eye stalks are very much reduced. Two of three optic ganglia, lamina ganglionaris and medulla externa, including the sensory cells have vanished. Only medulla interna and medulla terminalis may still be observed. The fact that the eye rudiments no longer show variability further demonstrates that this form is, phylogenetically, very old (for further details see Wilkens, 1973).

Both forms, *T. pearsei* and *C. morleyi*, deviate considerably from those whose ancestors lived in fresh water. *S. marmoratus* as well as *R. guatemalensis* could not live on the Yucatan before the caves were filled with fresh water. The different degree of reduction is probably due to a difference in salt-water sensitivity.

Whereas the synbranchidae are secondary fresh water fish which can live in salty or brackish water for some time, the pimelodidae are primary fresh water fish, which means that they cannot exist under such conditions (Myers, 1938, 1966). Consequently, *S. marmoratus* has been found on the islands surrounding the Yucatan as well as on Cuba, contrary to *R. guatemalensis* and all other ostariophysian species (Hubbs, 1938).

In the same way *S. marmoratus* might have lived in the caves before the Yucatan elevation was completed and when the cave water was still brackish to some extent - and thus considerably before *R. guatemalensis*.

With regard to their age the troglobites of marine derivation were - according to our present knowledge - isolated at earliest at the beginning of the Pleistocene Age. Possibly all isolations have been invalidated by submergences which followed each glaciation. In this case the cavernicolous evolution of the troglobites of marine derivation at latest started at the beginning of the last glaciation. The development of "*F. infernalis*" and especially that of *R. guatemalensis* probably only began in post-Pleistocene times.

References

Barr, T.C. (1968): Cave ecology and the evolution of troglobites. In: Evolutionary biology, Vol. 2. Eds. Th. Dobzhansky, M.K. Hecht, W.S. Steere. Amsterdam: North Holland Publishing Co.

Eigenmann, K. (1909): Cave vertebrates of America, a study of degenerative evolution. Carnegie Inst. Publ. 104, 1-241.

Hubbs, C.L. (1938): Fishes from the caves of Yucatan. Carnegie Inst. Publ. 491, 261-287.

Kosswig, C. (1940): Die Variabilität bei *Asellus aquaticus*, unter besonderer Berücksichtigung der Variabilität in isolierten unter- und oberirdischen Populationen. Rev. Fac. Sc. Univ. Instanbul (B) 13, 78-132.

Kosswig, C. (1948): Genetische Beiträge zur Präadaptionstheorie. Rev. Fac. Sci. Univ. Istanbul (B) 13, 176-209.

Kosswig, C. (1960): Darwin und die degenerative Evolution. Abhandl. u. Verh. Naturw. Verein Hamburg, N.F. 4, 21-42.

Kosswig, C. (1963): Genetische Analyse konstruktiver und degenerativer Evolutionsprozesse. Z. zool. Syst. Evolut.-forsch. 1, 205-239.

Mitchell, R.W. (1969): A comparison of temperate and tropical cave communities. South-western Natur. 14, 73-88.

Myers, G.S. (1938): Fresh-water fishes and West Indian zoogeography. Ann. Rep. Smith. Inst. for 1937, Wash., D.C., 339-364.

Parzefall, J., Wilkens, H. (1972).: Artbildung bei Höhlenfischen. Untersuchungen an zwei amerikanischen Synbranchiden. Z. Morph. Tiere 73, 63-79.

Pearse, A.S. (1936): The cenotes of Yucatan. Carnegie Inst. Publ. 457, 1-29.

Peters, N., Peters, G. (1968): Zur genetischen Interpretation morphologischer Gesetzmäßigkeiten der degenerativen Evolution. Z. Morph. Tiere 62, 211-244.

Vandel, A. (1964): Biospéléologie. La Biologie des animaux cavernicoles. Paris: Gauthiers-Villars.

Weyl, R. (1964): Die paläogeographische Entwicklung des mittelamerikanisch-westindischen Raumes. Geol. Rdsch. 54, 1213-1240.

Weyl, R. (1970a): Mittelamerika. Zbl. Geol. Paläont. 7/8, 243-291.

Weyl, R. (1970b): Mittelamerika. Umschau 10, 295-299.

Wilkens, H. (1971): Genetic interpretation of regressive evolutionary processes: Studies on hybrid eyes of two *Astyanax* cave populations (Characidae, Pisces). Evolution 25, 530-544.

Wilkens, H. (1973): Über das phylogenetische Alter von Höhlentieren. Untersuchungen über die cavernicole Süßwasserfauna Yucatans. Z. zool. Evolut.-forsch. 11, 49-60.

Origin and Synthesis of a Unisexual Fish

R. J. Schultz

Prior to 1932 the scientific community had already accepted the fact that invertebrates of all sorts reproduce parthenogenetically; none the less, it came as a surprise when Hubbs and Hubbs (1932) announced the existence of a unisexual fish, the now famous Amazon molly, *Poecilia formosa.* After more than 12 years of experimenting with this unique vertebrate, the Hubbses concluded that it was of hybrid origin probably involving *Poecilia sphenops (sensu lato)* and *P. latipinna* (Hubbs, 1955). They found that either of the presumed parental species provides sperm to the unisexual form but that its function is merely to stimulate the ova into development (gynogenesis). Offspring, thus, are characteristically females and identical to their mothers. Biochemical (Abramoff et al. 1968) and cytological studies (Drewry, 1964; Schultz and Kallman, 1968), and tissue transplant experiments (Kallman, 1962) subsequently confirmed that *P. formosa* is a gynogenetic hybrid; and, furthermore, that it is diploid and apparently produces diploid eggs.

Until the end of 1959, this fish stood as the only naturally occurring vertebrate "species" known; then, suddenly there was a "bloom" of discoveries of unisexual vertebrates, a bloom which has not yet ceased. Represented among these recent discoveries were not only additional fishes (Miller and Schultz, 1959; Schultz, 1967, 1969) but also amphibians (Uzzell, 1963) and reptiles (Darevsky, 1958; Maslin, 1962; Wright and Lowe, 1968). Today practically all who have worked with unisexual vertebrates conclude that they are of hybrid origin (see reviews by Maslin, 1971 and Schultz, 1971); yet, until recently, no one has successfully put together the presumed parents of these hybrids and formed a unisexual "species".

Now, after some 15 years of experimentation, a unisexual "species" has been produced in the laboratory (Schultz, 1973). It was accomplished by crossing two bisexual (gonochoristic) species of *Poeciliopsis*, a livebearing poeciliid from the same family as the Amazon molly. Like the molly it also lives mainly in Mexico; but whereas *Poecilia formosa* inhabits the rivers of northeastern Mexico, wild unisexual forms of *Poeciliopsis* live in the northwestern part of the country, ranging from the border of the United States south to the Rio Mocorito in Sinaloa.

The "species" to be dealt with principally in this discussion originally was designated as *Poeciliopsis* Cx (Miller and Schultz, 1959; Schultz, 1961) because its affinities at that time were uncertain. This single diploid form is known to have given rise to two other diploid unisexuals and three triploid unisexuals (Schultz, 1969), thus, its identity provides a key to understanding an entire evolutionary sequence of unisexuality within the genus.

---

[1]Supported by NSF GB-33451X

Before discussing how the parental origin of *P.* Cx was determined and how the laboratory synthesis was accomplished, the mode of reproduction practiced by unisexual "species" of *Poeciliopsis* must first be considered since it differs from that of *formosa* and, in fact, from any other vertebrate known. In nature *Poeciliopsis* Cx relies upon males of *Poeciliopsis lucida* for sperm. Mated to males of *lucida* in the laboratory, they consistantly produced all-female offspring for over 27 generations. In these matings or in matings to males of other species of *Poeciliopsis*, a true F_1 hybrid is formed which clearly possesses characteristics of both parents. Genetic evidence indicates, however, that random segregation does not occur during meiosis and that the paternal characteristics are not transmitted through the egg. By utilizing as a father a male from a species with dramatically different characters such as *P. latidens*, a very hybrid-looking F_1 can be produced; yet in a single mating back to *lucida* all *latidens* characters are lost (Schultz, 1966). How can such non-random segregation be explained cytologically? During the mitotic divisions just before meiosis, a unipolar spindle forms; one set of chromosomes becomes associated with it and migrates to the single pole. The other set remains in the cytoplasm and is excluded from the reconstituted nucleus and eventually lost - either by being absorbed or pinched off in a bud (Cimino, 1972). Whether the remaining set, then, undergoes an equational division is not known but the egg is haploid and matroclinous.

As a result of mating Cx to *latidens* several things were discovered regarding the origin of Cx plus the fact that it contains a certain amount of variation, suggestive of multiple origins. In the F_1 generation at least four different classes of progeny occur, each characteristic of the clone supplying the female: 1. both male and female offspring are produced which have spots inherited from *latidens*, 2. all males are produced which have spots but die before maturity, 3. all males occur which bear weak spots and a black wedge-shaped mark above the gonopodium, and 4. all males are produced but with bold spots and the black wedge.

The four classes of offspring from Cx x *latidens* matings proved to be exceedingly important in that the males produced provided us with the first clue as to the identity of the other parent involved in the hybrid origin of Cx; *lucida* was already assumed to be one of the parents since it appeared to be its natural mate. The black wedge-shaped mark above the gonopodium occurs in only two of the 16 or more species of *Poeciliopsis; P. viriosa* whose northern limit is the Rio Mocorito and *P. monacha* which occurs in the headwaters of the Rio Fuerte and has been taken occasionally from the Rio Sinaloa. After much difficulty, because of yolk-embryo size problems, we were able to produce viable F_1 offspring from Cx x *monacha* matings. Both sexes resulted which when backcrossed to *monacha* or were inbred with each other behaved and appeared like *monacha* in every respect.

Finally, now, we have successfully crossed *P. monacha* and *P. lucida.* The females used in this sythesis derive from all three Rio Fuerte locations known to contain *monacha*. Males of *lucida* have come from four different Fuerte sites. A total of 32 F_1 young were brought to maturity - all were females. These have since been backcrossed to *P. lucida* for five generations and a grand total of 334 young, all of which were females and look like any other diploid unisexual one might collect from the Rio Fuerte (Schultz, 1973).

This seemingly complex problem gathers simplicity with understanding. The fish we have called Cx would basically appear to be an F_1 *monacha* x *lucida* hybrid. It produces a *monacha*-like egg after expelling all *lucida* chromosomes during oogenesis.

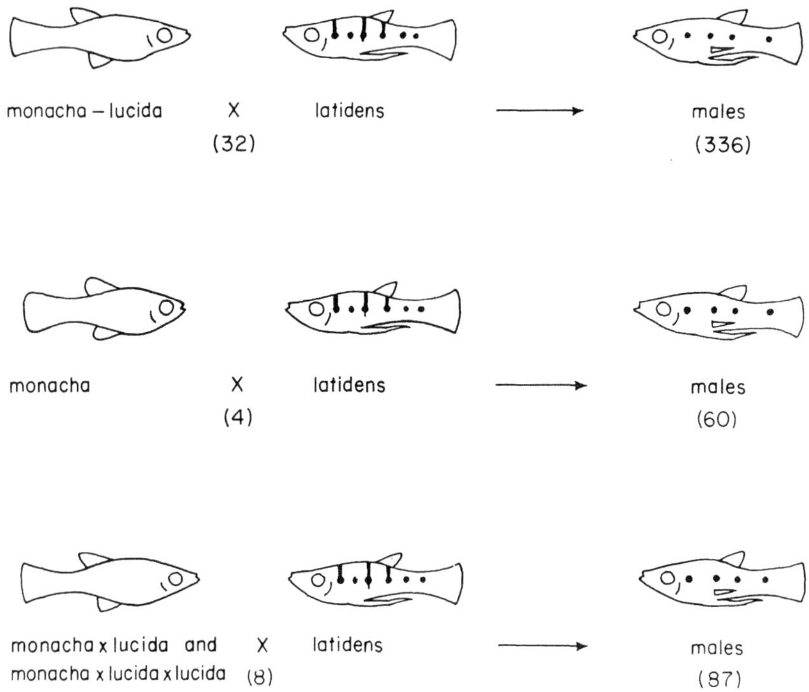

Fig. 1. The unisexual fish, *Poeciliopsis monacha-lucida* and hybrids of *P. monacha* x *P. lucida* both produce eggs that are genetically the same as *P. monacha*; these result in the same kind of all-male offspring when combined with sperm from *P. latidens*

If indeed we have synthesized a diploid unisexual equivalent to those in nature and if both forms have meiotic mechanisms that eliminate the paternal chromosomes and produce *monacha*-like eggs, then Cx x *latidens* is genetically the same mating as *monacha* x *latidens*. The reality of our circumstantial evidence can be verified by one last set of experiments (Fig. 1). When *monacha* is crossed with *latidens*, all males are produced; these have spots which they get from their *latidens* fathers and a black wedge above the gonopodium, characteristic of *monacha* males. When Cx females from the Rio Fuerte are crossed with *latidens*, they produce the same kinds of spotted males. Synthesized strains of *monacha-lucida* mated to *latidens* also produce all-male offspring, again with spots and the black wedge, like those from the other two crosses.

The laboratory synthesis of a unisexual "species" might appear to be a fitting climax for these studies, perhaps even bringing them to conclusion. Instead, however, it opens the way to a vast area of inquiry that is virtually unexplored. With the evidence that unisexual females can be synthesized in the laboratory, there is no reason to believe that they are not still being generated in nature. Narrow regions of overlap occur between the ranges of *P. monacha*, which lives in the mountain tributaries, and *P. lucida*, which lives further downstream. Here, the potential exists for *monacha* and *lucida* to continue hybridizing. This being true, it is imperative to ask such questions as: 1. How many clones of *monacha-lucida* have actually become established in nature?

2. What competitive relationship exists between clones of unisexuals, on the one hand, and between unisexuals and bisexuals, on the other? 3. Can a heterogeneous environment support more clones of unisexuals than one that is less so? 4. Do these hybrids show heterosis or does survival depend on production of two females for every one produced by bisexuals? and 5. Do unisexuals undergo evolution as a result of mutations and natural selection like other species, even though they have abandoned recombination? Students of the fossil record become impressed by the great time span through which evolution works. The dynamic processes going on in unisexual-bisexual populations of *Poeciliopsis* suggest that not all evolution proceeds at an imperceptible pace.

References

Abramoff, P., Darnell, R.M., Balsano, J.S. (1968): Electrophoretic demonstration of the hybrid origin of the gynogenetic teleost *Poecilia formosa*. Am. Naturalist 102, 555-558.

Cimino, M.C. (1972): Egg production, polyploidization, and evolution in a diploid all-female fish of the genus *Poeciliopsis*. Evolution 26, 294-306.

Darevsky, I.S. (1958): Natural parthenogenesis in certain subspecies of rocky lizard, *Lacerta saxicola* Eversmann. Dokl. Biol. Sci. Sect. 122, 877-879.

Drewry, G.E. (1964): *In* Interactions between a bisexual fish species and its gynogenetic sexual parasite. Bull. Texas Mem. Mus. 8, Appendix 1, 67.

Hubbs, C.L. (1955): Hybridization between fish species in nature. Syst. Zool. 4, 1-20.

Hubbs, C.L., Hubbs, L.C. (1932): Apparent parthenogenesis in nature, in a form of fish of hybrid origin. Science 76, 628-630.

Kallman, K.D. (1962): Gynogenesis in the teleost, *Mollienesia formosa* (Girard), with a discussion of the detection of parthenogenesis in vertebrates by tissue transplantation. J. Genet. 58, 7-21.

Maslin, T.P. (1962): All-female species of the lizard genus *Cnemidophorus*, Teiidae. Science 135, 212-213.

Maslin, T.P. (1971): Parthenogenesis in reptiles. Amer. Zoologist 11, 361-380.

Miller, R.R., Schultz, R.J. (1959): All-female strains of the teleost fishes of the genus *Poeciliopsis*. Science 130, 1656-1657.

Schultz, R.J. (1961): Reproductive mechanism of unisexual and bisexual strains of the viviparous fish *Poeciliopsis*. Evolution 15, 302-325.

Schultz, R.J. (1966): Hybridization experiments with an all-female fish of the genus *Poeciliopsis*. Biol. Bull. 130, 415-429.

Schultz, R.J. (1967): Gynogenesis and triploidy in the viviparous fish *Poeciliopsis*. Science 157, 1564-1567.

Schultz, R.J. (1969): Hybridization, unisexuality, and polyploidy in the teleost *Poeciliopsis* (Poeciliidae) and other vertebrates. Amer. Naturalist 103, 605-619.

Schultz, R.J. (1971): Special adaptive problems associated with unisexual fishes. Amer. Zool. 11, 351-360.

Schultz, R.J. (1973): Unisexual fish: laboratory synthesis of a "species". Science 179, 180-181.

Schultz, R.J., Kallman, K.D. (1968): Triploid hybrids between the all-female teleost *Poecilia formosa* and *Poecilia sphenops*. Nature 219, 280-282.

Uzzell, T.M. (1963): Natural triploidy in salamanders related to *Ambystoma jeffersonianum*. Science 39, 113-115.
Wright, J.W., Lowe, C.H. (1968): Weeds, polyploids, parthenogenesis, and the geographical and ecological distribution of all-female species of *Cnemidophorus*. Copeia 1967, 128-138.

Fish and Nature's Extensive Experiments with Gene Duplication

S. Ohno

When one attempts to look at the overall picture of vertebrate evolution, it soon becomes evident that the genetic foundation which assured the eventual emergence of mammals and of man was laid at the very beginning; i.e., at the stage of fish. For example, most relevant information on the origin of hemoglobin genes as well as immunoglobulin genes are found neither in reptiles nor in amphibians but in cyclostomes and elasmobranchs. This is the very reason that fish as a group constitutes the most fascinating subject for the student of vertebrate evolution.

I have already written two books on this subject (Ohno, 1970; Ohno, in press). Furthermore, Dr. Engel, a representative of the Freiburg group, is going to cover the topic of gene duplication by polyploid evolution immediately after me and Dr. Whitt has already touched upon gene duplication to a considerable extent in his talk. Therefore, I will confine myself to a few pertinent remarks on the nature of genomic changes and evolution.

1. Polymorphic Gene Loci, Chromosomal Changes and the Conservative Nature of Natural Selection

In a sense, it is unfortunate that Darwin's genius enabled him to conceive of the idea of evolution by natural selection long before the advent of modern genetics and molecular biology. The result has been that when our knowledge of genetics was still meager and limited to morphological traits, we embraced natural selection with religious fervor believing it to be the omnipotent advocator of progressive changes. Even in this age of modern biology, we tend to imagine the force of all powerful natural selection behind every genetic polymorphism which we encounter regardless of whether it is interspecific, interracial or intrapopulational.

The cold fact is that natural selection can actively favor a particular genetic trait in a population only by simultaneously disfavoring certain other traits in the same population. Thus, a population pays a considerable price for each gene locus which is actively subjected to natural selection. Let us imagine a simple polymorphism involving two alleles at a single gene locus, and assume that natural selection is favoring the heterozygous type at this locus. The inevitable corollary is that a fraction of homzygous individuals must be suffering genetic death (not bearing progeny) at each generation. Even if this fraction amounts to only 1% of the population, it becomes clear that similarly mild natural selection simultaneously operating on 99 other unlinked loci would completely exterminate a population within one generation, for every individual would be disfavored by natural selection at one gene locus or another.

The point to be made is clear - at any given time, natural selection can play its role as an advocator of progressive changes only on a very small fraction of the total gene loci contained in an organisms genome. So far as the vast majority of gene loci are concerned, the role natural selection plays is that of an extreme conservative.

Any gene product with a biological function contains an active site or sites within the molecule. In the case of an enzyme molecule, an active site represents that stretch of amino acid sequence which recognizes and binds to a substrate and a coenzyme. Any mutational amino acid substitution occurring within the active site would render a mutant enzyme delinquent in the performance of its assigned function. Thus, such a mutation is deleterious to an individual which bears it. Consequently, natural selection eliminates these bearers as unfit, and herein lies the extremely conservative nature of natural selection. Witness the fact that histone IV (110 amino acid residues) of cattle and garden peas differ from each other only by two amino acid substitutions (DeLange and Fambrough, 1968; Hunt, 1969). It appears that the entire molecule of histone IV represents a functionally critical active site, so that natural selection has eliminated almost all the mutations which from time to time affected this gene locus since plants and animals shared a common ancestor.

The cytochrome C gene has not been conserved as stringently. Of 103 amino acid residues, human and carp cytochrome C differ from each other by 17 substitutions. Obviously, amino acid substitutions at some parts of the cytochrome C molecule are not deleterious. Yet, evolutional amino acid substitutions have occurred at the rate of only 3 substitutions every 100 million years (Margoliash and Fitch, 1967). This reveals that natural selection has also been ruthlessly eliminating most mutants sustained by this locus.

As the truly conservative nature of the influence natural selection exercises over most of the functional gene loci becomes more evident, we are forced to conclude that only that part of the genome which contributes little or nothing to the well-being of an organism would be undergoing extremely rapid evolutional changes. Indeed, fibrinopeptides A (15 to 19 amino acid residues) and B (14 - 21 residues) have undergone evolutional changes (Blomback et al., 1965) at a rate 1500 times faster than that of histone IV. During blood clot formation, fibrinopeptides A and B are split off from the inert fibrinogen molecule by the trypsin-like action of thrombin. As a result, fibrinogen is converted to the active fibrin. Their role being passive, natural selection has ignored many amino acid substituting mutations during the course of evolution as they are harmless.

Up to 30% of the genomic DNA of vertebrates is made of millions of tandemly repeated copies of a short, apparently meaningless, base sequence. Such repetitious DNA is not even transcribed to RNA. It is this part of the genomic DNA which appears to be undergoing evolutional changes at the fastest rate (Southern, 1971).

Speciation has more often than not been accompanied by a chromosomal change or changes. In the past, this has been interpreted to mean that such changes are of adaptive value, therefore, they have been actively selected for by natural selection. In view of the above discussion, however, a more likely explanation appears to be that chromosomal changes are of trivial significance provided that such changes have not disturbed a functional part of the genome. It is for this reason that such changes have largely been ignored by natural selection. Intraspecific chromosomal polymorphisms which have begun to be found in increasing instances can be interpreted in the same manner, and so are allelic polymorphisms involving electrophoretic variants of enzymatic and nonenzymatic proteins. Such electrophoretic variants probably represent amino acid substitutions at functionally less critical sites of the polypeptide chain.

Even if apparent evidence of selection (excess of a certain genotype or genotypes in the population accompanied by proportional deficiency of certain other genotypes) is found with regard to a particular polymorphic enzyme locus, chances are great that selection is actually not operating upon that marker gene locus but rather on a chromosomal region adjacent to that locus. For example, if an allele, *a*, of the marker gene locus is closely linked to a deletion involving a few other unknown gene loci, there would be a distinct deficiency of *a/a*-homozygotes, not because *a*-variant enzyme is defective, but because *a/a*-homozygotes simultaneously became homozygous for a deletion.

All in all, it is a futile attempt to attribute adaptive significance to every genetic polymorphism one sees.

2. Polygenic Traits and Pleiotropic Effects of many Genes

Whenever a genetic analysis is attempted on body color and other easily visible heritable traits, the involvement of a number of gene loci rather than a single gene locus is usually found. Thus, morphological traits in general are said to be under polygenic control, and, from such observations, extreme complexity of vertebrate genetic systems is implied. The apparent complexity, however, is more imaginary than real.

For reasons which are not very clear to me, many investigators apparently start with the assumption that one particular somatic cell type, such as melanocytes, must ideally be under the control of a single regulatory system. The very fact that the type of mutations which eliminates one particular somatic cell type without affecting other types has seldom, if ever, been found in any form of multicellular organism reveals that such an assumption is based upon a fundamental misunderstanding of genetic regulatory systems.

Liver cells for example are not under the control of a single regulatory system. Within this cell type, several genetic regulatory systems which respond to different peptide and steroid hormones coexist. Hormones to which these systems respond include insulin, glucagon, hydrocortisone, estrogen and progesterone. Each genetic regulatory system in turn exercises its control over not one cell type but several different cell types. For example, not only liver cells but also fat cells and muscle cells among others respond to insulin. In short, cataloging the number of gene loci which affect one particular somatic cell type is not the way to analyze the genetic regulatory systems of multicellular organisms. Such attempts are a source of confusion with regard to the complexity of vertebrate regulatory systems.

The above discussion brings up the subject of pleiotropic effects of many gene loci. Not only regulatory gene loci but also structural loci are, as a rule, expressed in more than one cell type. Therefore, most mutations show pleiotropic effects, and this simple fact puts further constraint on natural selection's freedom of operation as an advocator of progressively adaptive changes.

The body color of vertebrates immensely influences the survival value of species, for it can be used as camouflage or as a warning. So far as the gene loci concerned with coloring due to production and distribution of melanin pigments are concerned, the tyrosinase or C-locus appears to be the one with the least pleiotropic effect. Thus, mutational deficiencies to various degrees of tyrosinase serve to lighten body color without deleterious side effects. It is apparently for this reason that such defective mutations of the C-locus as *chinchilla* and *himalaya* have recurrently been utilized by various mammalian species (Searle, 1968). If tyrosinase specified by the C-locus is concerned not only with synthesis of melanin but also with synthesis of related but more vital products such as epinephrine, defective mutations of this locus would show side effects too deleterious to be utilized for adaptive changes. Color spots on the body surface can be caused by mutations at the W-locus. However, since this locus governs not only the migration of melanoblasts from neural crests to various cutaneous areas of the body, but also the migration of primordial germ cells from the yolk sac to gonads, mutations of this locus are associated with sterility. Accordingly, W-locus mutations are not likely to be utilized for adaptive changes. Indeed, only these mutational changes occurring on a limited number of gene loci with relatively little pleiotropic effect can readily be utilized for adaptive changes and speciation. This is yet another reason which makes me believe that so far as the majority of functional gene loci in the genome are concerned, the role natural selection plays is that of an extreme conservative.

3. Gene Duplication, Genome Size Increase and the Number of Gene Loci in the Genome

Once we understand that natural selection more often than not functions as an avid advocator of status quo, it becomes evident that evolution on a grand scale such as the eventual emergence of mammals from tunicate-like sessile forms of some 500 million years ago was possible only because periodical escape from relentless surveillance by natural selection was, in the past, provided for individual gene loci. The mechanism of gene duplication provides this shelter for individual gene loci; thus, gene duplication emerges as the most important factor in evolution.

Evolution toward progressively more complex body forms would have required the continuous acquisition by the genome of new gene loci with hitherto nonexistent functions. Only the gene product with a new active site sequence can perform a new function; yet, so long as a particular function is assigned to a single gene locus in the genome, mutations affecting the active site are ruthlessly eliminated by natural selection. If two copies of the same gene become incorporated into the genome by gene duplication, what then? One of the two becomes redundant, and this redundant copy would be ignored by natural selection. While being ignored, it is free to accumulate all manner of mutations. Random accumulation of mutations by a redundant copy would most often result in total degeneration; a redundant copy becoming a functionless, nonsense DNA base sequence. Every now and then, however, random mutations by chance would result in modification of the old active site in such a way that a redundant copy would emerge triumphant as a new gene locus with a hitherto nonexistent active site, therefore, with a new function.

Indeed, one can trace the origin of most modern genes to their respective ancestry. For example, a set of gene loci for light- (220 amino acid residues) and heavy- (550 residues) chains of immunoglobulin have apparently been created by a series of gene duplications from a single ancestral gene locus which was specifying a polypeptide chain 110 amino acid residues long (Lennox and Cohn, 1967). It appears that the gene loci for various peptide hormones of vertebrates have also been created by gene duplication from a single ancestral locus. According to Adelson (1971), this ancestral locus was specifying a digestive enzyme in more primitive forms.

One expects that progressive increase in the number of functional gene loci would be reflected as progressive increase in genome sizes. Indeed, vertebrates seem to have started their evolution with a relatively small genome; the genome size of tunicates is only 6% of that of mammals. Subsequently, there appears to have been a very rapid increase in genome size for the genome size of amphioxus is 17% of that of mammals, and some of the cyclostomes representing the most primitive jawless state of vertebrate evolution already possess a genome almost as large as that of mammals (Atkin and Ohno, 1967).

However, it should be pointed out that a very large fraction of the genomic DNA of vertebrates appears to consist of nonsense base sequences which do not specify biologically meaningful products. Even the tunicate genome is 45 times larger than the genome of *E. coli*. Since *E. coli* is endowed with between 2 000 and 4 000 gene loci, the tunicate genome has room for 90 000 to 180 000 gene loci and the mammalian genome which is 750 times the size of the *E. coli* genome can contain 1.5 to 3 million gene loci. The fact is that there is a finite upper limit to the number of functional gene loci an organism can afford to maintain, and this limit is set by the spontaneous deleterious mutation rate per locus per organism generation. As already pointed out, any meaningful gene product contains a functionally critical active site within, and a mutational change affecting the active site would deprive a mutant gene product from the performance of its assigned function. Inevitably, a fraction of mutations randomly sustained by a functional gene locus would be deleterious to individuals that bear them. This is what Haldane (1957) meant by "the cost of natural selection".

For some gene loci such as that for histone IV, all mutations are apparently deleterious, while for other gene loci some mutations would be harmless. Nevertheless, the average spontaneous deleterious mutation rate is of the order of 10^{-5} per locus per organism generation. It is for this reason that, if the mammalian genome contained 3 million gene loci, mammals would have been exterminated from an unbearable mutation load a long time ago, for every individual would have been born with 30 newly sustained deleterious mutations. Since the observed overall deleterious mutation rate for mammals appears to be about 0.5, we can conclude that the mammalian genome contains only 40 000 to 50 000 functional gene loci (Müller, 1967; Crow and Kimura, 1970; Kimura and Ohta, 1971). This many functional gene loci for various peptide chains plus *ribosomal* and *transfer* RNA genes which exist in multiple copies put together would account for, at the most, only 3% of the genomic DNA of mammals. It then follows that more than 90% of the mammalian genomic DNA consists of nonsense base sequences of no apparent biological function. What is the nature of these nonsense DNA base sequences? The so-called highly repetitious DNA which are tens of millions of tandemly repeated copies of a very short basic sequence (9 base pairs or so) comprises 10 to 30% of the genomic DNA. Such repeated sequences are not even transcribed, and, even if they are transcribed, they can be translated only to nonsense oligopeptides (Southern, 1970). An additional 60 to 80% of the genomic DNA is made of so-called intermediately repetitious DNA. A more recent interpretation by Southern is that this class of repeated sequences is in fact the older, highly repetitious DNA which has diverged its sequences by random accumulation of mutations. Since these DNA base sequences have no apparent function, natural selection would permit all manner of mutations to accumulate in this large, nonfunctional fraction of the genome. Such nonfunctional DNA base sequences apparently occupied a large fraction of the genome from the very beginning of vertebrate evolution, since 30% of the tunicate genomic DNA has been found to consist of highly repetitious DNA (Lambert and Laird, 1971).

All in all, it appears that, despite the fact that only a very small fraction of the genomic DNA actually represents functional gene loci, the larger genome size by and large indicates a proportionally greater number of functional gene loci. But, there can surely be exceptions. The genome of lungfish is 23 to 35 times greater than the mammalian genome. Yet, it is inconceivable that lungfish would have that many times more functional gene loci than mammals.

There are two ways to progressively increase the genome size. One is by a continuous series of tandem duplications involving one small segment of a particular chromosome at a time. Indeed, there is little doubt that tandem duplications played a rather important role in vertebrate evolution. For example, a group of immunoglobulin heavy-chain gene loci such as the one for γ-chain of IgG class, another for α-chain of IgA class and a third for μ-chain of IgM class are closely linked to each other in all vertebrates species in which genetic analysis has been possible. Yet, with the vertebrate genome containing far more nonsense DNA base sequences than functional gene loci, one expects that a series of tandem duplications more often than not would merely result in increasing the nonfunctional part of the genome without creating redundant copies of functional genes. Thus, exclusive reliance on tandem duplications may ultimately prove to be deleterious in that they may result in saddling the genome with an unbearably large amount of nonsense DNA base sequences. Indeed, this appears to have happened in fish, for we see secondary decrease in the genome size in more specialized members of actinopterygian fish. While primitive fish appear to have increased their genome size almost to the level of mammals, more specialized members of the advanced teleosts have apparently regained smaller genome sizes by secondary decrease, for flatfish of *Heterosomata*, sea horses of *Solenichthyes* and puffers of *Tetraodontiformes* have genomes as small as amphioxus (Ohno and Atkin, 1967; Hinegardner, 1968).

The other way to increase the genome size is by a series of polyploidizations. By becoming tetraploid, not only are nonsense base sequences doubled, but so is every functional gene locus. This is the very reason that makes me believe that polyploid evolution occurring at the stage of fish or amphibians laid the foundation for the eventual emergence of mammals and of man (Ohno, 1970; Ohno, in press). The mammalian gene loci which apparently arose by gene duplication at the stage of fish are indeed unlinked to each other in all known cases. Hemoglobin α-chain gene for example is not linked to the β-, δ- and γ-chain gene complex. Similarly, immunoglobulin light-chain genes are not linked to the immunoglobulin heavy-chain gene complex. This is exactly what we expect if such gene duplications were the result of tetraploid evolution.

Based on a large amount of independent evidence, we have concluded that, of those teleost fish belonging to *Clupeiformes*, members of *Salmonidae*, *Coregonidae* and *Thymallidae* are tetraploid species which have almost completed the process of diploidization. Similarly, of those belonging to *Cypriniformes*, carp, goldfish and *Barbus barbus* are tetraploids which have long since completed the process of diploidization (Ohno et al., 1968). The genetic evidence of gene duplication by tetraploidy in these apparent tetraploid fish species is no doubt going to be discussed extensively by Dr. Engel in this symposium. Thus, I will conclude this talk by stating that, even in the human chromosome complement, there appears to be enough residual evidence left to indicate that we have indeed evolved from a tetraploid fish ancestor (Comings, 1972).

References

Adelson, J.W. (1971): Enterosecretory proteins. Nature 229, 321-325.
Atkins, N.B., Ohno, S. (1967): DNA values of four primitive chordates. Chromosoma 23, 10-13.
Blomback, B., Blomback, M., Grondahl, N.J. (1965): Studies on fibrinopeptides from animals. Acta Chem. Scand. 19, 1789-1791.
Comings, D.E. (1972): Evidence for ancient tetraploidy and conservation of linkage groups in mammalian chromosomes. Nature 238, 455-457.
Crow, F., Kimura, M. (1970): An introduction to population genetics theory. New York: Harper & Row.
DeLange, R.L., Fambrough, D.M. (1968): Identical COOH-terminal sequences of an arginine-rich histone from calf and pea. Fed. Proc. 27, 392.
Haldane, J.B.S. (1957): The cost of natural selection. J. Genet. 55, 511-524.
Hinegardner, R. (1968): Evolution of cellular DNA content in teleost fishes. Amer. Nat. 102, 517-523.
Hunt, L.T. (1969): Histone IV - bovine and pea. In: Atlas of protein sequence and structure. Ed. by M.O. Dayhoff, Silver Springs, Maryland: National Biomedical Research Foundation, pp. 190-191.
Kimura, M., Ohta, T. (1971): Protein polymorphism as a phase of molecular evolution. Nature 229, 467-469.
Lambert, C.C., Laird, C.D. (1971): Molecular properties of tunicate DNA. Biochem. Biophys. Acta 240, 39-45.
Lennox, E., Cohn, M. (1967): Immunoglobulin genetics. Ann. Rev. Biochem. 36, 365-406.
Margoliash, E., Fitch, W.M. (1967): Phylogenetic trees based on a matrix of differences of amino acid sequences. Science 155, 279-284.
Müller, H.J. (1967): The gene material as the initiator and the organizing basis of life. In: Heritage from Mendel. Ed. by R.A. Brink, Madison: The University of Wisconsin Press, pp. 419-447.
Ohno, S., Atkin, N.B. (1966): Comparative DNA values and chromosome complements of eight species of fishes. Chromosoma 18, 455-466.
Ohno, S., Wolf, U., Atkin, N.B. (1968): Evolution from fish to mammals by gene duplication. Hereditas 59, 169-187.
Ohno, S. (1970): Evolution by gene duplication. Berlin-Heidelberg-New York: Springer Verlag.
Ohno, S. (1973): Cytogenetics of chordates, protochordates, cyclostomes and fishes. Gebruder-Borntraeger, in press.
Searle, A.G. (1968): Comparative genetics of coat colour in mammals. London: Logos Press, Ltd.
Southern, E.M. (1970): Base sequence and evolution of guinea-pig α-satellite DNA. Nature 227, 794-798.
Southern, E.M. (1971): Effects of sequence divergence on the reassociation properties of repetitive DNAs. Nature New Biology 232, 82-83.

7. Biochemistry

Biochemical Polymorphism and Microevolution Processes in Fish

V. S. Kirpichnikov

1. Variability for Blood Groups

The use of serological reactions (precipitation and hemoagglutination) makes it possible to reveal in fishes blood groups similar to those in man and mammals. Up to the present such groups have been detected in more than 50 species of fresh water and marine fishes. Variability for erythrocytic antigenes underlying the formation of various blood groups seems to be typical of all fish species. The results of blood group studies in fish are summarized in a few special reviews (Cushing, 1964; de Ligny, 1969; Altukhov, 1969). For each species one or several series of genes determining antigenic variability of erythrocytes have been found. By the character of inheritance these series not infrequently resemble a ABO system long ago detected in man. Such kind of systems have been revealed, for instance, in brown trout (Sanders, Wright, 1962), anchovy from the Azov Sea (Limansky, 1964; Altukhov et al., 1969a, b; Limansky, Pajusova, 1969), herring (Altukhov et al., 1968; Truveller, 1971), tuna (Fujino, 1970) and in other fishes. Simpler systems consisting of two or more codominant alleles occur more often, their number being sometimes as high as 12 (sockeye salmon, see Ridgway, 1969). In some instances it has been possible to check hypotheses concerning the inheritance of blood groups by means of hybridological analysis (Sanders, Wright, 1961 - rainbow trout; Slota et al., 1970 - carp, and oth.). In the majority of cases the validity criterion of genetic interpretation is a correspondence between frequencies of blood groups and those calculated from Hardy-Weinberg's equation

$$p^2(AA) + 2pq(AA') + q^2(A'A') = 1$$

where p and q are the frequencies of alleles A and A', while AA, AA' and A'A', are three possible genotypes with the two allelic system.

Numerous investigations of blood groups in fishes carried out over the past few years permit the following conclusion:

1. Variability of blood groups is widespread in fish no less than in other animals. Heterogeneity of blood groups in fish populations is very large. The hereditary basis of antigenic differences is formed by allelic (not infrequently multiple) systems of genes codominant or sometimes dominant toward each other. Many of the systems include so called "zero" alleles - genes the products of which are still undetected.

2. Populations within a species differ from others in the gene concentration of blood groups. Examples are found in works conducted on herring (Sindermann, Mairs, 1959, 1960; Altukhov et al., 1968), anchovy (Limansky, 1964; Altukhov et al., 1969a,b), sardine (Sprague, Vrooman, 1962; Vrooman, 1964) and on tuna (Suzuki, 1966; Fujino, 1969). Differentiation of populations by the frequencies of individual genes enable the investigators to analyse the species structure and detect true intraspecific groups. A good example is given by Pacific tunas making up probably three main reproductively isolated large populations -

Southern, Northern and Western ones. They are closely related morphologically, but nevertheless can be easily recognized by their blood group genes (Fujino, 1970).

3. The adaptive significance of variability in blood groups is obvious. By analogy with higher animals the diversity of erythrocyte antigenes in fish can be regarded as a good means of protection from infestation and infectious deseases. The larger and more variable the antigenic content the harder the process of adaptation for a microbe or parasite (Efroimson, 1971). In fact, there is direct evidence demonstrating the adaptive meaning of blood group variability. Thus, latitudinal (not longitudinal) variability has been found in 11 species of catostomids. The dispersion analysis shows that in this particular case about 42% of the whole variation is concerned with latitudinal changes (Koehn, 1969) and, most likely, with temperature environmental conditions. Likewise, the adaptive character of antigenic variability may be judged about by the evidence indicating its persistence in populations despite considerable inbreeding (Sanders, Wright, 1961). The frequencies of various blood groups in flatfish populations change with age, which seems to be a result of natural selection (de Ligny, 1967, 1969).

2. Polymorphism for Proteins

The data on variability of fishes with respect to their blood groups can give a general idea of its scope and adaptive significance. But more precise information of the biochemical polymorphism of fish populations has been obtained by another method. At the end of the 1950's and at the beginning of the 1960's a new fine technique of separating isozymes and isoallele forms of proteins by electrophoresis in starch and acrylamide gels was elaborated (Smithies, 1955; Hunter and Markert, 1957; Hubby and Lewontin, 1966; and oth.). In the majority of animal and plant populations 30-50 and even larger percent of proteins coded by genes proved to be polymorphous, whereas the mean heterozygosity level (relative number of loci of a given individual in heterozygous state) as a rule was as low as 7-15% (Lewontin and Hubby, 1966; and many oth. For review see Kirpichnikov, 1972).

Similar calculations were made for fishes. In brook trout the number of polymorphic protein loci was 38% (Wright, Atherton, 1970). Inland representatives of the genus *Astyanax* (*characidae*) 29-41% of all the genes tested was polymorphic when the mean heterozygosity level made up 11.2%. Cave relatives of these land dwellers proved to be much less variable - the portion of polymorphic genes in different cave populations, as well as in different species, ranged from 0 to 20% and the heterozygosity level, from 0 to 7.7% (Avise, Selander, 1972). Chum salmon was found to be comparatively stable. According to Altukhov et al. (1972), only 11-18% of its genes was polymorphic and the heterozygosity level made up only 1.9 - 3.2%. The authors are apt to ascribe such low variability in this case to the polyploidy of salmonids. In diploid herring the polymorphism attains 45%, and the heterozygosity, 10.6% (Altukhov et al., 1972). The small number of fish samples tested seems to diminish the significance of the variability calculations in question.

A great deal of data has been collected about the genetic variability of fish proteins. Protein heterogeneity is found to be associated with all fish species. Structural proteins of myogene type, hemoglobins, blood serum proteins as well as all enzymes in blood and some organs of fishes appear to be variable. Let us summarize briefly the available data on the genetic variability of various proteins.

Haemoglobins. Among polymorphic species are sprats (Wilkins, Iles, 1966; Naevdal, 1970), whitefish (Lindsey et al., 1969), tinch (Callegarini, Cucchi, 1968), eight species of gadoids (Sick, 1965a,b; Frydenberg et al., 1965; Möller, 1968; Möller, Naevdal, 1969; Wilkins, 1971), eels (Sick et al., 1967) etc. Salmonids are generally monomorphic (Altukhov et al., 1969), as also are wild and domesticated carps, and many other fishes (Altukhov, 1969). In the case of polymorphism with regard to hemoglobins an allele series, as a rule, consists of two "ordinary" forms and rare single alleles. The alleles are inherited codominantly, and in the majority of cases no "hybrid" product is found. As in mammals, hemoglobins in fish are tetramers consisting of four subunits grouped in two (e.g. $\alpha_2\beta_2$). Different subunits are coded by different genes. The number of genes varies between 2 and 4, but in some fishes, for instance, in salmonids, it attains 8 (Tsuyuki, Ronald, 1970, 1971; Altukhov et al., 1972). Such diversity is likely to be associated with the polyploid origin of salmonids.

Due to the monomorphism of hemoglobins in salmonids and many other fishes they may be used successfully for studies on fish taxonomy.

Transferrins which are contained in blood serum of fishes and transfer iron required for hemoglobin formation are most variable. The polymorphism of the transferrins is mainly of genetic character (although unstable conformational changes may occur). In fish, the number of alleles in one population most frequently makes up 3 or 4, but in some instances it can reach 8-10 or even 15. Not a single species is found to contain transferrin isoforms (analogous to different types of hemoglobin) which can be probably accounted for by its monomeric structure. Some authors associate the presence of numerous genetic variants of transferrin with its second "loading", i.e. with its bactericidal (protective) properties (Manwell, Baker, 1970).

Polymorphism for transferrins has been registered in more than 30 fish species, but it has received the most study in tunas, some gadoids and salmonids, flatfish and carp. In the majority of cases the inheritance is codominant being associated with one locus. As can be seen from electropherograms homozygotes have one disk and heterozygotes two weaker disks. There is no "hybrid" disk, but now and then there occur individuals with three or four transferrin zones. However, the cause of appearance of such additional zones is still uncertain. It is not unlikely that the multiplicity of rings may be explained by different numbers of silicic acid residues (1-4) which can be attached to the transferrin molecule (Hershberger, 1970). The relations between transferrin types satisfy completely the Hardy-Weinberg's equation, although reliable deviations (excess or deficiency of homozygotes) are possible. The cause of such errors will be discussed below. An example of "equilibrium" populations is offered by brook trout populations from American fish farms with the following gene frequencies (Wright, Atherton, 1970):

Frequencies	Genotypes						Total
	AA	AB	AC	BB	BC	CC	*(n)*
Found	2	21	36	54	240	249	602
Expected	2	19	39	57	237	248	

There is nearly a complete coincidence of theoretical and empirical frequencies for all the six genotypes studied. The same relations are found by the analysis of populations of cod (Möller, 1968), tunas (Fujino, Kang, 1968; Fujino, 1970) and Atlantic salmon (Möller, 1970).

Consequently, transferrins belong to a group of proteins with maximally pronounced polymorphism. The heterozygosity level with respect to transferrins is generally 30-50%. The bulk of species are heterogenic. However, there are statistically reliable differences in frequencies of transferrin alleles between populations. Thus, obvious variation is found between coastal and Arctic races of cod dwelling near the Norwegian coast (Möller, 1968). There are also differences between European and American eels (Drilhon, Fine, 1971) as well as between European and American Atlantic salmons (Nyman, 1967; Wilkins, 1971; Payne et al., 1971).

High concentrations of allele D(q = 0.64) are observed in the Far East carp, while in European populations of wild and domesticated carps this allele occurs rather seldom (Creyssel et al., 1966; Valenta, Kalal, 1968; Balakhnin and Romanov, 1971; Balakhnin and Galagan, 1972).

One of the American brook trout populations is found to contain an allele form of transferrin lethal when in homozygous state. In its primary structure this transferrin differs significantly from the others (Hershberger, 1970) and appears to be a result of mutation involving a considerable region of the gene.

<u>Fish serum proteins</u> (transferrins exclusive) are as yet imperfectly studied from the genetic point of view. Some species show variability in albumins and haptoglobins. Net infrequently polymorphic systems of these proteins include "zero" alleles. They are found, for example, in haptoglobin of the sea bass *Sebastes marinus* (Nefedov, 1969) and most likely in albumin of the sturgeon (Lukjanenko et al., 1971) and carp (Truveller et al., in print).

The general spectrum of serum proteins is successfully used in work on intraspecific systematics. Thus, there are accounts of differences between populations of Siberian sturgeon (Lukjanenko, Popov, 1969) and local stocks of the white bass *Roccus chrysops* (Wright, Hasler, 1967). As shown in the latter case on the basis of these differences, fishes return back to a river they leave to go down to the lake.

Serum proteins of parental forms are summed up as a rule in interspecific fish hybrids (Nyman, 1965; Haen, O'Rourke, 1968).

<u>Proteins of crystalline lens</u>, much like serum proteins, proved to be subjected to considerable individual variability (Eckroat et al., 1969; Eckroat, 1971; and oth.). Thus, differences between populations have been found in Arctic char (Saunders, McKenzie, 1971).

<u>Proteins of skeletal muscles</u> are polymorphic in a number of fishes - common sucker, hake, cuttlass fish, steel-head trout and some others (Tsujuki, 1966; Utter, Hodgins, 1971; Grag, McKenzie, 1970, and oth.). In many species muscle proteins, as well as haemoglobins, are monomorphic and species-specific and, therefore, can be used for evolution studies (Huntsman, 1970; McKenzie, 1973; and oth.).

A great body of information concerns enzymes. We shall mention here only the most essential of them.

Lactate dehydrogenase (LDH) has been studied most thoroughly. Four subunits constituting a LDH tetrameric molecule are usually coded by two genes the products of which make up a series from 3-5 isozymes. Most often products of "a" and "b" genes are combined. As a result, isozymes A_4, A_3B, A_2B_2, AB_3 and B_4 are formed. In some fishes combinations A_3B, AB_3 and sometimes A_2B_2 are absent. Likewise, other genes, for instance, gene "e" functioning efficiently in nervous cells (retina, brain) and gene "c" which manifests itself in gonads participate in the formation of LDH heteropolymers (Holmes, Markert, 1969; Markert, Ursprung, 1971). According to the recent data, whitefish from the genus *Prosobium* contains 10 LDH genes constituting 5 systems (a, a', b, b', d, d', e, e', f, f') each consisting of 3-5 isozymes and being related to a definite tissue (Massaro, 1972). Generally, the tissue specificity is typical of the majority of LDH isozymes detected in fishes.

It appears that isozymes from different "series" differ significantly since all the attempts to carry out molecular hybridization *"in vivo"* between them have failed (Massaro, 1972). Some authors, however, find no significant difference between subunits A, B and E (Wuntsch, Goldberg, 1968).

The smallest number of LDH isozymes is found in mackerel (3), fishes from the fam. *Soleidae* (2 - A_4 and B_4), some flatfishes and turbots (A_4 only) (Markert, Faulhaber, 1965). Salmonids contain the greatest number of isozymes of this enzyme (up to 27 and more) which is likely to be caused by their polyploidy.

The diversity of LDH forms in the fish organism is complicated by the genetic variability for individual loci, and genes "a", "b" and "e", in particular. Hagfish, *selachia*, many salmonids, herring, cod, haddock, carp, hake and some other species are polymorphic (Sachko, 1971). The genetic variability for LDH alleles is best known for the North-American brook trout (Wright, Atherton, 1970, and oth.). In this case the analysis of population goes along with special crosses. In the brook trout gene "b" is represented by three alleles, genes "a" and "e", by two. Genes "a" and "b" are inherited not quite independently. During hybridization of the brook trout (*S. fontinalis*) and the lake trout (*S. namaycush*) differing in alleles of the both genes progeny of one of the reciprocal crosses showed obvious deviation from independent segregation (Morrison, 1970).

♀aabb x♂ aa'bb' = 136 aabb' + 122 aa'bb + 495 aabb + 461 aa'bb'
(brook trout)
(hybrid F_1)

In reciprocal crosses the segregation was in compliance with the expected result (1:1:1:1), while subsequent generations yielded all kind of transitions, from crosses with the predominance of parental genes (ab and a'b') to those with large excess of crossovers.

These deviations may be explained in two ways (Morrison, 1970):

1. Remote relationship between two pairs of chromosomes carrying genes "a" and "b" which is connected with ancient polyploidization and cause non-random assortment of chromosomes during meiosis.

2. Association of two different chromosomes resulting from fusion of their centrosomes (Robertson translocation). Associations of such kind not infrequently occur in some salmonids (Ohno et al., 1965, and oth.).

Similarly, interpopulation differences in the concentration of LDH alleles are well expressed in fishes. Thus, frequences of gene "b" alleles proved to be different in six populations of the brook trout from Newfoundland (Goldberg et al., 1969). There are differences between populations of hake (Utter and Hodgins, 1971), crested blenny *Anoplarchus purpureus* (Johnson, 1971) and other species. In some fishes, for example in cod and haddock, the allelic frequencies of genes "a" and "b" are invariably constant within a species (Lush, 1970).

Malate dehydrogenase (MDH) is likely to be a dimer in one kind of fishes (e.g. soury) and consists, like LDH, of four polypeptide chains, in the other. The majority of fishes contain several MDH isozymes in their tissue. Their number sometimes exceeds 9 (Numachi, 1970a, b) and in Atlantic salmon even attains 15 (Wilkins, 1971). In the karyotype of salmon and whitefish the isozymes are coded by no less than six different genes (Massaro, 1972).

Polymorphism with regard to MDH is detected in populations of herring (Odense, Allen, 1971), chum and pink salmon (Slynko, 1971a, b; Altukhov et al., 1972), soury, hardtail, *Distrema temminki* (Numashi, 1970a, b), *Fundulus heteroclitus* (Whitt, 1970) and in some others. A great number of species, however, are monomorphic with respect to this enzyme.

In rainbow trout and chinook salmon no less than four MDH genes are found, instead of two detected in herring and flatfish (Bailey et al., 1970). These findings are in accord with the hypothesis of the polyploid origin of salmonids.

Esterases. This term is used to designate a few groups of enzymes varying in their functions. Aryl-, acetyl-, cholin-, and carboxyl esterases are known (Holmes, Whitt, 1970; Manwell and Baker, 1970). According to Nyman (1971), serum esterases probably belong to a type of carboxyl esterases, but in the overwhelming majority of cases no precise determination of enzymes has been carried out.

Esterases are inherited codominantly without formation of hybrid rings. As a rule, the esterase molecule is a homopolymer (probably a dimer), and, therefore, the number of isozymes is not considerable for this enzyme. In some fishes, however, it is large (Holmes, Whitt, 1970).

Polymorphism with respect to esterase genes has been discovered in representatives of clupeoids, salmonids, osmerids, cyprinids, catostomids, eels, gadoids, tunas, flatfishes and other families. By the frequency and range of variability esterases resemble transferrins. Not infrequently a population contains three-four or even six-seven alleles of one locus, and in capelin (*Mallotus villosus*) their number reaches 10 (Nyman, 1971). One and the same esterase is sometimes coded by several loci.

Let us give some examples of polymorphism regarding exterase genes. Stable polymorphism has been detected in the liver and heart esterases of the Atlantic herring *Clupea harengus*. It is determined by the three-allelic system with the concentrations of three genes in different populations amounting to 0.71-0.77, 0.22-0.28 and 0.005-0.016, respectively. The populations are rather alike and the distribution of genotypes is in good agreement with the Hardy-Weinberg's equation, at the same time there are small reliable differences in allelic concentration between local groups (Ridgway et al., 1970). It appears that these populations are reproductively isolated.

Three- and four-allelic equilibrium systems can also be found in some tunas (Sprague, 1967; Fujino, 1970). However, the excess of homozygotes in esterase genes has been described, for example in mackerel (Jamieson et al., 1969). In this case we are probably concerned with a mixture of two populations, but none the less a decrease in the viability of heterozygotes is not unlikely. In the Arctic char (*Salvelinus alpinus*) the esterase gene is found to have two alleles. The molecular weight of esterase coded by one of the alleles twice as that of other esterase molecule (Nyman, Shaw, 1971). In viviparous ling (*Zoarces*) the gene of "eye" esterase is presented by two (active and "zero") alleles. In this case the inheritance of genes could be verified by comparing females with their progeny (Simonsen, Frydenberg, 1972). Clear cut Mendelian inheritance of esterases is followed also in carp (Tscherbenok, 1973).

The population analysis shows that differences between populations can be easily defined by the allelic concentration of esterase genes. Such differences have been revealed in Atlantic salmon (European and American subspecies), tunas, hake, herring, etc. The presence of "hybrid" esterase permits to determine precisely the occurrence (4%) of *S. salar* - *S. trutta* hybrid in a salmon population near the shores of England (Payne et al., 1972).

Other enzymes in fish are as yet little known. The findings on the polymorphism of some enzymes are presented below. The genetic variability has been found in the following fish species:

<u>Isocitrat dehydrogenase</u> - in common carp, crucian carp and other cyprinids, herring, rainbow trout (Quiroz-Gutierres, Ohno, 1970; Wolf et al., 1970; Engel et al., 1971; Moon, Hochachka, 1972).

<u>Sorbitol dehydrogenase</u> - in herring and some cyprinids (Engel et al., 1970, 1971; Lin et al., 1969).

<u>6-phosphogluconat dehydrogenase</u> - in common and crucian carps, barbel, roach, ling (Bender, Ohno, 1968; Wolf et al., 1969; Klose et al., 1969, 1970; Yndgaard, 1972).

<u>Hexose-6-phosphat dehydrogenase</u> - in brook trout (Stegman, Goldberg, 1971).

<u>α-glycerophosphat dehydrogenase</u> - in brown and rainbow trouts, chum salmon, smelt, soury (Engel et al., 1971; Numachi, 1971; Altukhov et al., 1972).

<u>L-glycerophosphat dehydrogenase</u> - in one of the American whitefishes (Lindsey et al., 1970).

<u>Catalase</u> - in sea bass (Numachi, 1971).

<u>Peroxidase</u> - in the bullhead *Myoxocephalus quadricornis* (Nyman, Westin, 1969).

<u>Tetrazoliumoxidase</u> - in tunas, chinook salmon and rainbow trout (Edmunds, Sammons, 1971; Utter et al., 1973).

<u>Aspartat aminotranspherase</u> - in herring (Odense et al., 1966; Naevdal, 1969), in Nevada pupfish *Cyprinodon nevadensis* (Turner, 1973).

<u>Phosphoglucomutase</u> - in herring, rainbow trout, sockeye salmon, ling, Pacific perch, cod (Lush, 1969; Roberts et al., 1968; Tills et al., 1969; Utter, Hodgins, 1970; Hjorth, 1971; Johnson et al., 1971).

<u>Creatinkinase</u> - in common carp (Scopes, Gosselin-Rey, 1968).

<u>Phosphogluoisomerase</u> - in ling (Yndgaard, 1972).

The list will be by far increasing as well as the number of fishes with polymorphic protein systems.

Summing up the results of genetic studies on fish proteins we must stress out the following:

1. Protein polymorphism is widespread in fishes. Transferrins, LDH and esterases appear to be most variable proteins.

2. There are clearly defined differences in the protein gene frequencies between fish populations and subspecies.

3. Genetic variability for proteins frequently goes along in fishes with the presence in the same individual of a larger or lesser number of isozymes or allozymes - products of different genes and combinations of these products (heteropolymers).

4. In the majority of cases "protein" genes are inherited codominantly in strong conformity with the Mendel's laws. Sometimes "hybrid" zones are formed in heterozygotes. In certain instances complete dominance may be observed associated, as a rule, with "zero" alleles which yield no protein products.

3. Adaptive Significance of Biochemical Polymorphism in Fishes

The adaptive meaning of biochemical polymorphism in animals and plants becomes more and more obvious. There are three groups of facts speaking about the adaptive significance of protein polymorphism in fish. Among them are the "clines" in frequencies of some genes, one-gene heterosis (increased fitness of heterozygotes as compared with homozygotes) and functional differences between isoalleles (and isozymes) often related to the environmental conditions.

Clines. Clinal variability of hemoglobins has been found in cod from the Norwegian and Baltic seas. The frequencies of two alleles change from south-west to north-east (Frydenberg et al., 1965; Sick, 1965a, b; and oth.). The presence of a cline cannot be accounted for only by the mixture of two forms (Arctic and coastal) with different allelic concentrations (Möller, 1968, 1971) since, despite the absence of Arctic cod from the Baltic sea, the cline there has the same direction as at the Norwegian coast.

The selective mechanism of formation of the hemoglobin cline is supported by other evidence, for instance such as persistent retention of definite gene concentration in each local population over many years (Frydenberg et al., 1968; Wilkins, 1971).

Clinal variability for hemoglobin alleles is found in the whiting *Merlangius merlangus* along the shores of Scotland (Wilkins, 1971).

The cline in allelic frequencies of the transferrin locus is typical of salmon from the Atlantic coasts of Canada and U.S.A. The gene concentration increases from 0.07-0.24 to 0.55-0.60 toward the south (Möller, 1970).

The crested blenny (*Anoplarchus purpureus*) shows latitudinal variability in LDH alleles - the frequency of LDH-1 gene increases southwards from 0.02 to 0.26. A closely related species *A. insignis* lives in colder waters and, correspondingly, is monomorphic with respect to the "north" $LDH\text{-}1^1$ allele (Johnson, 1971). In the sucker *Catostomus clarki* there is also a latitudinal cline in esterase. The concentration of gene Est-1 increases constantly over a distance of 525 miles from 0.18 in the south to 1.00 in the north. Closely related species of Catostomids dwelling in the north or ascending rivers higher above the sea level have only one "north" allele of this gene (Koehn, Rasmussen, 1967; Koehn, 1968, 1971). There are species containing two esterases revealed electrophoretically - i.e. they exhibit as if a "fixed" heterozygosity.

A similar cline has been detected in the Arctic char with the frequency of one of esterase gene alleles increasing southwards from 0.17 to 1.00 (Nyman, Shaw, 1971).

Functional differences between alleles and isozymes. Functional differences between LDH isozymes are undeniable. This fact being supported by numerous experiments on salmonids, *Fundulus heteroclitus*, etc. (Hochachka, 1967, 1968; Somero, Hochachka, 1969; Massaro, Book, 1972, and oth.). Markert has repeatedly emphasized in his works that the presence of numerous LDH isozymes provides normal functioning of this enzyme in different organs and tissues under sharply alterating external and internal conditions (Markert, Whitt, 1968; Markert, Ursprung, 1971, and oth.).

Isocitrat dehydrogenase in rainbow trout functious most perfectly (shows more stabilized reaction) when represented in the organism by several isozymes (Moon, Hochachka, 1972). The presence of a few hemoglobin types has by far an adaptive meaning. Thus, in anadromous eel two hemoglobin components are in conformity with two environments where the eel is living - fresh water (strong affinity to oxygen) and sea water (weakened affinity to oxygen, strong Bohr effect) (Poluhowich, 1972). In catostomids hemoglobins are found to differ adaptively for two groups of species (Powers, 1972).

Functional peculiarities of allelic forms of one and the same protein are also sharply defined. Thus, in brook trout isoallelic forms of LDH coded by genes B and B' differ drastically in the temperatures of enzyme inhibition by lactate (Wuntch, Goldberg, 1970). In largemouth bass x smallmouth bass hybrids of the second and third generations there is a selection in favour of allele E' characteristic of the smallmouth black bass (Whitt et al., 1971). In crested blenny the fry with LDH A and LDH A' genes exhibit different heat resistance (Johnson, 1971).

In lake trout one of the genetic types of transferrin is connected evidently with increased growth of the fish (Morrison, 1970), and gene A seems to be lethal in homozygous state (Hershberger, 1970). Decreased survival percent is also typical of transferrin genotype CC in carp (Balakhnin and Galagan, 1972). In flatfish various alleles of transferrin locus differ in viability, their concentration changes with age (de Ligny, 1967). Striking constancy in the polymorphism for transferrin in brook trout populations isolated over 100 years (Wright, Atherton, 1970) and in introduced population of the buffalo fish *Ictiobus cyprinellus* (see Koehn, Johnson, 1967) cannot be explained without assumption that in nature there is a genetic equilibrium of alleles maintained by selection. Different mortality of fishes with different alleles of the hemoglobin locus has been observed in the second generations of the smallmouth x largemouth black bass hybrids (Manwell et al., 1963).

Substrate specificity of esterase allelic forms and their different resistance to inhibitors have been revealed in *Fundulus heteroclitus* (Holmes, Whitt, 1970). Difference between the activities of alleles (rather, their products) are very significant in *Catostomus clarki* (Koehn, 1968, 1970) and *Notropis stramineus* (Koehn et al., 1971). In the both cases the relation of the activity to temperature alters. The frequencies of esterase alleles arranged in accordance with increasing electrophoretic mobility are characterized by well defined binomial distribution in sand launce populations (*Ammodytes*). Alleles of intermediate mobility seem to correlate most perfectly with the environment (Leung, Odense, 1973). Changes in the frequencies of esterase genes with age have been found in bleak *Alburnus alburnus* (Handford, 1971).

No doubt, allelic differences will be discovered, if not in all, at any rate in most cases of fish protein polymorphism. The same conclusion has been recently drawn by Harris relating to human polymorphic systems (Harris, 1971).

Increased fitness of heterozygotes. A good deal of data has been collected about obvious advantages of heterozygotes over homozygotes for protein alleles. In certain cases heterosis at the molecular level has been shown. Thus, in the hybrids *Lepomis cyanellus* x *Chaenobrittus gulosus* hemoglobin has an increased ability for oxygen transport (Manwell et al., 1963; Manwell, Baker, 1970). In the brook trout heterozygous transferrins AB, AC and BC show greater affinity to iron than homozygous combinations (Hershberger, 1970). In common carp heterozygotes relevant to transferrin grow more rapidly than homozygotes (Tscherbenok, 1973). As has already been shown there is a cline in esterase alleles in *Catostomus clarki*. At intermediate temperatures (about 10-20°) "hybrid" esterase (ab) shows the highest activity (Koehn, 1960, 1970). Increased activity at high temperatures is characteristic of esterase isolated from heterozygotes in sand shiner (*Notropis stramineus*). This case is of particular interest since in heterozygotes for gene Est-2 the activity of esterase-1 changes. It seems that a close functional relation exists between the two loci. The advantage of the heterozygotes in this case leads to seasonal selection - the number of heterozygous individuals increases in summer and decreases in winter (Koehn et al., 1971).

The advantage of heterozygotes is also proved by their increased numbers in natural populations or in hybrids. The excess of heterozygotes for hemoglobin genes is observed in the same populations of the cod, catfish *Ictaburus nebulosus*, as well as in hybrids *Lepomis* x *Chaenobrittus* (Sick, 1965a; Raunich et al., 1966; Manwell, Baker, 1970). The increased number of heterozygotes for transferrins is found in young individuals of skipjack tuna (Fujino, Kang, 1968), for LDH, in *Micropterus* hybrids (Whitt et al., 1971), for sorbitol dehydrogenase, in gold fish from the Lake Erie (Lin et al., 1969), for esterase, in bleak and in crappie hybrids *Pomoxis* (Handford, 1971; Metcalf et al., 1972).

The maintenance of protein variability in brook trout during long inbreeding and after domestication is an indirect evidence in favour of increased fitness of heterozygotes (Wright, Atherton, 1970). These facts allow, however, another interpretation.

It can be suggested that in the near future a great deal of new information will be gained about the existence of heterosis for separate protein genes in fish.

Heterosis plays by far an important role in the maintenance of biochemical polymorphism in fish populations.

4. Conclusion

Polymorphism for blood groups and proteins is widely distributed in fish. In natural populations fishes are heterozygous for a great number of genes. The level of heterozygosity is very high, decreasing only under specific conditions - i.e. when fishes live in stable environment (cave fishes), in the case of polyploidy as well as in some other cases. Such facts as strongly pronounced clinal variability, functional differences between alleles and increased fitness of heterozygotes in separate genes speak about adaptive significance of biochemical polymorphism. Among mechanisms responsible for its origin and prolonged maintenance are (Timofeeff-Resovsky and Svirezhev, 1966):

1. Alternating adaptation of various allelic forms, change of allelic fitness in time and space owing to which selection acts at different time in different directions.

2. Increased fitness (in survival rate) of heterozygotes for protein genes and genes of blood groups. This increased fitness of heterozygotes manifests itself, occasionally only under certain conditions. It plays probably an important role for many genes. Consequently, of great significance should be the selection aimed at establishing a definite general "heterozygosity optimum" (Kirpichnikov, 1972).

There are some additional factors promoting the maintenance of polymorphic systems in populations. It is primarily the advantage of rare genotypes which is well-expressed, for instance, in *Drosophila* populations (Ayala, 1972; and many oth.). Of primary significance is the selection which acts in opposite directions for males and females. Sometimes (probably rather seldom) polymorphism is a result of current substitution of one allele for another or hybridization between two species or subspecies.

The main microevolution processes in populations are mutations, random variation in the frequency of allelic genes (drift) and selection. Recent data on the level of protein polymorphism in populations allows for two essential corrections in the concepts of evolution factors and the structure of populations. Firstly, in natural populations the heterozygosity level is much higher than it has been suggested. A larger percentage of variability in populations involves genes which are manifested in heterozygotes. Secondly, the role of random genic drift in the accumulation of deleterious recessive genes in populations and in the increase of heterogeneity has been evidently exaggerated. Direct selection is the main factor responsible for great variability of populations.

According to Koehn (1971) the main cause of polymorphism is the necessity for organisms to adapt of ever altering environmental conditions. The "strategy of adaptation" in such changing environment consists in the formation of labile genetic systems capable of responding adaptively to a wide range of variable environmental factors.

New data on biochemical polymorphism of fishes and other organisms requires the revision of all our concepts relevant to factors of evolution and genetic structure of populations and species.

5. Summary

Polymorphism for blood groups and proteins, enzymes inclusively, occur widely in fish populations. The heterozygosity level of fishes is as high as that of other organisms. Cases of regular clinal variability, functional differences between isozymes and allelic forms of the same protein and facts of one-gene heterosis for protein suggest adaptive significance of biochemical polymorphism.

As has been recently established numerous genes which are expressed in heterozygote and produce no deleterious effect in homozygote accumulate in populations. This fact requires the revision of many concepts of population genetics concerning some microevolution factors.

References

Altukhov, Ju.P. (1969): On the immunogenetic point of view on the problem of intraspecific differentiation of fishes (russ.). Usp. sovrem. genet. Moscow, Nauka 2, 161-195.

Altukhov, Ju.P. (1969): On the ratio of monomorphic and polymorphic hemoglobins in the microevolution of fishes (russ.). Dokl. Acad. Nauk 189(5), 1115-1117.

Altukhov, Ju.P., Truveller, K.A., Zenkin, V.C., Gladkova, N.S. (1968): The A-system of blood groups in Atlantic herring (*Clupea harengus L.*) (russ.). Genetica 4(2), 155-167.

Altukhov, Ju.P., Limansky, V.V., Pajusova, A.N., Truveller, K.A. (1969): Immunogenetic analysis of the intraspecific differentiation in the European anchovy (*Engraulis encrasicholus*) inhabiting the Black Sea and the Sea of Azov. I. The blood groups and the hypothetical mechanism of the genic control. The heterogeneity of the Azov race (russ.). Genetica 5(4), 50-64.

Altukhov, Ju.P., Limansky, V.V., Pajusova, A.N., Truveller, K.A. (1969): Ibid. II. Elementary populations of anchovy and their connection with the genetic-populational structure of the species (russ.). Genetica 5(5), 81-94.

Altukhov, Ju.P., Salmenkova, E.A., Omelchenko, V.T., Sachko, T.D., Slynko, V.I. (1972): Number of mono- and polymorphic loci in the population of tetraploid salmon species *Oncorhynchus keta* (russ.). Genetica 8(2), 67-75.

Avise, J.C., Selander, R.K. (1972): Evolutionary genetics of cave dwelling fishes of the genus *Astyanax*. Evolution 26(1), 1-19.

Ayala, F.J. (1972): Frequency-dependent mating advantage in *Drosophila*. Behav. Genet. 2(1), 85-91.

Bailey, G.S., Wilson, A.C., Halver, J., Johnson, C. (1970): Multiple forms of supernatant malate dehydrogenase in salmonid fishes: biochemical, immunological and genetic studies. J. Biol. Chem. 245, 5927-5940.

Balachnin, I.A., Romanov, L.M. (1971): Gene frequency of the different transferrin types in the low-bred domesticated carp and Amur wild carp (russ.). Hydrobiol. J. 7(3), 84-86.

Balachnin, I.A., Galagan, N.P. (1972): Distribution and surviving of the individuals with different transferrin types in the various carp crossings (russ.). Hydrobiol. J. 8(3), 56-61.

Bender, K., Ohno, S. (1968): Duplication of the autosomally inherited 6-phosphogluconate dehydrogenase gene locus in tetraploid species of cyprinid fish. Bioch. Genet. 2(1), 101-107.

Callegarini, C., Cucchi, C. (1968): Intraspecific polymorphism of hemoglobin in *Tinca tinca*. Biochem. Biophys. Acta 160(2), 264-266.

Creyssel, R., Richard, G., Silberzahn, P. (1966): Transferrin variants in carp serum. Nature 212, 1362.

Cushing, J.E. (1964): The blood groups of marine animals. Adv. Mar. Biol., N.Y. Acad. Press 2, 85-131.

Drilhon, A., Fine, I.M. (1971): Les groupes de transferrines dans le genre *Anguilla L.* Rapp. Proc.-Verb. Reun. 161, 122-125.

Eckroat, L.R. (1971): Lens protein polymorphisms in hatchery and natural populations of brook trout, *Salvelinus fontinalis* (Mitchill). Trans. Amer. Fish. Soc. 100(3), 527-536.

Eckroat, L.R., Wright, J.E. (1969): Genetic analysis of soluble lens protein polymorphism in brook trout (*Salvelinus fontinalis*). Copeia 3, 466-473.

Edmunds, P.H., Sammons, J.I. (1971): Genetic polymorphism of tetrazolium oxidase in bluefin tuna, *Thunnus thynnus*, from the Western North Atlantic. J. Fish. Res. Bd. Canada 28(7), 1053-1055.

Efroimson, V.P. (1971): Immunogenetics (russ.). Moscow, Medicina.
Engel, W., Op't Hof, J., Wolf, U. (1970): Genduplikation durch polyploide Evolution: die Isoenzyme der Sorbitdehydrogenase bei herings- und lachsartigen Fischen (*Isospondyli*). Humangenetik 9(2), 157-163.
Engel, W., Schmidtke, J., Wolf, U. (1971): Genetic variation of L-glycerophosphate-dehydrogenase isoenzymes in clupeoid and salmonoid fish. Experientia 27(12), 1489-1491.
Frydenberg, O., Møller, D., Naevdal, G., Sick, K. (1965): Hemoglobin polymorphism in Norwegian cod populations. Hereditas 53, 255-271.
Frydenberg, O., Nielsen, J.T., Simonsen, V. (1968): The maintenance of the haemoglobin polymorphism of the cod. Proc. 12th Int. Congr. Gen. 2, 152.
Fujino, K. (1969): Atlantic skipjack tunas genetically distinct from Pacific specimens. Copeia 3, 626-628.
Fujino, K. (1970): Immunological and biochemical genetics of tunas. Trans. Amer. Fish. Soc. 99(1), 152-178.
Fujino, K., Kang, T. (1968): Transferrin groups of tunas. Genetics 59(1), 79-91.
Goldberg, E., Kerekes, J., Cuerrier, J.P. (1971): Lactate dehydrogenase polymorphism in wild populations of brook trout from Newfoundland. Rapp. Proc.-Verb. Reun. 161, 97-99.
Grag, R.W., McKenzie, J.A. (1970): Muscle protein electrophoresis in the genus Salmo of eastern Canada. J. Fish. Res. Board Canada 27(11), 2109-2112.
Haen, P.J., O'Rourke, F.J. (1968): Proteins and hemoglobins of salmon-trout hybrids. Nature 217, 65-67.
Handford, P.T. (1971): An esterase polymorphism in the bleak, *Alburnus alburnus*. Ecological Genetics and Evolution, Oxford a. Edinburgh, Blackwell Sci. Publ., 289-297.
Harris, H. (1971): Protein polymorphism in man. Canad. J. Genet. Cytol. 13(3), 381-396.
Hershberger, W.K. (1970): Some physicochemical properties of transferrins in brook trout. Trans. Amer. Fish. Soc. 99(1), 207-218.
Hjorth, J.P. (1971): Genetics of *Zoarces* populations I. Three loci determining the phosphoglucomutase isoenzymes in brain tissue. Hereditas 69(2), 233-241.
Hochachka, P.V. (1967): Organization of metabolism during temperature compensation. Molecul. Mechan. of Temper. Adaptation. New York: Acad. Press.
Hochachka, P.V. (1968): The nature of thermal optimum for lungfish lactate dehydrogenase. Compar. Bioch. Phys. 27(4), 609-612.
Holmes, R.S., Markert, C.L. (1969): Immunochemical homologies among subunits of trout lactate dehydrogenase isozymes. Proc. Nat. Acad. Sci. U.S.A. 64(1), 205-210.
Holmes, R.S., Whitt, G.S. (1970): Developmental genetics of the esterase isozymes of *Fundulus heteroclitus*. Bioch. Genet. 4(4), 471-480.
Hubby, J.L., Lewontin, R.C. (1966): A molecular approach to the study of genic heterozygosity in natural populations I. The number of alleles at different loci in *Drosophila pseudoobscura*. Genetics 54(2), 577-594.
Hunter, R.L., Markert, C.L. (1957): Histochemical demonstration of enzymes separated by zone electrophoresis in starch gels. Science 125, 1295.
Huntsman, G.R. (1970): Disc gel electrophoresis of blood sera and muscle extracts from some catostomid fishes. Copeia 3, 457-467.
Jamieson, A., de Ligny, W., Naevdal, G. (1971): Serum esterases in mackerel *Scomber scombrus*. Rapp. Proc.-Verb. Reun. 161, 109-117.

Johnson, A.C., Utter, F.M., Hodgins, H.O. (1971): Phosphoglucomutase polymorphism in Pacific Ocean perch, *Sebastodes alutus*. Comp. Bioch. Phys. 39(2), 285-290.
Johnson, M.S. (1971): Adaptive lactate dehydrogenase variation in the crested blenny, *Anoplarchus*. Heredity 27(2), 205-226.
Kirpichnikov, V.S. (1972): Biochemical polymorphism and the problem of so-called non-darwinian evolution (russ.). Usp. sovr. biol. 74/ 2(5), 231-246.
Klose, J., Wolf, U., Hitzeroth, H., Ritter, H., Ohno, S. (1969): Polyploidization in the fish family Cyprinidae order Cypriniformes II. Duplication of the gene loci coding for LDH (E.C. 1.1.1.27) and 6-PGDH (E.C. 1.1.1.44) in various species of *Cyprinidae*. Humangenetik 7(3), 245-250.
Klose, J., Wolf, U. (1970): Transitional hemizygosity of the maternally derived allele at the 6-PGD locus during early development of the cyprinid fish *Rutilus rutilus*. Bioch. Genet. 4(1), 87-92.
Koehn, R.K. (1968): The component of selection in the maintenance of a serum esterase polymorphism. Proc. 12th Int. Congr. Gen. 1, 1227.
Koehn, R.K. (1969): Esterase heterogeneity: dynamics of a polymorphism. Science 163, 943-944.
Koehn, R.K. (1970): Functional and evolutionary dynamics of polymorphic esterases in catostomid fishes. Trans. Amer. Fish. Soc. 99(1), 219-228.
Koehn, R.K. (1971): Biochemical polymorphism: a population strategy. Rapp. Proc.-Verb. Reun. 161, 147-153.
Koehn, R.K., Johnson, D.W. (1967): Serum transferrin and serum esterase polymorphisms in an introduced population of the bigmouth buffalofish, *Ictiobus cyprinellus*. Copeia 4, 805-809.
Koehn, R.K., Peretz, J.E., Merritt, R.B. (1971): Esterase enzyme function and genetical structure of populations of the freshwater fish, *Notropis stramineus*. Amer. Natur. 105(941), 51-69.
Koehn, R.K., Rasmussen, D.J. (1967): Polymorphic and monomorphic serum esterase heterogeneity in catostoned fish populations. Bioch. Genet. 1(2), 131-144.
Leung, T.E., Odense, P.H. (1972): Enzyme polymorphism in the herring and sand launce. Abstract, Ichthyological Symposium on Genetics and Mutagenesis, Neuherberg nr. Munich, 1972.
Lewontin, R.C., Hubby, J.L. (1966): A molecular approach to the study of genic heterozygosity in natural populations. II. Amount of variation and degree of heterozygosity in natural populations of *Drosophila pseudoobscura*. Genetics 54(2), 595-609.
De Ligny, W. (1967): Polymorphism of serum transferrins in plaice. Proc. Xth Eur. Conf. Anim. Blood Groups, 373-378.
De Ligny, W. (1969): Serological and biochemical studies on fish populations. "Oceanogr. a. Marine Biol.", Ann. Rev. 7, 411-513.
Limansky, V.V. (1964): Analysis of the intraspecific differentiation in some fishes of Black Sea and Sea of Azov by means of precipitation reaction (russ.). Vopr. physiol. ryb Chern. i Azovck. morei, M., Nauka, 31-39.
Limansky, V.V., Pajusova, A.N. (1969): Immunogenetic differences between elementary populations of anchovy (russ.). Genetica 5(6), 109-118.
Lin, C.C., Schipmann, G., Kittrell, W.A., Ohno, S. (1969): The predominance of heterozygotes found in wild goldfish of Lake Erie at the gene locus for sorbitol dehydrogenase. Bioch. Genet. 3(6), 1603-1607.
Lindsey, C.C., Clayton, J.W., Franzin, W.G. (1970): Zoogeographic problems and protein variation in the *Coregonus clupeaformis* whitefish species complex. Biol. of Coreg. Fishes, Winnipeg, 127-146.

Lukjanenko, V.I., Popov, A.V. (1969): Proteins of the serum in two allopatric populations of the Siberian sturgeon *Acipenser baeri Br.* (russ.). Dokl. Acad. Nauk 186(1), 233-235.
Lukjanenko, V.I., Popov, A.V., Mishin, E.A. (1971): Heterogeneity and polymorphism of the fish serum albumins (russ.). Dokl. Acad. Nauk 201(3), 737-740.
Lush, J.E. (1969): Polymorphism of a phosphoglucomutase isoenzymes in the herring (*Clupea harengus*). Compar. Bioch. Phys. 30(2), 391-397.
Lush, J.E. (1970): Lactate dehydrogenase isoenzymes and their genetic variation in coalfish (*Gadus vireus*) and cod (*G. morrhua*). Compar. Bioch. Phys. 32(1), 23-32.
Manwell, C., Baker, C.M.A., Childers, W. (1963): The genetics of hemoglobin in hybrids. I. A molecular basis for hybrid vigor. Compar. Bioch. Phys. 10(1), 103-120.
Manwell, C., Baker, C.M.A. (1970): Molecular biology and the origin of species. Heterosis, protein polymorphism and animal breeding. Sidgwick a. Jackson, L.
Markert, C.L., Faulhaber, J. (1965): Lactate dehydrogenase isozyme patterns of fish. J. Exper. Zool. 159(2), 319-332.
Markert, C.L., Ursprung, H. (1971): Developmental genetics. New-Jersey: Prentice-Hall, Inc.
Markert, C.L., Whitt, G.S. (1968): Molecular varieties of isoenzymes. Experientia 24(10), 977-991.
Massaro, E.J. (1972): Isozyme patterns of coregonine fishes: evidence for multiple cistrons for lactate and malate dehydrogenases and achromatic bands in the tissues of *Prosobium cylindraceum* (Pallas) and *P. coulteri* (Eigenmann a. Eigenmann). J. Exper. Zool. 179(2), 247-262.
Massaro, E.J., Booke, H.E. (1972): A mutant A-type lactate dehydrogenase subunit in *Fundulus heteroclitus* (Pisces: *Cyprinodontidae*). Copeia 2, 298-302.
McKenzie, J.A. (1973): Comparative electrophoresis of tissues from blueback herring *Alosa aestivalis* (Mitchill) and caspareau, *Alosa pseudoharengus* (Wilson). Compar. Bioch. Phys. 44(IB), 65-68.
Metcalf, R.A., Whitt, G.S., Childers, W.F., Metcalf, R.L. (1972): A comparative analysis of the tissue esterases of the white crappie (*Pomoxis annularis Rafinesque*) and black crappie (*P. nigromaculatus Lesneur*) by electrophoresis and selective inhibitors. Compar. Bioch. Phys. 41B(1), 27-38.
Møller, D. (1968): Genetic diversity in spawning cod along the Norwegian cost. Hereditas 60(1), 1-32.
Møller, D. (1969): The relationship between Arctic and coastal cod in their immature stages illustrated by frequencies of genetic characters. FiskeriDir. Skr., Ser. HavUnders 15, 220-233.
Møller, D. (1970): Transferrin polymorphism in Atlantic salmon (*Salmo salar*). J. Fish. Res. Bd. Canada 27, 1617-1625.
Møller, D., Naevdal, G. (1969): Studies on hemoglobins of some gadoid fishes. FiskeriDir. Skr., Ser. HavUnders. 15(2), 91-97.
Moon, T.W., Hochachka, P.W. (1972): Temperature and the kinetic analysis of trout isocitrat dehydrogenases. Compar. Bioch. Physiol., 42B(4), 725-730.
Morrison, W.J. (1970): Non-random segregation of two lactate dehydrogenase subunits loci in trout. Trans. Amer. Fish. Soc. 99(I), 193-206.
Naevdal, G. (1969): Further studies on blood protein polymorphism in sprat. FiskeriDir. Skr., Ser. HavUnders. 15, 555-564.
Naevdal, G. (1969): Distributions of multiple forms of LDH, AAT and serum Est in herring from Norwegian waters. FiskeriDir. Skr., Ser. HavUnders. 15, 565-572.
Nefyodov, G.N. (1969): Serum haptoglobins of the sea redfishes *Sebastes* (russ.). Vestnik Mosc. Univ. 1, 104-107.

Numachi, K. (1970): Lactate and malate dehydrogenase isozyme patterns in fish and marine mammals. Bull. Jap. Soc. Sci. Fish. 36(10), 1067-1075.
Numachi, K. (1970): Polymorphism of malate dehydrogenase and genetic structure of juvenile population in soury *Cololabis saira*. Bull. Jap. Soc. Sci. Fish. 36(12), 1235-1241.
Numachi, K. (1971): Electrophoretic variants of catalase in the black rockfish, *Sebastes inermis*. Bull. Jap. Soc. Sci. Fish. 37(12), 1177-1181.
Numachi, K. (1971): Genetic polymorphism of α-glycerophosphate dehydrogenase in soury, *Cololabis saira*. I. Seven variants forms and genetic control. Bull. Jap. Soc. Sci. Fish. 37(6), 755-760.
Nyman, L. (1965): Variation of proteins in hybrids and parental species of fishes. Rep. Swed. Salm. Res. Inst. 13, 1-11.
Nyman, L. (1967): Protein variations in *Salmonidae*. Rep. Inst. Freshwater. Res. Drottningholm 47, 5-38.
Nyman, L. (1971): Plasma esterases of some marine and anadromous teleosts and their application in biochemical systematics. Rep. Inst. Reshwater Res. Drottningholm 51, 109-123.
Nyman, O.L., Shaw, D.H. (1971): Molecular weight heterogeneity of serum esterases in four species of salmonid fish. Compar. Bioch. Phys. 40(2B), 563-566.
Nyman, L., Westin, L. (1969): Blood protein systematics of *Cottidae* in the Baltic drainage area. Rep. Inst. Freshwater Res. Drottningholm 49, 164-174.
Odense, P.H., Allen, T.M., Leung, T.C. (1966): Multiple forms of lactate dehydrogenase and aspartate aminotransferase in herring (*Clupea harengus L.*). Canad. J. Biochem. 44, 1319-1324.
Odense, P.H., Allen, T.M. (1971): A biochemical comparison of some Atlantic herring populations. Rapp. Proc.-Verb. Reun. 161, 26.
Ohno, S., Stenius, C., Faisst, E., Zenzes, M.T. (1965): Post-zygotic chromosomal rearrangements in rainbow trout (*Salmo irideus Gibbons*). Cytogenetics 4(2), 117-129.
Payne, R.H., Child, A.R., Forrest, A. (1971): Geographical variation in the Atlantic salmon. Nature 231, 250-252.
Payne, R.H., Child, A.R., Forrest, A. (1972): The existence of natural hybrids between the European trout and the Atlantic salmon. J. Fisher. Biol. 4(2), 233-236.
Poluhowich, J.J. (1972): Adaptive significance of eel multiple hemoglobin. Physiol. Zool. 45(3), 215-222.
Powers, D.A. (1972): Hemoglobin adaptation for fast and slow habitats in sympatric catostomid fishes. Science 177, 360-362.
Quiroz-Gutierrez, Ohno, S. (1970): The evidence of gene duplication for S-form NADP-linked isocitrate dehydrogenase in carp and goldfish. Bioch. Genet. 4(1), 93-99.
Raunich, L., Callegarini, C., Cavicchioli, G. (1966): Polymorfismo emoglobinico e caratteri sistematici, del genere *Ictalurus* del Italia settentrionale. Arch. Zool. Ital. 51, 497-510.
Ridgway, G.J. (1971): Problems in the application of serological methods to population studies on fish. Rapp. Proc.-Verb. Reun. 161, 10-14.
Ridgway, G.J., Sherburne, S.W., Lewis, R.D. (1970): Polymorphism in the esterase of Atlantic herring. Trans. Amer. Fish. Soc. 99(1), 147-151.
Roberts, F.L., Wohnus, J.F., Ohno, S. (1969): Phosphoglucomutase polymorphism in the rainbow trout *S. gairdneri*. Experientia 25, 1109-1110.
Sachko, G.D. (1971): On the degree of polymorphism of the lactate dehydrogenase genes in fishes (russ.). Nauchn. soob. In-ta biol. morja. Vladyvostock 2, 190-195.

Sanders, B.G., Wright, J.E. (1961): Inheritance of a major blood group system in rainbow trout. Genetics 46(8), 895.
Sanders, B.G., Wright, J.E. (1962): Immunogenetic studies in two trout species of the genus *Salmo*. Ann. N.Y. Acad. Sci. 97, 116-130.
Saunders, L.H., McKenzie, J.A. (1971): Comparative electrophoresis of Arctic char. Compar. Bioch. Phys. 38(3), 487-491.
Scopes, R.K., Gosselin-Rey, C. (1968): Polymorphism in carp muscle creatine kinase. J. Fish. Res. Bd. Canada 25, 2715.
Shcherbenok, Ju.I. (1973): Polymorphic systems of esterases and transferrins and their connection with economic-important carp characters (russ.). Mater. Soveshch. po bioch. genet. ryb (in press). Leningrad.
Sick, K. (1965a): Hemoglobin polymorphism of cod in the Baltic and the Danish Belt Sea. Hereditas 54(1), 19-48.
Sick, K. (1965b): Hemoglobin polymorphism of cod in the North Sea and the North Atlantic Ocean. Hereditas 54(1), 49-73.
Sick, K., Bahn, E., Frydenberg, O., Nielsen, J.T., von Wettstein, D. (1967): Hemoglobin polymorphism of the American fresh water eel *Anguilla*. Nature 214, 1141-1142.
Simonsen, V., Frydenberg, O. (1972): Genetics of *Zoarces* populations. II. Three loci determining esterase isozymes in eye and brain tissue. Hereditas 70(2), 235-242.
Sindermann, C.J., Mairs, D.F. (1959): A major blood group system in Atlantic sea herring. Copeia 3, 228-232.
Sindermann, C.J. (1962): Serological studies of Atlantic redfish. Fish. Bull. US. Fish. Wildl. Serv. 191, 351-354.
Slota, E., Rapacz, J., Stefan, L. (1970): Wstepne bodania ned grupanti krwi u karpia (*Cyprinus carpio*). Zesz. probl. postep. nauk rol. 104, 71-78.
Slynko, V.I. (1971): Analysis of the malate dehydrogenase gene frequencies in the populations of Pacific salmon in the Sakhalin rivers (russ.). Nauchn. soob. In-ta biol. morja. Vladyvostock 2, 212-214.
Slynko, V.I. (1971): Polymorphism of the malate dehydrogenase isozymes in Pacific salmons (russ.). Nauchn. soob. In-ta biol. morja, Vladyvostock 2, 207-211.
Smithies, O. (1955): Zone electrophoresis in starch-gel: group variations in the serum proteins of normal human adults. Bioch. J. 61, 629-641.
Somero, G.N., Hochachka, P.W. (1969): Isoenzymes and short-term temperature compensation in poikilotherms: activation of lactate dehydrogenase isoenzymes by temperature decreases. Nature 223, 194-195. 194-195.
Sprague, L.M. (1967): Multiple molecular forms of serum esterase in three tuna species from the Pacific Ocean. Hereditas 57(1-2), 198-204.
Sprague, L.M., Vrooman, A.M. (1962): A racial analysis of the Pacific sardina (*Sardinops caerulea*) based on studies of erythrocyte antigenes. Ann. N.Y. Acad. Sci. 97, 131-138.
Stegeman, J.J., Goldberg, E. (1971): Distribution and characterization of hexose 6-phosphate dehydrogenase in trout. Bioch. Genet. 5(6), 579-589.
Suzuki, A. (1967): Blood type of fish. Bull. Jap. Soc. Sci. Fish. 33(4), 372-381.
Tills, D., Mourant, A.E., Jamieson, A. (1971): Red-cell enzyme variants of Icelandic and North Sea cod (*Gadus morrhua*). Rapp. Proc.-Verb. Reun. 161, 73-74.
Timoffeef-Ressovsky, N.V., Svirezhev, Ju.M. (1966): Adaptive polymorphism in the populations of *Adalia bipunctata L.* (russ.). Probl. kibernetiki 16, 137-146.
Truveller, C.A. (1971): A study of blood groups in herring (*Cl. harengus L.*) from the North Sea in connection with the problem of race differentiation. Rapp. Proc.-Verb. Reun. 161, 33-39.

Tsuyuki, H. (1966): Comparative electropherograms of haemoglobins and other fish protein. Proceed. 11th Pacif. Sci. Congr. 7, 8.

Tsuyuki, H., Ronald, A.P. (1970): Existence in salmonid haemoglobins of molecular species with three and four different polypeptides. J. Fish. Res. Bd. Canada 27, 1325-1328.

Tsuyuki, H., Ronald, A.P. (1971): Molecular basis for multiplicity of Pacific salmon haemoglobins: evidence for in vivo existence of molecular species with up to four different polypeptides. Compar. Bioch. Phys. 39(3B), 503-522.

Turner, B.J. (1973): Genetic variation of mitochondrial aspartate aminotransferase in the teleost *Cyprinodon nevadensis*. Comp. Bioch. Phys. 44(IB)., 89-92.

Utter, F.M., Hodgins, H.O. (1970): Phosphoglucomutase polymorphism in sockeye salmon. Compar. Bioch. Phys. 36(1), 195-200.

Utter, F.M., Hodgins, H.O. (1971): Biochemical polymorphisms in the Pacific hake (*Merluccius productus*). Rapp. Proc.-Verb. Reun. 161, 87-89.

Utter, F.M., Hodgins, H.O., Allendorf, F.W., Mighell, J.L., Johnson, A.G. (1973): Inheritance of biochemical variants in salmonid species and their application to population studies. This Symposium.

Valenta, M., Kâlal, L. (1968): Polymorfismus sérových transferinu u karpa (*Cyprinus carpio L.*) a lina (*Tinca tinca L.*). Sbornik VSZ, Praha, B, 93-103.

Vrooman, A.M. (1964): Serologically differentiated subpopulations of the Pacific sardine, *Sardinops caerulea*. J. Fish. Res. Bd. Canada 21(4), 691-701.

Whitt, G.S. (1970): Genetic variation of supernatant and mitochondrial malate dehydrogenase isoenzymes in the teleost *Fundulus heteroclitus*. Experientia 26, 734-736.

Whitt, G.S., Childers, W.F., Wheat, T.E. (1971): The inheritance of tissue-specific LDH isozymes in interspecific bass (*Micropterus*) hybrids. Bioch. Genet. 5(3), 257-273.

Wilkins, N.P. (1971): Hemoglobin polymorphism in cod, whiting and pollack in Scottish Waters. Rapp. Proc.-Verb. Reun. 161, 60-63.

Wilkins, N.P. (1971): Biochemical and serological studies on Atlantic salmon (*Salmo salar L.*). Rapp. Proc.-Verb. Reun. 161, 91-95.

Wilkins, N.P., Iles, T.D. (1966): Hemoglobin polymorphism and its ontogeny in herring (*Clupea harengus*) and sprat (*Sprattus sprattus*). Compar. Bioch. Phys. 17, 1141-1158.

Wolf, U., Ritter, H., Atkin, N.B., Ohno, S. (1969): Polyploidization in the fish family *Cyprinidae* order *Cypriniformes*. I. DNA-content and chromosome sets in various species of *Cyprinidae*. Humangenetik 7(3), 240-244.

Wolf, U., Engel, W., Faust, J. (1970): Zum Mechanismus der Diploidisierung in der Wirbeltierevolution: Koexistenz von tetrasomen und disomen Genloci der Isocitrat-Dehydrogenasen bei der Regenbogenforelle (*Salmo irideus*). Humangenetik 9(2), 150-156.

Wright, J.E., Atherton, L.M. (1970): Polymorphism for LDH and transferrin loci in brook trout populations. Trans. Amer. Fish. Soc. 99(1), 179-192.

Wright, T.D., Hasler, A.D. (1967): An electrophoretic analysis of the effects of isolation and homing behavior upon the serum proteins of the white bass (*Roccus chrysops*) in Wisconsin. Amer. Natur. 101, 921, 401-413.

Wuntch, T., Goldberg, E. (1970): A comparative physicochemical characterization of lactate dehydrogenase isozymes in brook trout, lake trout and their hybrid splake trout. J. Exp. Zool. 174(3), 233-252.

Yndgaard, C.F. (1972): Genetically determined electrophoretic variants of phosphoglucose isomerase and 6-phosphogluconate dehydrogenase in *Zoarces viviparus L.* Hereditas 71(1), 151-154.

Added in Proof

The paper being in preparation some new interesting data proving the adaptive nature of biochemical polymorphism in fish have been published. In stickleback and in ling the clinal variability on esterase has been discovered (Raunich et al., 1972; Frydenberg et al., 1973). In addition in ling the intensive natural selection for esterase genes has been shown (Christiansen et al., 1973). In American eel populations there is a distinct cline for alleles of ADH, Sorbitol DH, Est and PXI genes (Williams et al., 1973). Excess of heterozygotes for Hp locus has been found in crucian carp (Polyakovsky et al., 1973).

Of great importance is the pioneer discovery in food-fish of linked genes coding some enzymes: αGPDH and 6GPDH in Lepomis (Wheat et al., 1972), G6PDH and 6GPDH in swordtail hybrids (Wright, personal communication). Many papers report new materials on biochemical polymorphism in various fish species. These data concern the genetical variability of transferrins, hemoglobins, haptoglobins and albumins, muscle myogens and many enzymes (LDH, MDH, IDH, 6PGDH, G6PDH, αGPDH, Pep, TO, P6M, KK).

Christiansen, F.B., Frydenberg, O., Simonsen, V. (1973): Genetics of Zoarces populations. IV. Selection component analysis of an esterase polymorphism using population samples including mother-offspring combinations. Hereditas 73(2), 291-304.
Frydenberg, O., Gyldenholm, A.O., Hjorth, J.P., Simonsen, V. (1973): Ibid, III. Geographic variations in the esterase polymorphism Est. III. Hereditas 73(2), 233-238.
Polyakovsky, V.I., Papkovskaya, A.A., Bogdanov, L.V. (1973): Biochemical polymorphism in crucian carp (*Carassius auratus gibelio*) from the Sudoble lake (Belorussia)(russ.). Bioch. Gen. Fish, Leningrad, 161-166.
Raunich, L., Callegarini, C., Cucchi, C. (1972): Ecological aspects of haemoglobin polymorphism in *Gasterosteus aculeatus* (Teleostea). 5th Eur. Mar. Biol. Symp., Padowa, Piccin Ed., 153-162.
Wheat, T.E., Whitt, G.S., Childers, W.F. (1972): Linkage relationships among six enzymes in interspecific sunfish hybrids. Genetics 71(3), 2, 67-68.
Williams, G.C., Koehn, R.K., Mitton, J.B. (1973): Genetic differentiation without isolation in the american eel, *Anguilla rostrata.* Evolution 27(2), 192-204.

Developmental and Biochemical Genetics of Lactate Dehydrogenase Isozymes in Fishes

G. S. Whitt, E. T. Miller, and J. B. Shaklee

1. Introduction

The differential regulation of gene function is a fundamental aspect of cellular differentiation. Isozymes (multiple molecular forms of an enzyme, Markert and Møller, 1959) are ideal gene products for the analysis of gene activation during embryogenesis.

As the most extensively studied isozyme system, L-lactate dehydrogenase (E.C. 1.1.1.27) has proven to be an excellent gene marker for examining differential gene activation during development (Markert, 1962; Cahn, et al., 1962; Markert and Ursprung, 1962; Vesell and Philip, 1963; Auerbach and Brinster, 1967; Goldberg and Hawtrey, 1967; Wright and Moyer, 1968). The lactate dehydrogenase isozymes of fish are a particularly useful model system for studying developmental mechanisms because of the existence of three distinct and homologous lactate dehydrogenase genes whose regulation is tissue specific (Markert and Faulhaber, 1965; Ohno, 1969; Goldberg et al., 1969; Whitt, 1968, 1969, 1970a; Whitt et al., 1971, 1972).

The enzyme lactate dehydrogenase (LDH) occupies an important position in cell metabolism off the glycolytic pathway. During anaerobic periods when the Kreb's cycle is inoperative, pyruvate is converted to lactate by LDH with a concomitant production of NAD^+. This oxidation of NADH to NAD^+ allows glycolysis to continue, which in turn permits the continued production of ATP for energy. The enzyme lactate dehydrogenase is ubiquitous in vertebrates and exists in most tissues in several isozymic forms. In vertebrates, the polypeptide subunits of LDH are encoded in at least two genetic loci whose alleles are co-dominantly expressed (Shaw and Barto, 1963). Mammals and birds possess a third LDH locus, the C locus (Blanco and Zinkham, 1963; Goldberg, 1963) which is apparently restricted in its function to the primary spermatocytes. In many tissues there are five isozymes formed by the random association of the A and B subunits into all possible tetramers (A_4, A_3B_1, A_2B_2, A_1B_3 and B_4). These five isozymes are readily separable by electrophoresis because the two types of subunits differ in their net charge (Appella and Markert, 1961; Markert, 1962). However, in many species of fish, fewer than five isozymes are observed because of restrictions upon subunit association or tetrameric instability (Markert and Faulhaber, 1965; Whitt, 1970a, b).

The LDH isozyme patterns of a representative teleost, the green sunfish (*Lepomis cyanellus*) (Perciformes, Centrarchidae) are shown in Fig. 1. Three isozymes, A_4, A_2B_2, and B_4, are present in most tissues. The asymmetric heteropolymers A_3B_1 and A_1B_3 are absent, presumably due to a restriction of polypeptide assembly. In addition to the usual A and B subunits synthesis observed in most tissues, a unique isozyme is synthesized in the eye of the green sunfish. In most teleosts this retinal-specific LDH or "eye band" usually possesses a distinctive anodal mobility. This eye band was first observed in many species of fish by Markert and Faulhaber (1965) and was referred to as E_4 (Massaro and Markert, 1968) to emphasize its unique subunit composition.

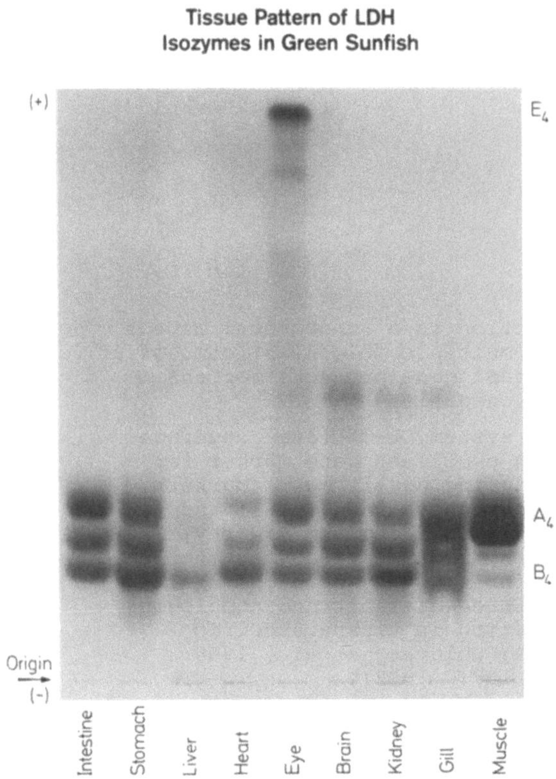

Fig. 1. Lactate dehydrogenase isozymes of the green sunfish (*Lepomis cyanellus*). The A_4 and B_4 isozymes show a reversed relative electrophoretic mobility compared to most other vertebrates. Note the highly anodal "eye band" or LDH E_4 isozyme in the eye

The different physiological roles proposed for these isozymes are based, in part, upon kinetic behavior and in part upon tissue specificity (Plagemann, et al., 1960a, b; Cahn, et al., 1962; Markert and Ursprung, 1962). The LDH A_4 is found predominantly in tissues subject to periods of anaerobiosis (e.g. skeletal muscle). This enzyme possesses a higher Km substrate, a higher substrate concentration optimum and a greater resistance to substrate inhibition. These are kinetic properties suited for the conversion of pyruvate to lactate when there are high substrate concentrations, as during anaerobic glycolysis. Tissues that are more highly oxygenated, such as the heart, contain relatively higher amounts of LDH B_4, an enzyme which has kinetic properties suitable for functioning at the lower substrate concentrations typical of more aerobic tissues. The relative significance of the different kinetic properties of the A_4 and B_4 isozymes is discussed by Kaplan (1964), Everse, et al.,(1970) and Wuntch, et al.,(1970a, b).

2. Genetic and Molecular Bases of the Retinal Specific LDH

The retinal-specific LDH isozyme is an excellent gene product for analyzing the genetic and epigenetic regulation of enzyme synthesis during cellular differentiation. In many teleosts the LDH E polypeptides are synthesized predominantly in regions of the nervous system concerned with vision (Nakano and Whiteley, 1965; Markert and Faulhaber, 1965; Morrison and Wright, 1966) particularly in the neural retina (Whitt, 1970a; Whitt and Booth, 1970).

The retinal-specific LDH has been demonstrated to be a homotetramer composed of subunits distinct from the other teleost LDH subunits (A and B) by in vitro molecular hybridization between the E_4 and the A_4 and B_4 homopolymers (Goldberg, 1966; Massaro and Markert, 1968; Whitt, 1970a, b; Whitt, et al., 1971). In vivo molecular hybridization also occurs between the LDH E and other LDH subunits in interspecific fish hybrids (Goldberg, 1966; Hitzeroth, et al., 1968; Ohno, 1969, Whitt, et al., 1971, 1972). In fact, the presence of the E subunit, in vivo and in vitro, can facilitate the association of A and B subunits in the same tetramer, even though in some fish the A and B subunits, by themselves, are not capable of associating with each other (Whitt, 1970a, b).

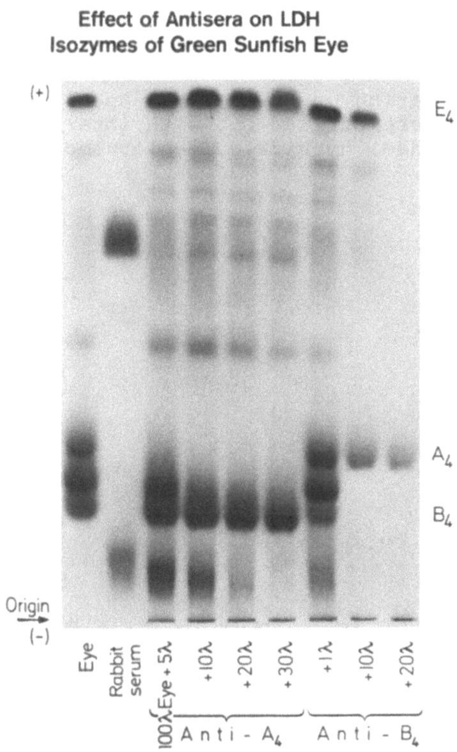

Fig. 2. Effects of antisera upon the eye lactate dehydrogenase isozymes of the green sunfish (*Lepomis cyanellus*). Antisera to weakfish (*Cynoscion regalis*) A_4 and B_4 homopolymers were each mixed with the eye isozymes and allowed to react for at least one hour prior to electrophoresis. The absence of LDH bands on the gel after treatment with antiserum is an indication of immunochemical homology

Immunochemical procedures have also proven very useful in assessing the distinctness of the retinal-specific LDH (Markert and Holmes, 1969; Whitt, 1969, 1970a, b; Whitt, et al., 1971; Horowitz and Whitt, 1972). The effect of antibodies, directed against purified A_4 and purified B_4, on the isozymes of the green sunfish, is shown in Fig. 2. The anti-A antiserum mixed with the eye extract, precipitates only those isozymes containing one or more A subunits. In low concentrations the anti-B antiserum precipitates only those isozymes containing B subunits. However, at higher concentrations the anti-B serum also precipitates the LDH E_4 isozyme. On the basis of this antigenic similarity, the LDH B and E subunits appear to be closely related (Whitt, 1969, 1970 a, b).

None of the previously discussed procedures have genetically demonstrated the existence of a distinct LDH E locus, for they are not capable of excluding the possibility that the retinal specific LDH is epigenetically derived from the other LDH isozymes. The fact that the E_4 isozyme is antigenically similar to the B_4 isozyme but <u>not</u> to the A_4 isozyme tends to exclude the possibility that the LDH E_4 is derived from the A_4 isozyme by post-translational modifications (Whitt, 1969). Genetic polymorphisms at the LDH B locus of *Fundulus heteroclitus* (Atheriniformes, Cyprinodontidae) have no effect on the E_4 isozyme mobility. In addition, rare electrophoretic variants of the E_4 isozyme were observed with no concomitant variation of the other LDH isozymes. These data provide <u>indirect</u> evidence for the genetic distinctness of the eye band (Whitt, 1969, 1970a). The LDH E_4 variants of *Fundulus* were too infrequent to analyze by genetic crosses.

In order to obtain direct genetic data bearing on the retinal-specific LDH of teleosts, genetic investigations were carried out with interspecific sunfish (Perciformes, Centrarchidae) hybrids. The largemouth

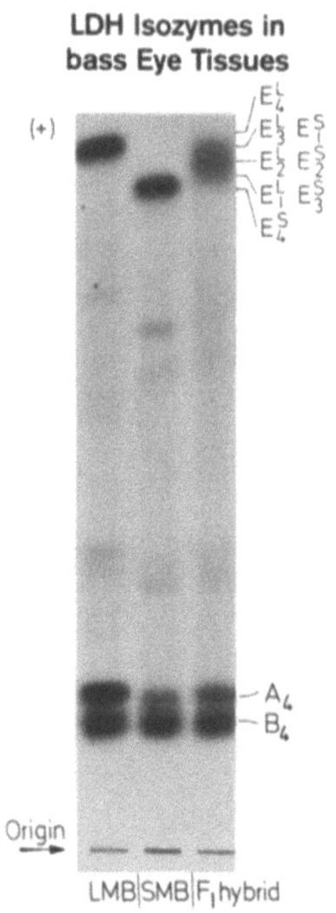

Fig. 3. Lactate dehydrogenase E_4 isozyme phenotypes of the eyes from largemouth bass (LMB) (*Micropterus salmoides*), smallmouth bass (SMB) (*M. dolomieui*) and their F_1 hybrid. The isozyme bands in the LDH E_4 region of the F_1 hybrid (LMB x SMB) are difficult to resolve because of their similar electrophoretic mobilities

bass (LMB)(*Micropterus salmoides*) was crossed with the smallmouth bass (SMB)(*M. dolomieui*) in the laboratory. The F_1 hybrids were placed in ponds to give rise to the F_2 and F_3 generations. In Fig. 3 are shown the LDH isozyme phenotypes of the eyes from the largemouth bass, smallmouth bass, and the F_1 hybrid. The E_4 isozymes of the smallmouth bass and the largemouth bass have different electrophoretic mobilities even though the A_4 homopolymers and the B_4 homopolymers of the two species are indistinguishable. The F_1 hybrid exhibits 5 eye-specific isozymes formed by the random association of the two parental types of E subunits. Both the F_2 and F_3 hybrid generations possessed all three phenotypes in the approximate ratio of 1 LMB : 2 F_1 hybrid : 1 SMB, which would be expected for equal frequencies of two co-dominant alleles at a single nuclear locus. These three LDH phenotypes were evenly distributed between both sexes suggesting that the E locus is autosomally located. This first direct genetic evidence for the LDH E_4 isozyme being encoded in a separate locus (Whitt, et al., 1971) has been confirmed by recent studies in our laboratory where the F_1 hybrids (LMB x SMB) have been back crossed to each parental species and have generated the expected 1:1 ratio of hybrid to parental phenotypes.

As mentioned above the tissue-specific regulation of the LDH isozyme synthesis in fishes suggests that these isozymes might have kinetic properties tailored for particular metabolic environments. The homopolymeric lactate dehydrogenases (A_4, B_4, and E_4) from the Atlantic mackerel (*Scomber scombrus*)(Perciformes, Scombridae) were isolated

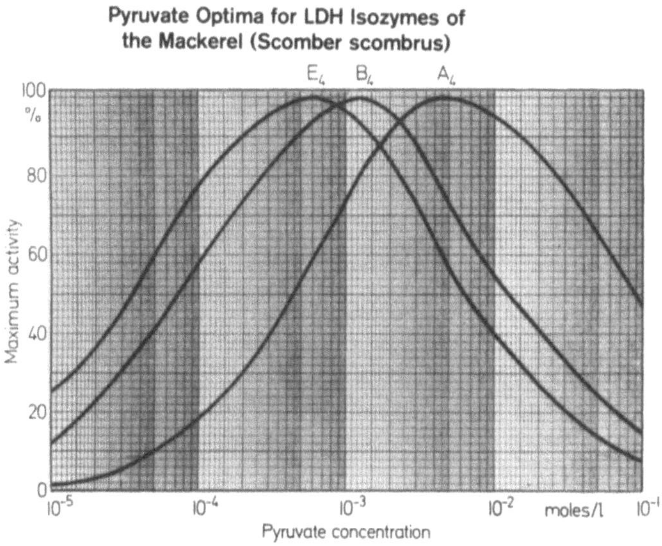

Fig. 4. Effect of pyruvate concentration on the three homopolymeric lactate dehydrogenase isozymes of the mackerel (*Scomber scombrus*). Conditions of kinetic analysis are given in Whitt (1970a)

and purified for kinetic analysis (Whitt, 1970a). The pyruvate concentration optima of these isozymes is shown in Fig. 4. The retinal specific LDH E_4 has the lowest substrate concentration optimum and the greatest susceptibility to inhibition by substrate. Thus, the retinal specific LDH behaves more like the LDH B_4 isozyme which is predominant in aerobic tissues than the A_4 isozyme which is predominant in anaerobic tissues.

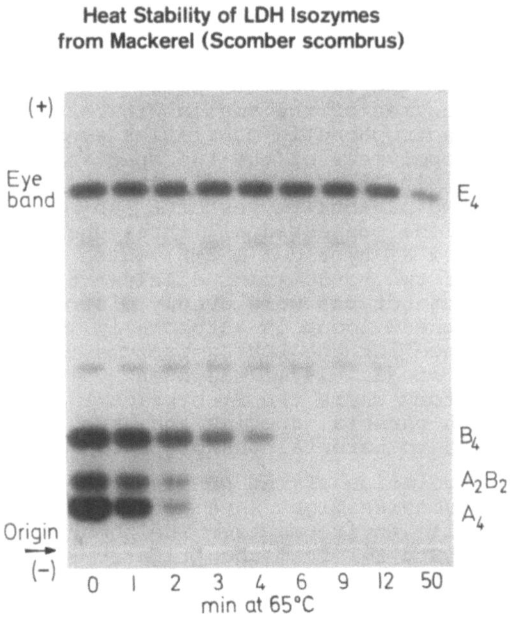

Fig. 5. Heat inactivation of the lactate dehydrogenase isozymes of the eye of the Atlantic mackerel (*Scomber scombrus*). Aliquots of the extract were placed in different test tubes and subjected to heating at 65°C for different lengths of time. After this heat treatment, the samples were cooled, centrifuged, and subjected to electrophoresis

This similarity of the LDH E_4 kinetic properties to those of the LDH B_4 is also observed for certain physical properties. The retinal specific LDH is considerably more stable to denaturation by heat and urea than the other isozymes (Whitt, 1970a). The effect of heat treatment on the eye isozymes of the Atlantic mackerel is shown in Fig. 5. The eye band is considerably more stable to heat inactivation than the B_4 isozyme which in turn is more stable than the A_4 isozyme. The possible roles for the distinctive kinetic and physical properties of the eye specific LDH will be discussed later.

The presence of three homologous LDH genes in many teleosts provides an excellent opportunity to investigate the specificity of gene activation within differentiated cells. In general, the E polypeptides are synthesized predominantly in derivatives of neural ectoderm. The E_4 isozyme is observed in highest levels in the neural retina (Whitt and Booth, 1970). E subunit synthesis is considerably lower in the brain, and occurs predominantly in the mesencephalon, diencephalon, and optic nerve (Whitt, 1970a). Because the relative level of E subunit synthesis is so low only heteropolymers containing E subunits are detected in brain tissues. The heteropolymers B_3E_1, B_2E_2, B_1E_3 as well as the E_4 homopolymers are synthesized in the retina. Although the differences in the levels of E polypeptide synthesis between the brain and retina probably reflects differential LDH E gene function, the isozyme patterns of vertebrate tissues may also be strongly influenced by preferential catabolism (Fritz et al., 1969, 1971).

In the retina of some fish, there is an apparent absence of heteropolymers containing E and A polypeptides (Whitt, 1970a). This phenomenon does not appear to be due to the inability of A and E subunits to assemble because the A, B and E subunits all readily assemble in vivo in the cells of the brain and readily assemble in vitro after the eye isozymes have been subjected to freeze-thaw molecular hybridization. Some possible interpretations of these data are that the LDH A subunits in the retina are either synthesized in different cells, different regions of the same cell, or at different times than the E subunits. This spatial or temporal isolation of A and E polypeptide synthesis in the retina presumably reflects a temporal or spatial

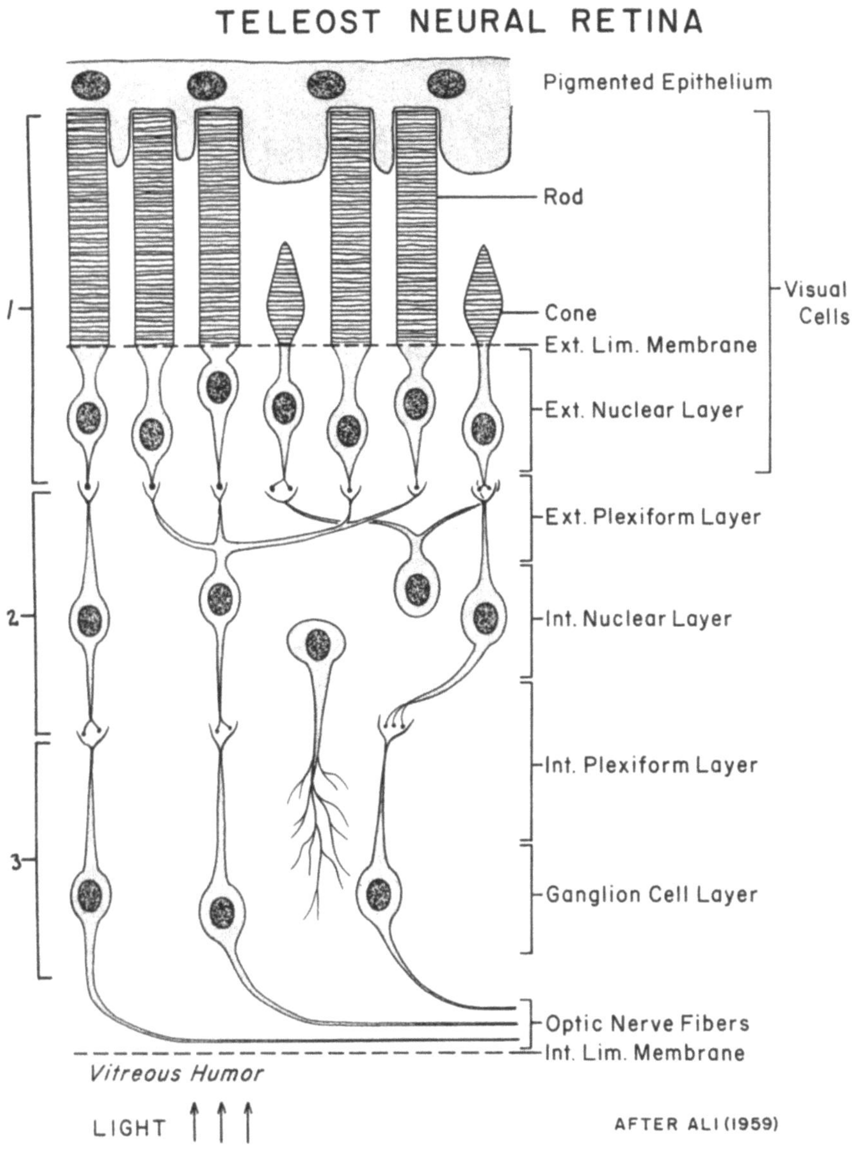

Fig. 6. Schematic representation of the cell layers in the differentiated retina of a teleost. Modified after Ali (1959). The cell layers are indicated on the left: (1) photoreceptor cells, (2) bipolar cells, and (3) ganglion cells

isolation of LDH gene function. Nevertheless, the B gene function appears to overlap the functions of the A and of the E genes in the brain and eye as shown by the presence of heteropolymers containing B and A polypeptides (B_2A_2) and those with B and E polypeptides (B_3E_1) (Whitt, 1970a, b). The LDH E gene tends to be accompanied in its activation by the more closely related LDH B gene, rather than the more distantly related LDH A gene. This system offers an excellent opportunity to investigate to what degree the physical relatedness of homologous structural genes determines the specificity of their activation.

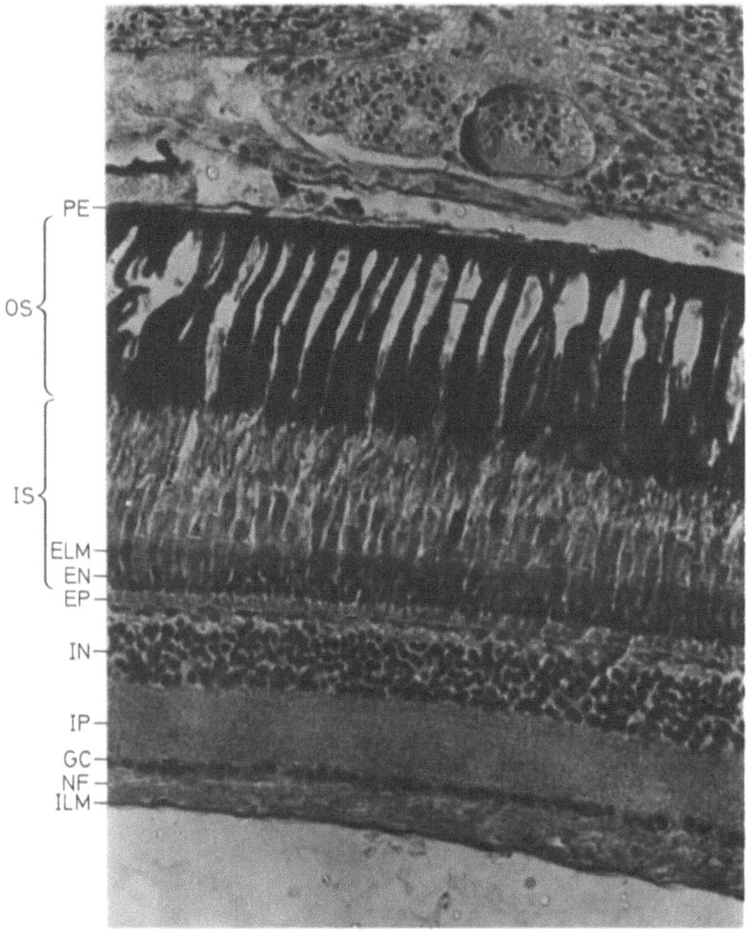

Fig. 7. Photomicrograph of the swordtail (*Xiphophorus sp.*) retina after hematoxylin staining. The retinal layers are designated after Ali (1959). PE - pigmented epithelium; OS - outer segment region of the photoreceptor cells; IS - inner segment region of the photoreceptor cells; ELM - external limiting membrane; EN - external nuclear layer; EP - external plexiform layer; IN - internal plexiform layer; GC - ganglion cell layer; NF - nerve fiber layer; ILM - internal limiting membrane. Procedures given in Whitt and Booth (1970)

3. LDH E Gene Function in the Differentiated Retina

Histochemical analyses of the LDH isozyme repertory of the teleost retina were initiated to determine which differentiated cells of the neural retina possess the LDH E gene function and whether the low level of heteropolymers (containing E and A polypeptides) in the retina is due to a temporal or to a spatial isolation of subunit synthesis.

A diagrammatic representation of a teleost neural retina is portrayed in Fig. 6 (after Ali, 1959). There are three main cell layers in the neural retina: 1. photoreceptor cells, 2. bipolar, amacrine, and horizontal cells and 3. ganglion cells. Fig. 7 is a cross section through the neural retina of a swordtail (*Xiphophorus sp.*)(Atheriniformes, Poeciliidae) after staining with hematoxylin. The nuclei of the three cell layers are readily visible after staining with hematoxylin, particularly those in the external and internal nuclear layers.

Once the anatomy of the swordtail retina was determined it was then possible to localize the LDH isozymes to specific cell layers (Whitt and Booth, 1970). Eight-micron sections of the frozen retina were

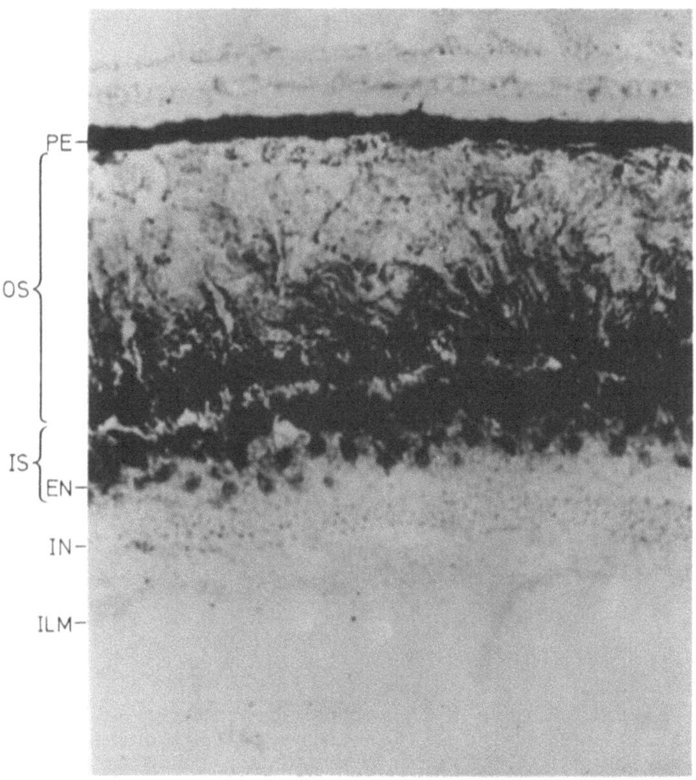

Fig. 8. Photomicrograph of the swordtail (*Xiphophorus sp.*) retina after staining for lactate dehydrogenase activity in the presence of 1.8M urea. The E_4 LDH is the only isozyme active at this concentration of urea. The LDH E_4 activity is most intense in the inner segment region closest to the outer segments (the ellipsoid region). There is a second but lighter layer of formazan deposition over the external nuclear layer. Procedures given in Whitt and Booth (1970)

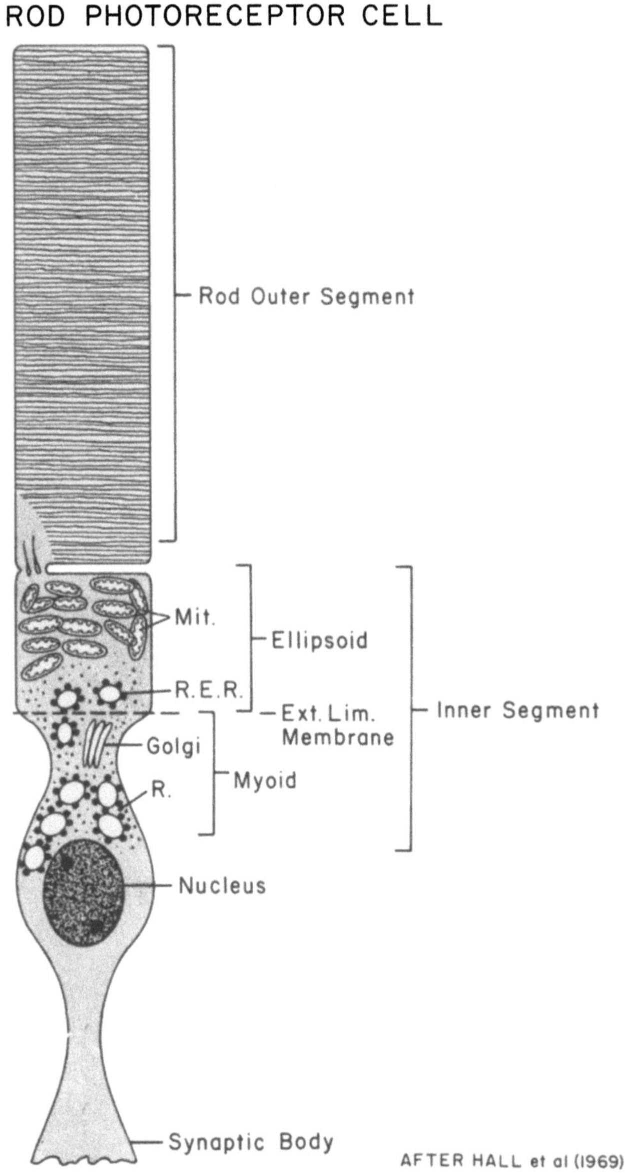

Fig. 9. Schematic representation of a rod photoreceptor cell. Modified after Hall et al.,(1969). The ellipsoid region possesses high concentrations of mitochondria. Mit. - mitochondria; R.E.R. - rough endoplasmic reticulum; R. - ribosome

incubated in LDH staining solutions similar to those used for gel staining. Fig. 8 shows a cross section of the swordtail retina incubated in LDH staining solution plus 1.8M urea. The urea inactivates the LDH A_4 and B_4 isozymes without affecting the activity of the retinal-specific LDH E_4. The predominant LDH E_4 activity, as indicated by heavy formazan dye deposition, is restricted to the photoreceptor cell layer, particularly the inner segment regions of these cells. Additional dye deposition in the external nuclear layer may result from LDH E_4 activity in other cell types, e.g. Müller cells. A schematic drawing of a rod photoreceptor cell is displayed in Fig. 9 (after Hall et al., 1969). As observed in the ellipsoid region of the amphibian rod, there are considerable numbers of mitochondria (Hall et al., 1969). It is concluded that the retinal specific LDH is primarily localized in the ellipsoid region of the inner segment of at least some photoreceptor cells (Whitt and Booth, 1970; Whitt, 1970a). This localization of the retinal specific LDH at the light microscope level has recently been confirmed by Kunz (1971) who employed the urea inhibition procedure on sections of the retina of the guppy (*Lebistes sp.*) (Atheriniformes, Poeciliidae). In this fish the retinal specific LDH is reported to be restricted to photoreceptor cells, specifically the cones.

The data from the histochemical study of the swordtail retina indicate that the A and E subunit syntheses are spatially isolated--their synthesis taking place in different cells of the retina, due probably to differential regulation of gene function (Whitt and Booth, 1970). The electrophoretic detection of heteropolymers with B and E subunits in the retina indicates that these subunits are both synthesized in

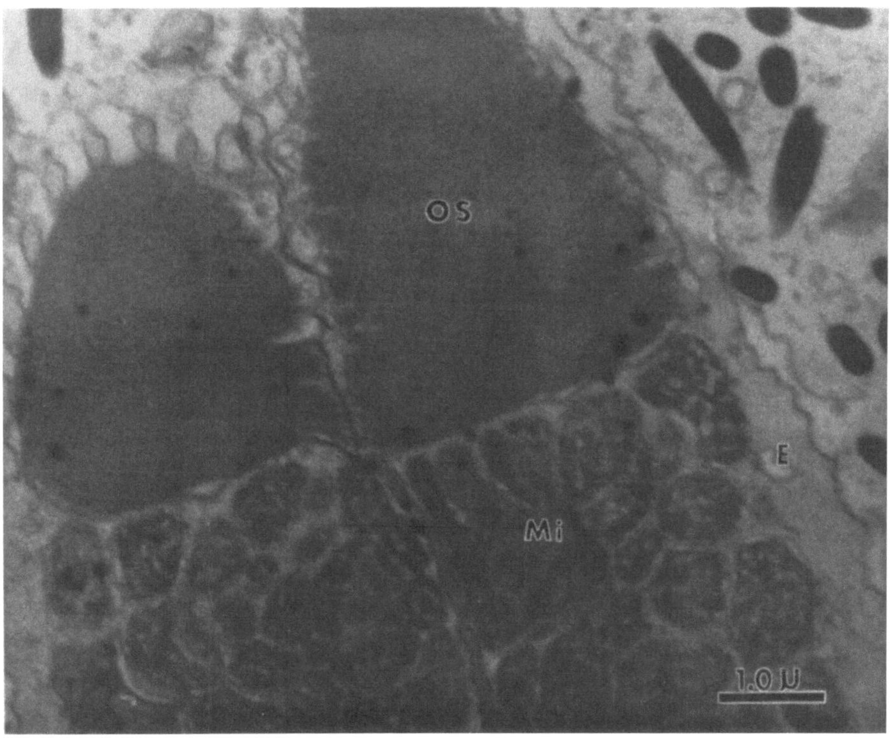

Fig. 10. Electron micrograph of the cone photoreceptor cell from the differentiated neural retina of the green sunfish (*Lepomis cyanellus*). OS - outer segment region; E (ellipsoid region is to the left of E); Mi - mitochondria. Note the density of mitochondria in the ellipsoid region

the photoreceptor cells (Whitt, 1970a). However, the low levels of these heteropolymers indicates that there may be temporal or spatial isolation of B and E subunit synthesis within the photoreceptor cells themselves.

As shown above, the retinal specific LDH isozyme is located in the inner segment and predominantly in the ellipsoid region of the photoreceptor cells. Examination of the fine structure of a typical green sunfish photoreceptor cell (in this case a cone cell) by electron microscopy reveals that the ellipsoid region of this cell possesses large numbers of mitochondria (Fig. 10).

A closer examination of this region reveals that the mitochondria are tightly clustered (Fig. 11). These mitochondria in the ellipsoid region appear to be arranged in a sort of rosette configuration. This concentration of mitochondria in the region of the photoreceptor cell containing the retinal specific LDH is particularly interesting in view of the fact that the kinetic properties of the E_4 LDH isozyme

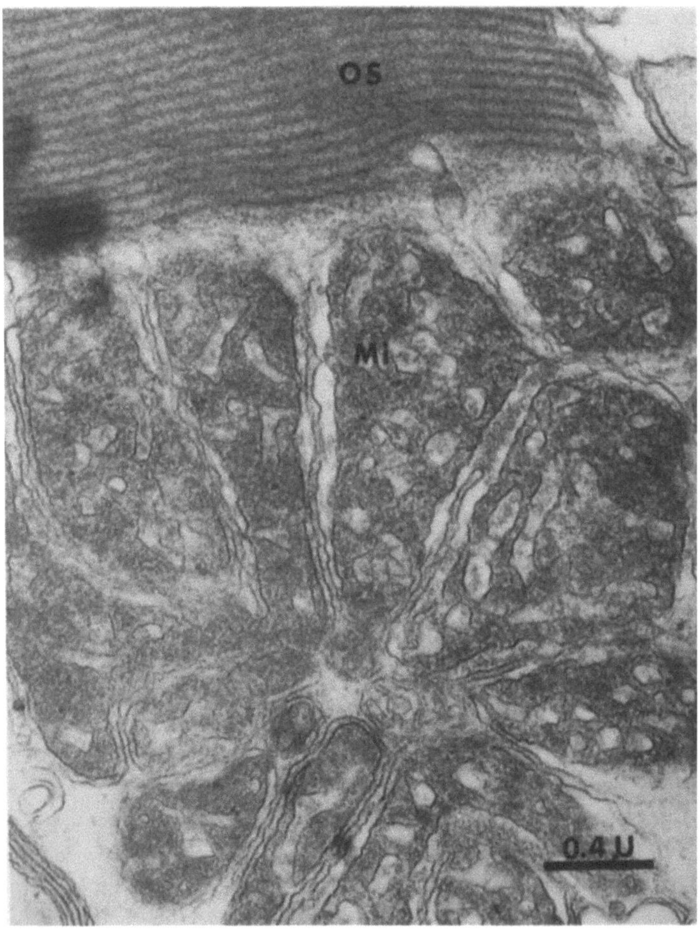

Fig. 11. High magnification electron micrograph of a cone cell. OS - outer segment region; Mi. - mitochondria. The mitochondria appear to be almost contiguous with one another and arranged in a rosette pattern

are consistent with an aerobic environment. A more specific localization of the LDH isozymes to organelles or regions of the photoreceptor cell is being undertaken to clarify the metabolic role of these isozymes.

4. Physiological Role of the Retinal Specific LDH Isozyme

The cellular and temporal-specificity of LDH E gene expression and the distinctive kinetic and physical properties of the E_4 LDH suggests that this enzyme may play an important role in the metabolism of the photoreceptor cells (Whitt, 1970a; Whitt and Booth, 1970).

As mentioned before, the kinetic behavior of the retinal specific E_4 isozyme is similar to that of the B_4 isozyme. The retinal E_4 isozyme of the Atlantic mackerel (Fig. 4) has a higher affinity for substrate and a lower substrate concentration optimum than the B_4 isozyme (Whitt, 1970a). These kinetic properties of the retinal specific LDH suggest that this isozyme might be especially suited for cells with a high constant aerobic metabolism. The photoreceptor cell has been shown to have a high degree of aerobic glycolysis (Cohen and Noell, 1965). In fact, the neural retina has perhaps the highest rate of oxygen utilization of any tissue of the body (White et al., 1968). Since much of the carbohydrate metabolism of the retina is directed toward the regeneration of rhodopsin, it is proposed that the LDH E_4 isozyme may play a role in the regeneration of visual pigments in the photoreceptor cells of teleosts (Whitt, 1970a).

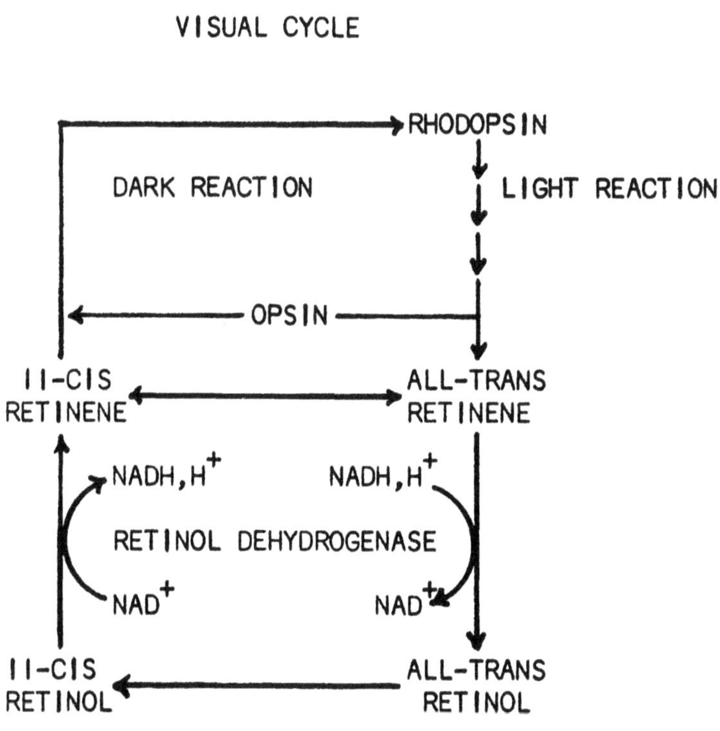

Fig. 12. Outline of the visual cycle. Modified after Wald (1968)

An outline of the metabolism of the visual cycle of vertebrates (modified after Wald, 1968; Bonting and Bangham, 1968) is shown in Fig. 12. The light converts rhodopsin to opsin + retinene. This trans-retinene is converted to trans-retinol (vitamin A) by retinol dehydrogenase (Koen and Shaw, 1966). The all-trans retinol is converted to 11-cis retinol. The retinol dehydrogenase acts as a catalyst for the conversion of 11-cis retinol to 11-cis retinene. Once the cis retinene is formed, it is bound up by opsin as fast as it appears.

In some fish and amphibians (Wald, 1949, 1968; Wald and Hubbard, 1949), as opposed to other vertebrates (Newhouse et al., 1972), NAD^+ is the coenzyme utilized by retinol dehydrogenase. This coenzyme, generated from retinal respiration and glycolysis, is required by the retinol dehydrogenase reaction to promote rhodopsin synthesis. Therefore, one of the limiting processes of rhodopsin generation is the oxidation of retinol to retinene.

How may the retinal specific lactate dehydrogenase participate in visual metabolism? If, during certain phases of teleost retinal metabolism, the amount of NAD^+ becomes limiting to the rate of rhodopsin regeneration, then an enzyme such as LDH which can convert NADH to NAD^+ would be important. If concentrations of substrate in the photoreceptor cells are relatively low, the affinity of the enzyme for substrate is of more physiological significance than its susceptibility to substrate inhibition. The high affinity of the E_4 isozyme for pyruvate would enable this enzyme to convert efficiently NADH to NAD^+ which would then be utilized by retinol dehydrogenase during the regeneration of rhodopsin.

There are a number of alternative hypotheses regarding the physiological role of the retinal specific isozyme which have not yet been experimentally tested. The higher susceptibility of the LDH E_4 isozyme to inhibition by lactate may help prevent the build-up of deleterious concentrations of lactate within the photoreceptor cells. Alternatively, the high affinity of the E_4 isozyme for substrate might enable the photoreceptor cells to utilize low levels of exogenous lactate efficiently as a source of energy.

Another possibility, is that the physiological significance of the retinal specific LDH may not depend solely upon its kinetic properties but also upon its distinctive physical attributes. The highly negative net charge of many LDH E_4 isozymes may be an indirect consequence of charge requirements determining specific intracellular location. In fact, there have been some reports of the preferential localization of mammalian LDH B_4 with the mitochondrial fraction (Agostoni et al., 1966; Guttler and Clausen, 1967). Furthermore, the high stability of the E_4 LDH to denaturation in vitro (Whitt, 1970a) may reflect a corresponding stability to intracellular catabolism. However, the investigations by Fritz et al.,(1969, 1971) indicate that mechanisms other than intrinsic enzyme stability also control LDH isozyme catabolism. It is not presently known why most teleosts, and not others exhibit a retinal specific LDH. One would assume that the absence of the E_4 LDH indicates that an alternative mechanism has evolved in place of the retinal specific LDH. Another interpretation is that the eyes of those fish lacking E_4 are not subjected to the same physiological conditions as the eyes of fishes possessing the E_4 isozyme.

No physiological or ecological variable has yet been correlated with the presence or absence of the retinal specific LDH in teleosts (Whitt, 1970a; Whitt and Maeda, 1970; Whitt and Horowitz, 1970; Whitt and Prosser, 1971). Comparative studies of retinal metabolism for fish with and without the E_4 isozyme might shed some light on this problem. In any event, the restriction of a distinctive LDH isozyme to differentiated photoreceptor cells certainly does suggest that it plays a very specialized role--a role unable to be fulfilled by the other LDH isozymes in these teleosts.

5. LDH E Gene Function in Non-Neural Tissues

Although the lactate dehydrogenase E gene usually functions predominantly in neural tissues, the E gene can be expressed to a limited extent in some non-neural tissues. The E_4 isozyme found in vitreous humor is probably the result of preferential secretion (or leakage) from the neural retina (Whitt, 1970a). In a few species of teleosts the E gene can function in the lenticular epithelial cells (Kusa, 1966; Whitt, 1968, 1970a) which is unexpected because the lens is noted for its low rate of oxygen utilization. There have been other reports for unexpected LDH isozyme patterns from the lens cells of a number of verte-

LDH E gene function in bass Tissues

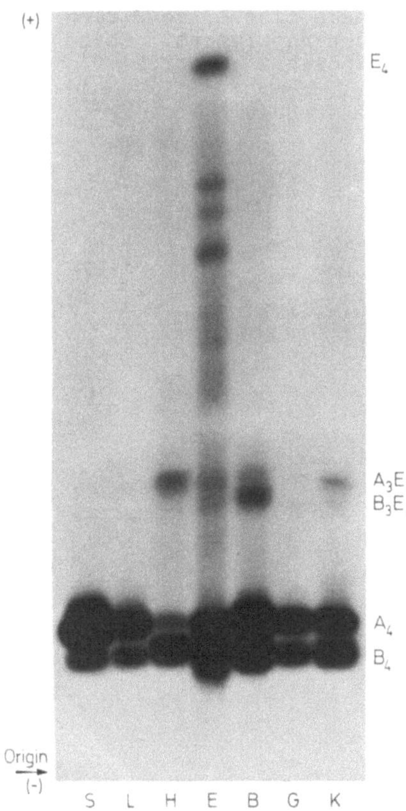

Fig. 13. Lactate dehydrogenase E gene function in the tissues of the largemouth bass (*Micropterus salmoides*). S - skeletal musclc; L - liver; H - heart; E - eye; B - brain; G - gill; and K - kidney. LDH E polypeptide synthesis is detected in heart, brain, and kidney in addition to its predominant synthesis in the eye

brates (Glaesser and Berg, 1966; Papaconstantinou, 1967; Vesell, 1965; Bernstein et al., 1966). The predominance of A subunits in some species and B subunits in other species indicates that the LDH isozyme pattern of terminally differentiated vertebrate lens cells may not be critical to their metabolism. Because of the close similarity of lens structure and function in all vertebrates, the striking species-specific difference in isozyme patterns does not appear to be due to differences in such common attributes as replicative states of the cells or degree of aerobiosis. The unexpected presence of the LDH E_4 isozyme in the lens cells of some species of fish may be due to an aberration of regulatory gene function analogous to the LDH regulatory gene mutant operating in the terminally differentiated red blood cell of some rodents (Shows and Ruddle, 1968). The presence of the LDH E_4 isozyme in the lenticular cells of some fish probably does not exert a deleterious effect upon the fate of these terminally differentiated cells.

The LDH E gene has also been demonstrated to be activated in tissues other than eye of some fishes (Whitt et al., 1971). The LDH isozyme patterns of the tissues in the largemouth bass are shown in Fig. 13. The LDH E subunit synthesis is most prominent in the retina--as shown by the presence of the E_4 isozyme in the eye. Tissues such as heart, brain, and kidney also possess E subunit synthesis. However, the E gene appears to function at a lower level in these tissues resulting in the detection of only the A_3E_1 and/or B_3E_1 heteropolymers. These tissues having low levels of E gene expression possess an aerobic

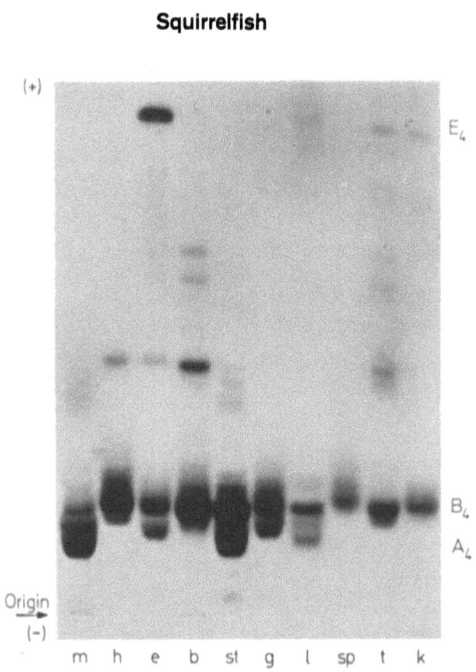

Fig. 14. Lactate dehydrogenase isozymes of the longspine squirrelfish (*Holocentrus rufus*). The LDH E_4 isozyme is predominantly synthesized in the eye but is observed in lesser amounts in liver, testis, and kidney. Heteropolymers containing E subunits are present in all ten tissues. M - skeletal muscle; h - heart muscle; e - eyes; b - brain; st - stomach; g - gill; l - liver; sp - spleen; t - testis; k - kidney

metabolism which suggest that the E subunits synthesized in these tissues may be of physiological importance.

In most of the fish species examined the E_4 homopolymer is electrophoretically detected only in the retina. However, in a few species of fish, the LDH E_4 isozyme is also detected in other tissues. Fig. 14 shows the LDH isozymes of the longspine squirrelfish (*Holocentrus rufus*) (Beryciformes, Holocentridae) tissues. The LDH E_4 isozyme is predominantly synthesized in the eye and to a lesser extent in testis, kidney and liver. E subunit synthesis was detected in all of the tissues of this fish that were examined.

The observations made on E gene expression in differentiated tissues indicate that LDH isozymes containing one or more E polypeptides may be especially suitable for function in highly aerobic tissues (usually retina and brain, but occasionally other tissues)(Whitt et al., 1971, 1972). Furthermore, this diversity of tissue-specific isozyme patterns among the fish groups does suggest there is considerable difference in the specificity of regulation of LDH E gene function.

In an attempt to study the specificity of regulation of the LDH E gene function, the expression of E subunit synthesis has been investigated in the tissues of interspecific sunfish hybrids (Whitt et al., 1972). Some sunfish species possess the LDH E gene function in the heart, e.g. warmouth (*Lepomis gulosus*), others lack the E gene function in their cardiac tissues, e.g. red-ear sunfish (*L. microlophus*)

Lactate Dehydrogenase Isozymes of the Heart

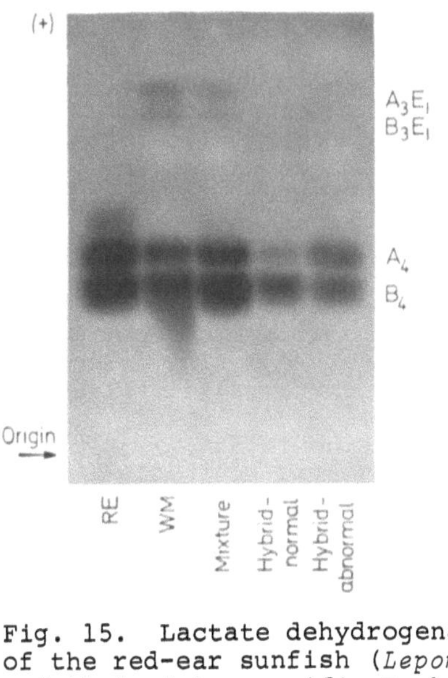

Fig. 15. Lactate dehydrogenase E gene expression in the heart muscle of the red-ear sunfish (*Lepomis microlophus*), warmouth (*L. gulosus*), and their interspecific F_1 hybrid (red-ear x warmouth). RE - red-ear sunfish; WM - warmouth; mixture - mixture of the enzyme extracts from the two parental species; Hybrid-normal - morphologically normal F_1 hybrids; Hybrid-abnormal - morphologically abnormal F_1 hybrids. The presence of heteropolymers containing E polypeptides (A_3E_1, B_3E_1) indicates the LDH E gene is expressed in the cardiac tissues of the warmouth (the maternal parent) while absence indicates the E gene does not function in the cardiac tissues of the red-ear (the paternal parent) or in the F_1 hybrid

(Fig. 15). A male red-ear sunfish was crossed with a female warmouth. The E gene function (present in the maternal parent) is absent in the cardiac tissue of the F_1 hybrids. This repression of maternal E subunit synthesis occurs in both morphologically normal and abnormal F_1 hybrids (Whitt et al., 1972). To gain further insight into the mechanisms of gene regulation, experiments are planned to determine whether the genes controlling this tissue specificity of E subunit synthesis are linked to the LDH E structural gene.

6. Expression of the LDH E Gene Function During Development

Nakano and Whiteley (1965) working with the medaka (*Oryzias latipes*) (Gasterosteiformes, Gasterosteidae) observed that the retinal specific LDH isozyme did not appear until abruptly at hatching. These researchers attributed the sudden appearance of the E_4 isozyme to the rapid change in metabolism at hatching, specifically the change-over from the anaerobic conditions of the egg to the aerobic conditions after hatching.

Because of this apparent dependence of the LDH E gene function at hatching, this hypothesis was tested by following the ontogeny of the LDH isozyme synthesis in a different species, *Fundulus heteroclitus* (Whitt, 1970a) a fish which has been subjected to numerous developmental analyses. In this fish the E_4 isozyme was first detected at least 24 hours prior to hatching which of course indicated that the LDH E gene was activated at least 24 hours prior to hatching (Whitt, 1968, 1970a). Thus, the act of hatching is not a required stimulus for E gene function. The onset of E polypeptide synthesis (and E gene activation) is more probably related to the state of differentiation of the neural retina (Whitt, 1968, 1970a).

A different pattern of E gene expression has been observed during development of fish in the family Salmonidae. The retinal-specific LDH isozyme appears considerably after hatching in trout (Hitzeroth et al., 1968; Goldberg et al., 1969). These results are also consistent with the postulate that hatching is not a direct stimulus to E gene activation. An examination of the literature on salmonid development (Ali, 1959; Woodhead, 1957) reveals that the larvae do not develop a mature visual apparatus, i.e. rods, retinomotor response, and visual acuity until 2-3 weeks after hatching which correlates well with the time of appearance of the retinal-specific LDH reported in trout. Because the appearance of the LDH E_4 isozyme during teleost development occurs at the time of structural and functional differentiation of the retina, the E locus activation is probably tightly coupled with the differentiation of the neural retina (Whitt, 1968, 1970a). This postulate of retinal specific LDH isozyme synthesis being dependent upon differentiation of the photoreceptor cells has been confirmed by the results of Kunz (1971) and Nakano and Hasegawa (1971). Nakano and Hasegawa (1971) have demonstrated that the differentiation of the medaka retinal layers precedes the appearance of the retinal specific LDH.

Our laboratory is focusing its efforts on investigating the onset of the LDH E_4 isozyme synthesis and correlating this with the onset of various morphological, functional, and enzymatic events during retinal differentiation. We have investigated the ontogeny of the LDH isozymes in the sunfishes (Perciformes, Centrarchidae). The sunfishes were chosen because of their ease of interspecific hybridization and their short developmental time. In Fig. 16 is shown the progression of LDH isozyme patterns observed during development of a typical sunfish (see Fig. 1 for adult tissue patterns). It takes from two to two and

LDH E_4 Ontogeny in Sunfish

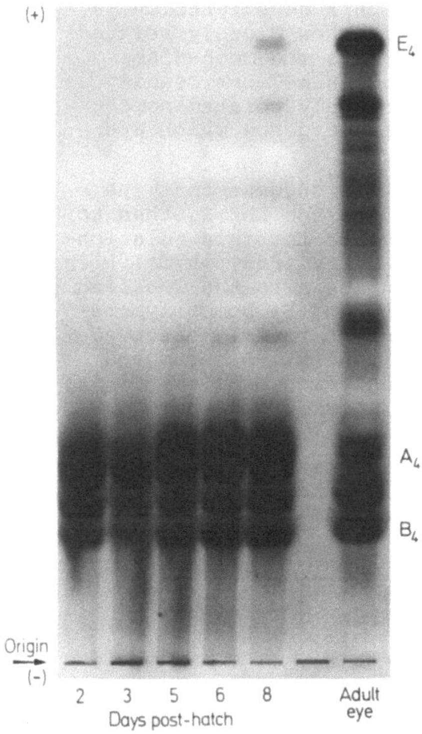

Fig. 16. Synthesis of the retinal specific lactate dehydrogenase isozyme (LDH E_4) during development of a typical sunfish. Hatching occurs three days after fertilization. Heteropolymers containing E subunits (B_3E_1) were first detected 5 days after hatching. The first time that the LDH E_4 isozyme was observed was 6 days after hatching

one half days from fertilization to hatching at 27°C. At hatching the eyes are poorly developed and possess no pigmentation. Two days after hatching the eyes are completely pigmented. The eyes appear completely differentiated 5 days after hatching, based on external morphology. No E polypeptide synthesis is detected until 5 days after hatching, when the B_3E_1 heteropolymer is first observed. The E_4 isozyme is first detected at 6 days after hatching. These results were observed during the ontogeny of the green sunfish, the bluegill sunfish (*Lepomis macrochirus*), and the F_1 hybrid formed from the bluegill X green sunfish cross.

We hope to determine eventually whether this relatively late appearance of the E_4 isozyme in the sunfish may represent a late transcriptional event or a delay of translational or post-translational events. Because it has been proposed that the LDH E_4 isozyme plays a role in the regeneration of rhodopsin, one might expect the onset of its synthesis to be coupled with the appearance of other enzymes (e.g. retinol dehydrogenase and acetylcholinesterase) associated with photoreceptor cell function. A concerted effort is presently underway to correlate this pattern of E_4 isozyme synthesis with the differentiation (biochemical, functional, and morphological) of the teleost photoreceptor cells.

7. LDH E Gene Homology

The cellular and temporal specificity of the LDH E gene function is so striking that it was of interest to determine how closely related the LDH E gene is to the other LDH genes. Perhaps the mechanisms determining the specificity of gene regulation can be more readily determined by examining the functions of the closely related homologous LDH loci than by examining the functions of genes which are only distantly related.

The kinetic, physical, and developmental analyses suggest that the retinal LDH E_4 isozyme is more closely related to the LDH B_4 than to the LDH A_4. However, the most persuasive evidence for this relationship is derived from immunochemical studies (Whitt, 1969, 1970a). In Fig. 17 is shown the inhibitory effects of antisera to the weakfish (*Cynoscion regalis*) (Perciformes, Sciaenidae) A_4 and B_4 homopolymers on the activity of the weakfish LDH E_4. As can be readily observed, there is no detectable cross-reaction between the anti-A serum and the E_4 isozyme, whereas the anti-B serum strongly inhibits the retinal-specific LDH which indicates there is considerable homology between the E_4 and B_4 isozymes (Whitt, 1969, 1970a).

Since the antigenic, physical, and kinetic properties of the retinal specific isozyme are similar to those of the B_4 isozyme, it has been proposed that the evolutionary relationships of the three main LDH

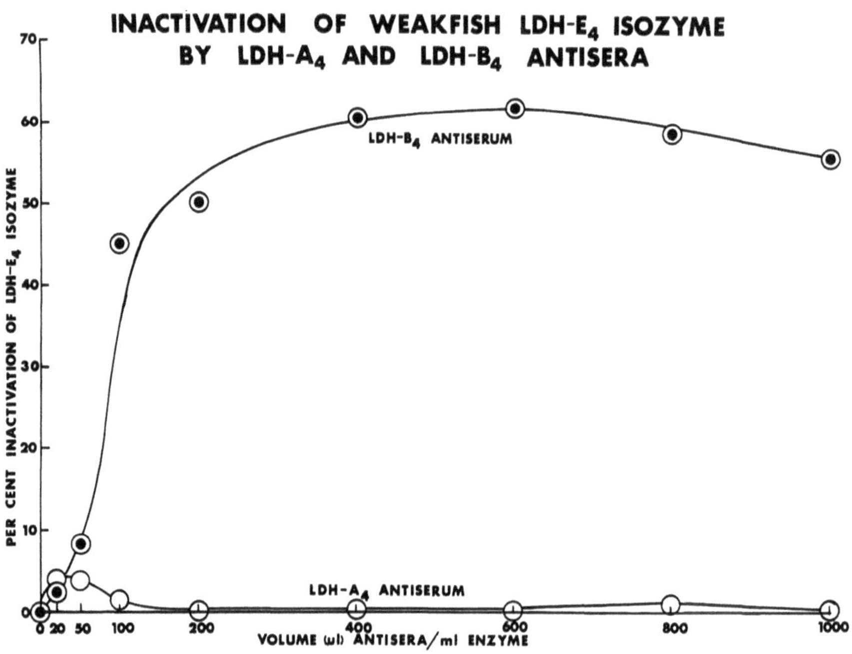

Fig. 17. Inactivation of weakfish (*Cynoscion regalis*) retinal specific lactate dehydrogenase isozyme (LDH E_4) by antisera against weakfish LDH A_4 and B_4. The anti-A_4 serum does not react with the LDH E_4 whereas the anti-B_4 serum reacts strongly with this retinal specific LDH. Procedures are given in Whitt (1970a)

genes of teleosts are as shown in Fig. 18 (Whitt, 1969, 1970a). The LDH E gene appears to have been derived from the LDH B gene by duplication and subsequent divergence. However, the LDH B and E loci have not diverged to the same extent as the A and B loci have diverged, presumably because of their more recent identity. More recent polyploidization resulting in the further duplication of some LDH genes appears to have occurred in the salmonids and some cyprinids (Klose et al., 1968, 1969; Massaro and Markert, 1968; Ohno, 1970). The LDH loci of salmonids duplicated by this more recent polyploidy are very closely related (Bailey and Wilson, 1968; Holmes and Markert, 1969). This postulate of LDH gene evolution (Fig. 18) is supported by a number of observations.

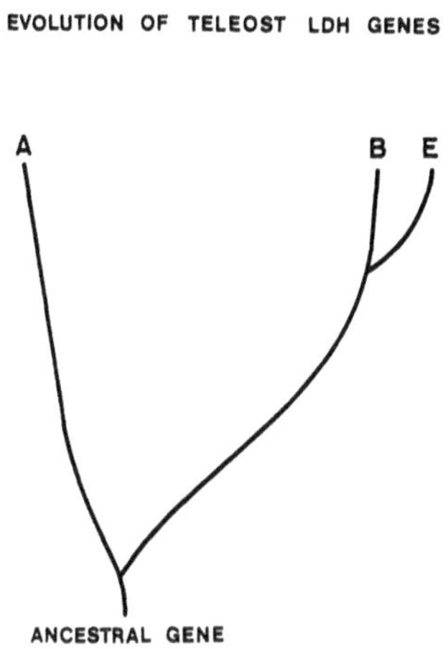

Fig. 18. Evolutionary relationships of the three lactate dehydrogenase genes of fishes. The LDH E gene is derived from the LDH B gene by duplication. Not shown in this figure are the more recent polyploidization events which have caused these loci to undergo further duplication in fishes in the orders Salmoniformes and Cypriniformes

A comparison of the LDH A and B gene products strongly suggests these two loci have arisen from a common ancestral gene. The LDH A and B polypeptides have very similar amino acid sequences in the region of the active site (Allison, 1968). Furthermore, the LDH A and B polypeptides derived from quite distant classes of vertebrates (Markert, 1965, 1968; Markert and Whitt, 1968) can be made to polymerize readily with each other in vitro to generate active hybrid enzymes. Although the A_4 and B_4 isozymes of most vertebrates are immunochemically distinct, these heterologous homopolymers possess a slight immunochemical similarity in some fishes (Holmes and Markert, 1969; Horowitz and Whitt, 1972) and may in fact share some antigenic determinants. These results are consistent with the hypothesis that the LDH A and B loci are derived from a common ancestral gene. However, the A and B loci have undergone extensive divergence as shown by the considerable differences between the A_4 and B_4 isozymes in their amino acid composition (Pesce et al., 1967) and immunochemical properties (Markert and Appella, 1963).

A recent evolutionary survey of fishes indicates that the retinal specific lactate dehydrogenase isozyme is not found in the non-tele-

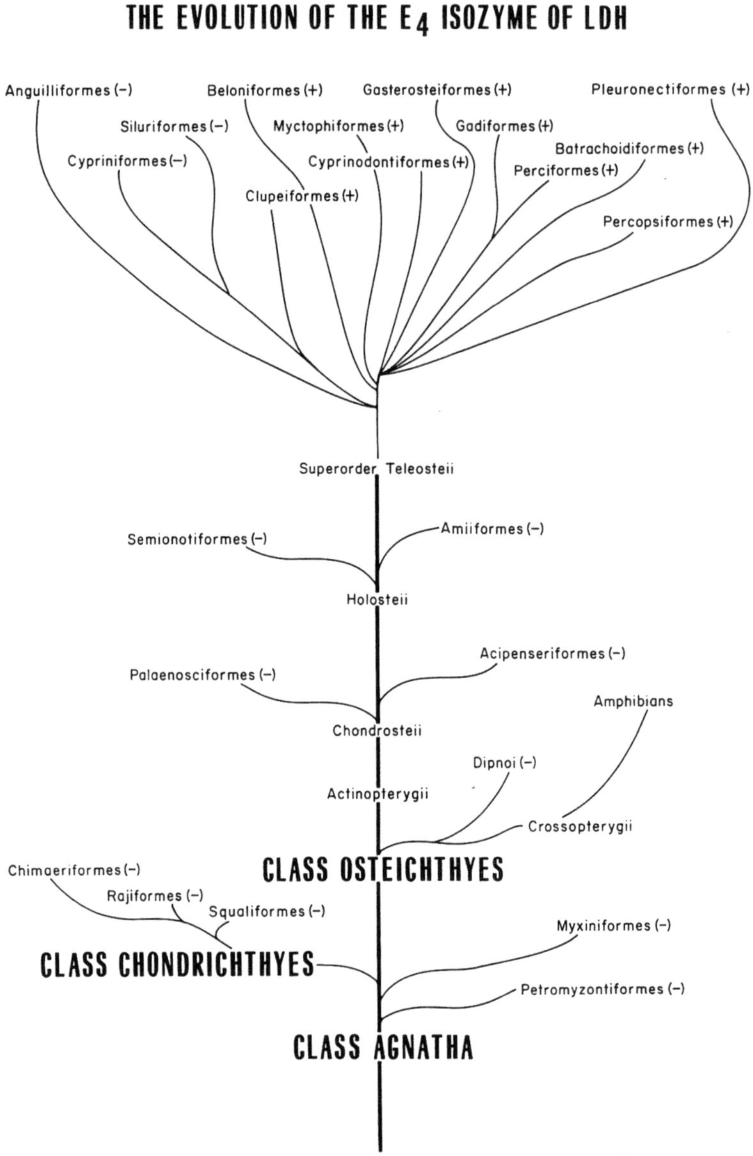

Fig. 19. The phylogenetic distribution of the retinal specific lactate dehydrogenase isozyme (LDH E_4) in fishes. A plus sign by an order indicates that at least one family in that order possesses species with an LDH E_4 isozyme. No retinal specific LDH was detected in any non-teleostean fishes. Most, but not all, of the orders of teleosts investigated have a retinal specific LDH. It is postulated, therefore, that the B gene duplication which gave rise to the E gene occurred prior to the adaptive radiation of the teleosts

ostean fishes (Whitt and Horowitz, 1970; Horowitz and Whitt, 1972) and is summarized in Fig. 19. A plus sign (+) by an order indicates that at least one family possesses species with a retinal specific LDH. No retinal specific LDH was detected in any group of fish other than teleosts. Most of the teleost orders examined possess the E_4 LDH. Therefore, it has been postulated that the B gene duplication which gave rise to the E locus occurred prior to the time of the adaptive radiation of the teleosts (Whitt and Horowitz, 1970; Horowitz and Whitt, 1972). However, some orders of teleosts such as Anguilliformes, Cypriniformes, Siluriformes and some families of Gadiformes and Clupeiformes lack the retinal specific LDH isozyme. However, the presence of a retinal specific LDH in many groups of fish suggests that there may be a significant difference in visual metabolism between some contemporary advanced teleosts and other vertebrates.

8. The Liver Specific LDH Gene Function

The absence of the retinal specific LDH isozymes in some orders of teleosts has stimulated this laboratory to more closely examine these groups of fish. One of the first studies on the Cypriniformes was the

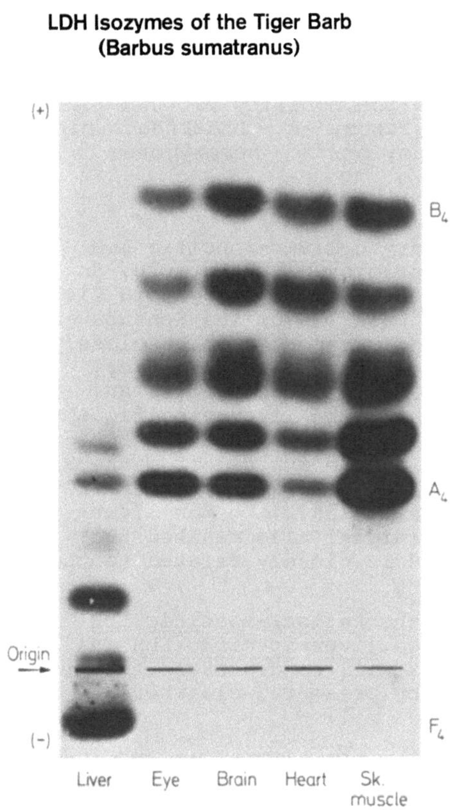

Fig. 20. Lactate dehydrogenase isozymes of the tiger barb (*Barbus sumatranus*). The cathodal isozyme specific to the liver is referred to as LDH F_4. Note the absence of a retinal specific LDH

Liver LDH Phenotypes of striped shiner

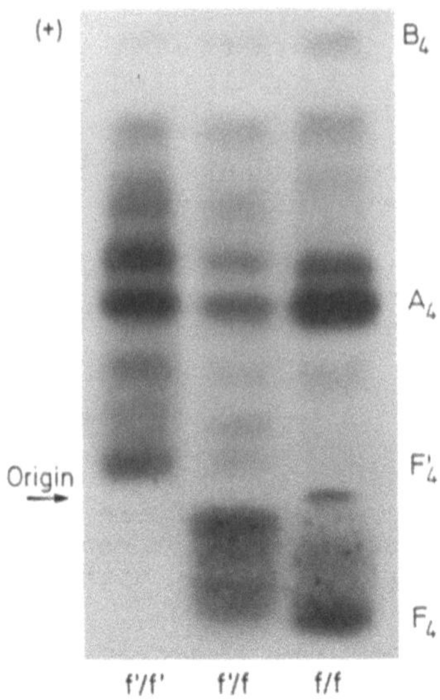

Fig. 21. Liver lactate dehydrogenase isozyme phenotypes of the striped shiner (*Notropis chrysocephalus*). The liver-specific LDH is the most cathodal isozyme. Genetic variation alters the mobility of this liver specific LDH without affecting the A_4 or B_4 isozymes. Underneath each isozyme phenotype is its postulated genotype: f'/f' - homozygote; f'/f - heterozygote; f/f - homozygote

work of Klose et al.,(1969) where they observed liver-specific LDH isozymes in both diploid and tetraploid cyprinids. Furthermore, an LDH subunit synthesized predominantly in the liver of the gadoid fish has been reported by Odense et al.,(1969), Lush (1970) and Sensabaugh and Kaplan (1972). This liver-specific LDH is reported to be closely related to the LDH B_4 on the basis of heat inactivation, kinetics, and immunochemistry (Sensabaugh and Kaplan, 1972). They found an individual with an electrophoretic variant for the liver-specific LDH and propose that this liver-specific LDH is encoded by a third LDH gene. These investigators propose that this gene evolved from the B gene by a duplication event separate from that giving rise to the E gene. A preliminary report by Kepes and Whitt (1972) on the liver-specific LDH of barbs (Cypriniformes, Cyprinidae) also reached the conclusion that the barb liver-specific LDH is closely related to the B_4 isozyme.

We have been investigating the absence of the retinal-specific LDH in some orders of fish and the presence of a liver-specific LDH in these same orders. A summary of these data will be discussed in this paper while a more detailed analysis will be presented elsewhere (Shaklee et al., 1973).

The LDH isozyme patterns of the tissues of the tiger barb (*Barbus sumatranus*)(Cypriniformes, Cyprinidae) are displayed in Fig. 20. Five main isozymes are synthesized in most tissues. These are formed by the random association of the A and B polypeptides. There is no eye-specific LDH, which appears to be the case for all cyprinids examined (Whitt and Maeda, 1970; Whitt and Horowitz, 1970; Horowitz

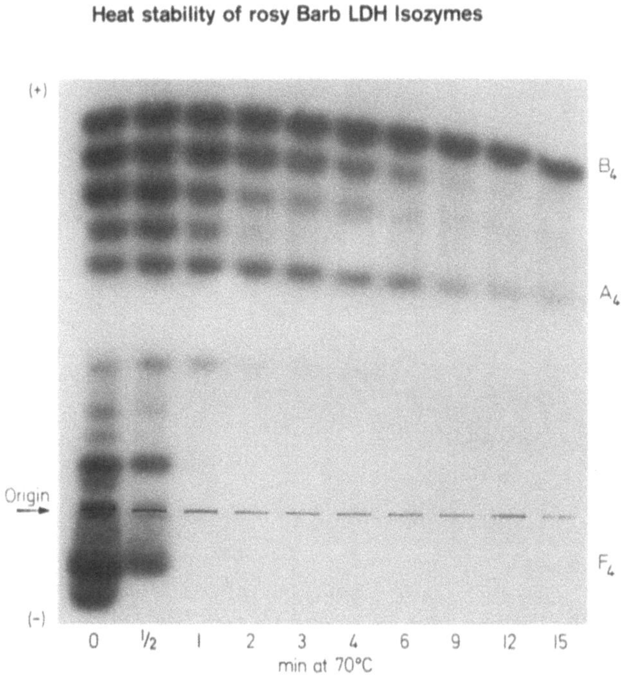

Fig. 22. Heat inactivation of the lactate dehydrogenase isozymes of the rosy barb (*Barbus conchonius*). Extracts of brain and liver were mixed together. Aliquots of this mixture were placed in separate centrifuge tubes and then heated at 70°C for different lengths of time. After the heat treatment, the samples were cooled to 4°C, centrifuged, and then subjected to electrophoresis

and Whitt, 1972). However, there is a unique isozyme synthesized in the liver of these fish.

For the purpose of this paper, we will refer to this cathodal liver specific isozyme as LDH $\underline{F}_4$ in order to distinguish it from the anodal retinal specific LDH $\underline{E}_4$. The results of freeze-thaw molecular hybridization of the liver LDH isozymes indicate that the liver specific LDH exists as a homotetramer whose subunits readily assemble with the other LDH subunits. The best evidence for the genetic distinctness of this liver specific LDH has been derived from the studies of W. Rainboth (unpublished) on the LDH isozyme polymorphisms in the liver of another cyprinid, the striped shiner (*Notropis chrysocephalus*) (Cypriniformes, Cyprinidae). The three main liver isozyme phenotypes observed in this population of shiners are shown in Fig. 21. The mobility of the liver specific LDH is altered by the genetic variation--without affecting the mobility of the A_4 or B_4 isozymes. This represents the first clear direct demonstration of the genetic distinctness of the liver-specific LDH.

Several approaches were taken to determine whether this liver-specific LDH possesses physical attributes similar to those of the retinal-specific LDH of other teleosts. The effect of heating upon a mixture of brain and liver LDH isozymes of the rosy barb (*B. conchonius*) is shown in Fig. 22. The F_4 isozyme of this fish is relatively most

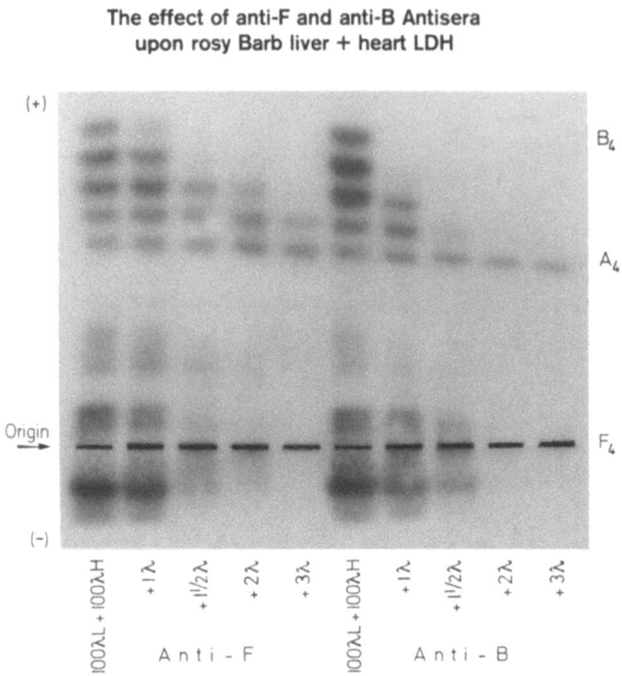

Fig. 23. Effects of antisera on the lactate dehydrogenase isozymes of the rosy barb (*Barbus conchonius*). The sample was a mixture of extracts from liver (L) and heart (H). The antiserum to LDH B_4 (weakfish) and the antiserum to LDH F_4 (Atlantic cod) were each mixed with separate aliquots of the sample. This mixture was allowed to react for at least one hour prior to electrophoresis. The amounts of antisera added to a constant volume of enzyme extract are indicated below each slot

unstable. These results are opposite to those reported for the liver-specific LDH of gadoids (Sensabaugh and Kaplan, 1972) and the eye-specific LDH (Whitt, 1970a). Although these differences in net charge and heat stability between the E_4 and F_4 isozymes presumably reflect differences in amino acid composition, a better estimate of enzyme relatedness is provided by our immunochemical studies.

The antisera to LDH A_4, B_4, and E_4 (retinal specific) of the weakfish and LDH F_4 (liver specific) of the Atlantic cod were prepared in rabbits according to the procedures described in Markert and Holmes (1969).

Fig. 23 shows the effect of antibodies upon the LDH isozymes of the rosy barb liver. The anti-LDH F_4 serum precipitates all those isozymes containing B or F polypeptides. The anti-B serum precipitates isozymes containing B subunits at lower concentrations and, in addition, precipitates the liver-specific LDH at higher concentrations. The A_4 isozyme is not precipitated by either anti-F or anti-B, even at high concentrations. These data indicate that there is considerable homology between the liver-specific LDH and the B_4 isozyme.

The tissue specific LDH isozyme patterns of gadoid fishes are somewhat similar to those of the barbs (Fig. 24). Although the F_4 LDH isozyme is predominant in the liver of the Atlantic cod (*Gadus morhua*)

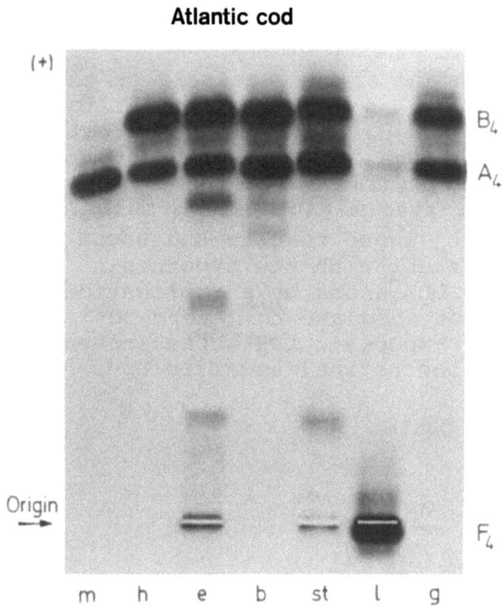

Fig. 24. Lactate dehydrogenase isozymes of the Atlantic cod (*Gadus morrhua*)

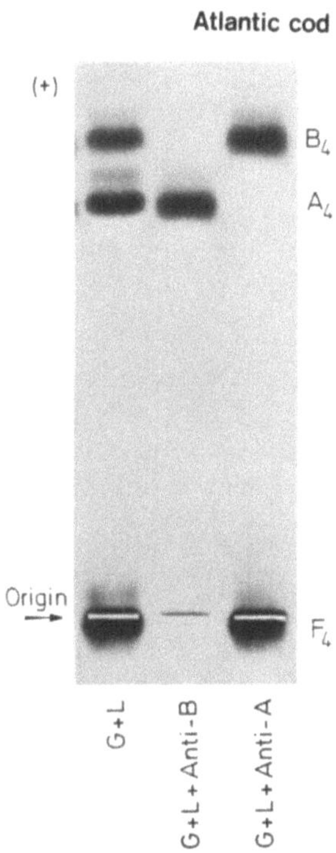

Fig. 25. Effects of antisera on the lactate dehydrogenase isozymes of the Atlantic cod (*Gadus morrhua*). The sample was a mixture of extracts from gills (G) and liver (L). The antisera to the LDH A_4 and the LDH B_4 of the weakfish were mixed with separate aliquots of the sample. This mixture was allowed to react for at least one hour prior to electrophoresis. Both the LDH B_4 and LDH F_4 isozymes are precipitated by the anti-B serum

(Gadiformes, Gadidae) it is also synthesized to a lesser extent in several other tissues such as eye, stomach, brain and gills.

In Fig. 25 are shown the effects of anti-A and anti-B antisera on the LDH isozymes of the Atlantic cod. These results are very similar to those observed for the barb LDH. Both the LDH B_4 and F_4 are unaffected by the anti-A serum whereas high concentrations of the anti-B serum precipitate both the LDH B_4 and the liver-specific LDH. These immunochemical data suggest again that the liver-predominant LDH is much more similar to the LDH B_4 than to the A_4. These results and those of Sensabaugh and Kaplan (1972) are consistent with the hypothesis that the gene encoding the liver-specific LDH arose by a duplication of the LDH B locus. But these data are not adequate to determine whether the LDH locus coding for the liver-specific LDH arose independently from the genetic locus coding for the retinal-specific LDH.

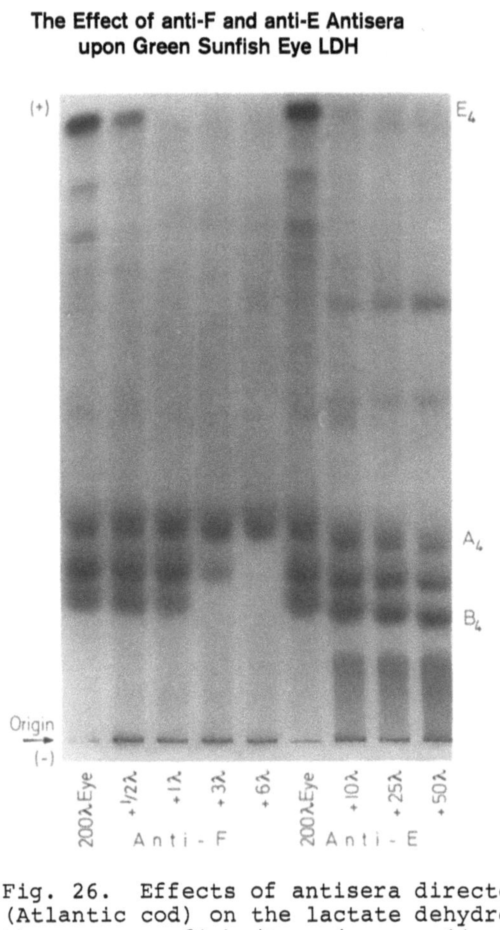

Fig. 26. Effects of antisera directed against the liver specific LDH (Atlantic cod) on the lactate dehydrogenase isozymes from the eye of the green sunfish (*Lepomis cyanellus*). The antiserum to LDH F_4 (cod) and the antiserum to LDH E_4 (weakfish) were separately mixed with aliquots of the extract of the sunfish eye. This mixture was allowed to react for at least one hour prior to being subjected to electrophoresis. The amounts of antisera added to a constant volume of enzyme extract are indicated below each slot. The most striking result is that the anti-LDH F antiserum exhibits greater cross-reaction with the retinal-specific E_4 isozyme than with the LDH B_4 isozyme

In order to determine the relative homology of the LDH B_4, E_4, and F_4 isozymes, the anti-F and anti-E antisera were reacted with the retinal specific LDHs of other fishes. The effects of the anti-LDH E_4 serum and the anti-LDH F_4 serum on the LDH isozymes of the green sunfish eye are shown in Fig. 26. On the right side is shown that the anti-E_4 serum precipitates primarily the E_4 isozyme with little effect upon the other LDH isozymes. But of particular interest is the effect of anti-liver LDH upon the eye isozymes--the LDH E_4 isozyme is precipitated by the anti-F serum at concentrations which have essentially <u>no effect</u> on the LDH B_4 isozyme. To summarize the various immunochemical studies, the teleost A_4 is immunochemically distinct from the B_4, E_4, and F_4. Both the retinal-specific LDH and the liver-specific LDH are immunochemically related to the B_4 isozyme. But most significantly, these two tissue specific LDHs appear more antigenically related to each other than either is to the B_4 LDH. While the E_4 and F_4 have similar physical characteristics and antigenic properties, they probably do not play identical physiological roles in such metabolically different tissues.

From the immunochemical and genetic data we have generated two alternative hypotheses for the evolution of these two tissue-specific LDH isozymes. Our main postulate is that only <u>one</u> LDH gene arose from the LDH B gene (probably by tandem duplication). This gene appears to have evolved to a liver-specific function in some orders of fish such as the Cypriniformes and Gadiformes. This probably involved a co-evolution of both gene regulation and kinetic properties of the isozyme. In most orders of teleosts, this gene evolved in its structure and regulation to become restricted to a retinal function. Perhaps the strongest evidence supporting this hypothesis is the mutually exclusive presence of the liver-specific and retinal-specific LDH in species of fish. In all of the teleosts we have examined there is either a retinal-predominant LDH or a liver-predominant LDH--but <u>never both</u>. This observation, plus the immunochemical data showing the closer relatedness of the E_4 and F_4 to each other than to the B_4, is consistent with only a single duplication of the B locus.

An alternative hypothesis, although more cumbersome, is also consistent with the data. Let us assume the presence of three LDH genes, e.g. A, B, and E, early in teleost evolution. Then assume the E locus duplicated and gave rise to another locus (the F locus). The E locus evolved to an eye-specific function and the F locus to a liver-specific function. One would also have to invoke that these two loci have since become mutually exclusive in their function--when one locus is active in a species, the other is inactive or absent. The second hypothesis is somewhat uneconomical, but it would be greatly favored if groups of fish are found which possess both a liver-specific LDH and a <u>different</u> eye-specific LDH. A detailed analysis of the LDH gene evolution in fishes and other vertebrates will be presented elsewhere (Markert et al., 1973).

The presence of these two tissue-specific LDHs encoded in the same homologous locus offers an unusual opportunity to study the evolution of the specificity of gene regulation. Ultimately, we may be able to determine the extent that identical genes have had to diverge in their nucleotide sequences before they can come under separate genetic control. Multigenic enzyme systems exhibiting tissue specific expression such as the isozymes of lactate dehydrogenase provide excellent model systems for investigating the specificity of gene function during cellular differentiation.

9. Summary

The L-lactate dehydrogenase (LDH) isozymes of fish are useful gene markers for investigating gene action during embryogenesis and the mechanisms of evolution of homologous genes. Teleosts possess at least three genetic loci coding for lactate dehydrogenase subunits.

The main purpose of this presentation is to discuss the evidence for, and the significance of, LDH isozymes whose synthesis is usually restricted to highly differentiated tissues--LDH E_4 in the eye, and LDH F_4 in the liver. The retinal specific LDH E_4 is encoded in a third LDH locus and is more closely related in its antigenic, physical and kinetic properties to the LDH B_4 isozyme (in aerobic tissues) than to the LDH A_4 (in anaerobic tissues). The retinal specific LDH (with kinetic properties suitable for aerobic conditions) is localized primarily in the ellipsoid region of the photoreceptor cells--containing high concentrations of mitochondria. It is proposed that the LDH E_4 isozyme may play a role in the visual metabolism of some fish, perhaps in rhodopsin regeneration. The activation of this gene during teleost ontogeny is correlated with the structural and functional differentiation of the retina--not the changeover from anaerobic to aerobic conditions at hatching.

A retinal specific LDH E_4 isozyme is found only in the teleosts, not in non-teleostean fishes. However, in some orders of teleosts, e.g. Anguilliformes, Cypriniformes, Gadiformes and Siluriformes, the retinal specific LDH is lacking. Immunochemical and phylogenetic data suggest that the LDH E gene resulted from a duplication of the LDH B gene (by tandem duplication) prior to the adaptive radiation of the teleosts.

Recent investigations of the LDH isozyme patterns of the Cypriniformes and Gadiformes reveal that these fishes, which lack the anodal retinal specific LDH, possess a cathodal LDH isozyme (LDH F_4) predominantly synthesized in the liver. The LDH F_4 is encoded in a third locus. The liver specific isozyme differs from the retinal specific isozyme in its electrophoretic mobility and its lability to heat denaturation. Immunochemical studies indicate that the F_4 is closely related to the B_4. However, the LDH F_4 and E_4 appear to be more closely related to each other than to the LDH B_4.

On the basis of recent investigations, it is proposed that only a single LDH gene arose from the B gene by duplication. In the group of fish leading to the Cypriniformes, etc., this third LDH evolved in its regulation and its kinetic properties to an isozyme (LDH F_4) synthesized predominantly in the liver. In the other line of fish, leading to most orders of teleosts, this third gene evolved in its structure and regulation to become restricted to a retinal function.

10. Acknowledgments

The original research reported here was supported by National Science Foundation grants GB 16425 and GB 5440X. Some of the research reported here represents work toward the Ph.D. dissertations of E.T.M. (University Fellowship, Cell Biology, University of Illinois, Urbana) and J.B.S. (National Science Foundation Fellowship, Biology, Yale University). Some of the research with the cyprinids was carried out by K. Kepes. All common and scientific names of fishes discussed are based on American Fishery Society recommendations of 1970.

References

Agostoni, A., Vergoni, C., Villa, L. (1966): Intracellular distribution of the different forms of lactic dehydrogenase. Nature 209, 1024-1025.

Ali, M.A. (1959): The ocular structure, retinomotor, and photobehavioral responses of juvenile Pacific salmon. Can. J. Zool. 37, 965-996.

Allison, W.S. (1968): Structure and evolution of triosephosphate and lactate dehydrogenases. Ann. N.Y. Acad. Sci. 151, 180-189.

American Fisheries Society (1970): A list of common and scientific names of fishes from the United States and Canada. (Special Publication No. 6), Wash. D.C.

Appella, E., Markert, C.L. (1961): Dissociation of lactate dehydrogenase into subunits with guanidine hydrochloride. Biochem. Biophys. Res. Comm. 6, 171-176.

Auerbach, S., Brinster, R.L. (1967): Lactate dehydrogenase isozymes in the early mouse embryo. Exp. Cell Res. 46, 89-92.

Bailey, G.S., Wilson, A.C. (1968): Homologies between isoenzymes of fishes and those of higher vertebrates: evidence for multiple H_4 lactate dehydrogenase in trout. J. Biol. Chem. 243, 5843-5853.

Bernstein, L., Kerrigan, M., Maisel, H. (1966): Lactic dehydrogenase isozymes in lens and cornea. Exp. Eye Research 5, 309-314.

Blanco, A., Zinkham, W.H. (1963): Lactate dehydrogenase in human testes. Science 139, 601-602.

Bonting, S.L., Bangham, A.D. (1968): On the biochemical mechanism of the visual process in: Biochemistry of the eye. pp. 493-513. Basel: Karger.

Cahn, R.D., Kaplan, N.O., Levine, L., Zwilling, E. (1962): Nature and development of lactic dehydrogenase. Science 136, 962-969.

Cohen, L.H., Noell, W.K. (1965): Relationships between visual function and metabolism. In: Biochemistry of the retina, Graymore, C.N., pp. 36-50. New York: Academic Press, 172 p.

Everse, J., Berger, R.L., Kaplan, N.O. (1970): Physiological concentrations of lactate dehydrogenases and substrate inhibition. Science 168, 1236-1238.

Fritz, P.J., Vesell, E.S., White, E.L., Pruitt, K.M. (1969): The roles of synthesis and degradation in determining tissue concentrations of lactate dehydrogenase-5. Proc. Nat'l Acad. Sci. U.S. 62, 558-565.

Fritz, P.J., White, E.L., Vesell, E.S., Pruitt, K.M. (1971): New theory of the control of protein concentrations in animal cells. Nature (New Biology) 230, 119-122.

Glässer, D., Berg, V. (1966): Lactate dehydrogenase isoenzyme pattern in the lens of cattle eyes as a function of growth and age. Hoppe-Seyler's Z. Physiol. Chem. 345, 61-64.

Goldberg, E. (1963): Lactic and malic dehydrogenase in human spermatozoa. Science 139, 602-603.

Goldberg, E. (1966): Lactate dehydrogenase of trout: hybridization in vivo and in vitro. Science 151, 1091-1093.

Goldberg, E., Cuerrier, J.P., Ward, J.C. (1969): Lactate dehydrogenase ontogeny, paternal gene activation and tetramer assembly in embryos of brook trout, lake trout, and their hybrids. Biochemical Genetics 2, 335-350.

Goldberg, E., Hawtrey, C. (1967): The ontogeny of sperm specific lactate dehydrogenase in mice. J. Exp. Zool. 164, 309-316.

Güttler, F., Clausen, J. (1967): Cellular compartmentalization of lactate dehydrogenase isoenzymes. Enzymol. Biol. Clin. 8, 456-470.

Hall, M.O., Bok, D., Baharoch, A.D.E. (1969): Biosynthesis and assembly of the rod outer segment membrane system. Formation and fate of visual pigment in the frog retina. J. Mol. Biol. 45, 397-406.

Hitzeroth, H., Klose, J., Ohno, S., Wolf, V. (1968): Asynchronous activation of parental alleles at the tissue-specific gene loci observed in hybrid trout during early development. Biochemical Genetics 1, 287-300.

Holmes, R.S., Markert, C.L. (1969): Immunochemical homologies among subunits of trout lactate dehydrogenase isozymes. Proc. Nat'l Acad. Sci. U.S. 64, 205-210.

Horowitz, J.J., Whitt, G.S. (1972): Evolution of a nervous system specific lactate dehydrogenase isozyme in fish. J. Exp. Zool. 180, 13-32.

Kaplan, N.O. (1964): Lactate dehydrogenase--structure and function, pp. 131-153. In: Subunit structure of proteins. Brookhaven Symposia in Biology 17.

Kepes, K.L., Whitt, G.S. (1972): Specific lactate dehydrogenase gene function in the differentiated liver of cyprinid fish. Genetics 71, (no. 3/part 2) 529.

Klose, J., Wolf, V., Hitzeroth, H., Ritter, H., Atkin, N.B., Ohno, S. (1968): Duplication of the LDH gene loci by polyploidization in the fish order Clupeiformes. Humangenetik 5, 190-196.

Klose, J., Wolf, V., Hitzeroth, H., Ritter, H., Ohno, S. (1969): Polyploidization in the fish family Cyprinidae, order Cypriniformes II. Humangenetik 7, 245-250.

Koen, A.L., Shaw, C.R. (1966): Retinol and alcohol dehydrogenase in retina and liver. Biochem. Biophys. Acta 128, 48-54.

Kunz, Y. (1971): Distribution of lactate dehydrogenase (and its E-isozymes) in the developing and adult retina of the guppy (*Lebistes reticulatus*). Revue Suisse de Zoologie 78, 761-776.

Kusa, M. (1966): Lactate dehydrogenase isozyme patterns of the Stickleback, *Pungitius pungitius*. Proceedings of the Japan Academy 42, 146-150.

Lush, I.E. (1970): Lactate dehydrogenase isoenzymes and their genetic variation in coalfish (*Gadus virens*) and cod (*Gadus morhua*). Comp. Biochem. Physiol. 32, 23-32.

Markert, C.L. (1962): Isozymes in kidney development. In: Hereditary, developmental, and immunologic aspects of kidney disease. J. Metcoff, ed. Northwestern University Press.

Markert, C.L. (1965): Developmental genetics. In: The Harvey Lectures. Series 59. New York: Academic Press, Inc., pp. 187-218.

Markert, C.L. (1968): The molecular basis for isozymes. Ann. N.Y. Acad. Sci. 151, 14-40.

Markert, C.L., Appella, E. (1963): Immunochemical properties of lactate dehydrogenase isozymes. In: Antibodies to enzymes: a three component system. Ann. N.Y. Acad. Sci. 103, 915-929.

Markert, C.L., Faulhaber, I. (1965): Lactate dehydrogenase isozyme patterns of fish. J. Exp. Zool. 159, 319-332.

Markert, C.L., Holmes, R.S. (1969): Lactate dehydrogenase isozymes of the flatfish, Pleuronectiformes: Kinetic, molecular, and immunochemical analysis. J. Exp. Zool. 171, 85-104.

Markert, C.L., Møller, F. (1959): Multiple forms of enzymes: Tissue, ontogenetic, and species specific patterns. Proc. Nat'l Acad. Sci. 45, 753-763.

Markert, C.L., Ursprung, H. (1962): The ontogeny of isozyme patterns of lactate dehydrogenase in the mouse. Dev. Biol. 15, 363-381.

Markert, C.L., Whitt, G.S. (1968): Molecular varieties of isozymes. Experientia 24, 977-991.

Markert, C.L., Shaklee, J.B., Whitt, G.S. (1974): The evolution of a gene. (In preparation).

Massaro, E.J., Markert, C.L. (1968): Isozyme patterns of salmonid fishes: Evidence for multiple cistrons for lactate dehydrogenase polypeptides. J. Exp. Zool. 168, 223-238.

Morrison, W.J., Wright, J.E. (1966): Genetic analysis of three lactate dehydrogenase isozyme systems in trout: Evidence for linkage of genes coding for subunits A and B. J. Exp. Zool. 163, 259-270.
Nakano, E., Hosegawa, M. (1971): Differentiation of the retina and retinal lactate dehydrogenase isoenzymes in the teleost, *Oryzias latipes*. Development, Growth and Differentiation 13, 351-357.
Nakano, E., Whiteley, A.H. (1965): Differentiation of multiple molecular forms of four dehydrogenases in the teleost, *Oryzias latipes*, studied by disc electrophoresis. J. Exp. Zool. 159, 167-179.
Newhouse, J.P., Graymore, C.N., Kissun, R.D. (1972): Co-enzyme dependence of retinal dehydrogenase in the retina of adult rat. Exp. Eye Research 14, 82-83.
Odense, P.H., Leung, T.C., Allen, T.M., Parker, E. (1969): Multiple forms of lactate dehydrogenase in the cod *Gadus morhua* L. Biochemical Genetics 3, 317-334.
Ohno, S. (1969): The preferential activation of maternally derived alleles in development of interspecific hybrids. Wistar Institute Symposium Monograph 9, 137-150.
Ohno, S. (1970): Evolution by gene duplication. Berlin-Heidelberg-New York: Springer-Verlag.
Papaconstantinou, J. (1967): Molecular aspects of lens cell differentiation. Science 156, 338-346.
Pesce, A., Fondy, T.P., Stolzenbach, F., Castillo, F., Kaplan, N.O. (1967): The comparative enzymology of lactic dehydrogenases. III. Properties of the H_4 and M_4 enzymes from a number of vertebrates. J. Biol. Chem. 242, 2151-2167.
Plagemann, P.G.W., Gregory, K.F., Wroblewski, F. (1960a): The electrophoretically distinct forms of mammalian lactic dehydrogenase. I. Distribution of lactic dehydrogenase in rabbit and human tissues. J. Biol. Chem. 235, 2282-2287.
Plagemann, P.G.W., Gregory, K.F., Wroblewski, F. (1960b): The electrophoretically distinct forms of mammalian lactic dehydrogenase. II. Properties and interrelationships of rabbit and human lactic dehydrogenase isoenzymes. J. Biol. Chem. 235, 2288-2293.
Sensabaugh, G.F. Jr., Kaplan, N.O. (1972): A lactate dehydrogenase specific to the liver of gadoid fish. J. Biol. Chem. 247, 585-593.
Shaklee, J.B., Kepes, K., Whitt, G.S. (1973): Specialized lactate dehydrogenase isozymes: The molecular and genetic basis for the unique eye and liver LDHs of teleost fish. J. Exp. Zool. 185, 217-240.
Shaw, C., Barto, E. (1963): Genetic evidence for the subunit structure of lactate dehydrogenase isozymes. Proc. Nat'l Acad. Sci. U.S. 50, 211-214.
Shows, T.B., Ruddle, F.H. (1968): Function of the lactate dehydrogenase B gene in mouse erythrocytes: Evidence for control by a regulatory gene. Proc. Nat'l Acad. Sci. 61, 574-581.
Vesell, E.S. (1965): Lactate dehydrogenase isozyme patterns of human platelets and bovine lens fibers. Science 150, 1735-1737.
Vesell, E.S., Philip, J. (1963): Isozymes of lactate dehydrogenase: sequential alterations during development. Ann. N.Y. Acad. Sci. 111, 243-257.
Wald, G. (1949): The enzymatic reduction of the retinenes to the vitamins A. Science 109, 482-483.
Wald, G. (1968): Molecular basis of visual excitation. Science 162, 230-239.
Wald, G., Hubbard, R. (1949): The reduction of retinene, to vitamin A, in vitro. J. Gen. Physiol. 32, 367-389.
Wheat, T.E., Childers, W.F., Miller, E.T., Whitt, G.S. (1971): Genetic and in vitro molecular hybridization of malate dehydrogenase isozymes in interspecific bass (*Micropterus*) hybrids. Anim. Blood Grps. Biochem. Genet. 2, 3-14.

White, A., Handler, P., Smith, E.L. (1968): Principles of Biochemistry. New York: McGraw-Hill Book Co., 1187 pp.
Whitt, G.S. (1968): Developmental genetics of lactate dehydrogenase isozymes unique to the eye and brain of teleosts. Genetics 60, 237.
Whitt, G.S. (1969): Homology of lactate dehydrogenase genes: E gene function in the teleost nervous system. Science 166, 1156-1158.
Whitt, G.S. (1970a): Developmental genetics of the lactate dehydrogenase isozymes of fish. J. Exp. Zool. 175, 1-36.
Whitt, G.S. (1970b): Directed assembly of polypeptides of the isozymes of lactate dehydrogenase. Arch. Biochem. Biophys. 138, 352-354.
Whitt, G.S., Booth, G.M. (1970): Localization of lactate dehydrogenase activity in the cells of the fish (*Xiphophorus helleri*) eye. J. Exp. Zool. 174, 215-224.
Whitt, G.S., Childers, W.F., Wheat, T.E. (1971): The inheritance of tissue specific lactate dehydrogenase isozymes in interspecific bass (*Micropterus*) hybrids. Biochemical Genetics 5, 257-273.
Whitt, G.S., Cho, P.L., Childers, W.F. (1972): Preferential inhibition of allelic isozyme synthesis in an interspecific sunfish hybrid. J. Exp. Zool. 179, 271-282.
Whitt, G.S., Horowitz, J.J. (1970): Evolution of a retinal specific lactate dehydrogenase isozyme in teleosts. Experientia 26, 1302-1304.
Whitt, G.S., Maeda, F.S. (1970): Lactate dehydrogenase gene function in the blind cave fish, *Anoptichthys jordani*, and other characins. Biochemical Genetics 4, 727-741.
Whitt, G.S., Prosser, C.L. (1971): Lactate dehydrogenase isozymes, cytochrome oxidase activity, and muscle ions of the rattail (*Coryphaenoides* sp). Am. Zoologist 11, 503-511.
Woodhead, P.M.J. (1957): Reactions of salmonid larvae to light. J. Exptl. Biology 34, 402-417.
Wright, D.A., Moyer, F.H. (1968): Inheritance of frog lactate dehydrogenase patterns and the persistence of maternal isozymes during development. J. Exp. Zool. 167, 197-206.
Wuntch, T., Chen, R.F., Vesell, E.S. (1970a): Lactate dehydrogenase isozymes: kinetic properties at high enzyme concentrations. Science 167, 63-65.
Wuntch, T., Chen, R.F., Vesell, E.S. (1970b): Lactate dehydrogenase isozymes: further kinetic studies at high enzyme concentrations. Science 169, 480-481.

Biochemical Evolution in the Genus *Xiphophorus* (Poeciliidae, Teleostei)

A. Scholl

1. Introduction

In recent years electrophoretic techniques are increasingly used to demonstrate allelic variation of proteins in conspecific animal populations. An unexpected abundance of enzyme polymorphisms has been disclosed (Stone et al., 1967; Prakash, 1969; Selander et al., 1970; Selander et al., 1971; Ayala et al., 1971; Avise and Selander, 1972). Considerable structural variation of homologous enzyme proteins is usually observed beyond the species level (Ayala et al., 1970; Johnson and Selander, 1971; Johnson et al., 1972).

Taking into account the extent of biochemical divergence which is generally seen in congeneric animal species, we have suggested (Scholl, 1973a) that fishes of the genus *Xiphophorus* might be potentially adventagous for studies in biochemical genetics, since interspecific hybrids are frequently obtained in this genus (Gerschler, 1914; Bellamy, 1922; Kosswig, 1927). Structurally different homologous proteins might serve as biochemical markers in studies on the organization and regulation of genetic material in these fishes. Since a greater portion of homologous proteins of congeneric animal species has been found to differ species-specifically with respect to electrophoretic mobilities, one might expect to find easily marker proteins in platyfish and swordtails, which will allow for such studies in biochemical genetics. We have previously surveyed the extent of structural variation of several enzyme proteins in inbred strains of platyfish and swordtails (Scholl, 1973a) and based on these investigations we have been able to investigate the inheritance of structurally different homologous proteins in interspecific platyfish hybrids (Scholl, 1973b). But is was observed that the number of suitable marker proteins for such studies is unexpectedly low in the genus *Xiphophorus*. This might indicate that platyfish and swordtails have biochemically less diverged than would be expected, based on their taxonomic classification into various species. In fact, some scientists maintain, that fishes of the genus *Xiphophorus* should be subdivided into two genera, *Xiphophorus* and *Platypoecilus* respectively (Zander, 1967), as in former systematics.

Conventional systematics usually depend to a larger extent on morphological characters. But morphological differences in congeneric animal species probably depend on relatively few genes. It can therefore not be ruled out that special selection pressure might have been exerted predominantly on this small fraction of the genome only, which mainly determined morphological characters. Morphological dissimilarities therefore may falsely imply in some cases considerable evolutionary divergence. Conversely it has already been shown that slight morphological and/or ecological dissimilarities such as exist between sibling species cannot be taken as evidence of little genetic differentiation. Sibling species of the *Drosophila willistoni* group are morphologically very similar but genetically very different (Ayala et al., 1970).

From this point of view it appears of interest to compare a randomly selected sample of loci in *Xiphophorus* species, and in particular structural variation of homologous enzyme proteins, as revealed by electrophoretic techniques appears very suitable for comparative purposes. It is of interest to comment on the extent of biochemical evolution in the genus *Xiphophorus* as compared to the extent of biochemical differentiation in other conspecific animal populations and congeneric vertebrate species.

2. Material

Nine inbred strains of platyfish and swordtails have been used. These fish originate from the collections of Gordon and have been maintained in closed stocks since their capture. Kosswig obtained them from Gordon and since 1959 they have been bred by Anders. The locality where the original material was collected is indicated for each strain (see also Kallman and Atz, 1966). The year of capture is also indicated and the approximate number of generations of inbreeding:

Platyfish[1]:

1. *Xiphophorus xiphidium* (Gordon, 1932) from Rio Purification/Rio Soto la marina, Mexico; collected in 1939, inbred for 55 generations.
2. *Xiphophorus variatus* (Meek, 1904) probably from the Panuco river system, Mexico. No precise information available about year of capture and number of generations of inbreeding.
3. *Xiphophorus maculatus* (Guenther, 1866) from Rio Jamapa, Mexico; collected in 1939, inbred for 55 generations.
4. *Xiphophorus maculatus* (Guenther, 1866) from Belize River, British Honduras; collected in 1949, inbred for 40 generations.

Swordtails:

5. *Xiphophorus helleri strigatus* (Regan, 1907) from Rio Papaloapan, Mexico; collected in 1939, inbred for 35 generations.
6. *Xiphophorus helleri guentheri* (Jordan and Evermann, 1896) from Belize River, British Honduras; collected in 1949, inbred for 25 generations.
7. *Xiphophorus helleri guentheri* (Jordan and Evermann, 1896) from Rio Lancetilla, Honduras; collected in 1951, inbred for 25 generations.
8. *Xiphophorus helleri helleri* (Haeckel, 1848) from Rio Jamapa, Mexico; collected in 1949, inbred for 25 generations.
9. *Xiphophorus montezumae cortezi* (Rosen, 1960) from the Rio Axtla/ Rio Panuco-System, Mexico; probably collected by Gordon in 1939, inbred for 35 generations.

All investigations were carried out on adult fish. Electrophoresis and enzyme staining were carried out according to routine procedures (Shaw and Prasad, 1970). However, the optimal conditions of electro-

[1]In the literature of platyfish species are also referred to as *Platypoecilus xiphidium, variatus* and *maculatus* respectively. *X. xiphidium* (Gordon, 1932) and *X. variatus xiphidium* (Rosen, 1960) are identical. For systematics of the genus *Xiphophorus* see Zander (1967).

phoresis and specific enzyme staining for this material were investigated and are indicated below.

3. Methods

Homogenization

The tissues of the fish were immediately removed upon sacrifice of the animals and were hand homogenized in three volumes of buffer (tris 0.1 M, pH 8.0 at 4°C). Homogenates were subjected to centrifugation at 20.000 g for 15 min. at 4°C. The final supernatants were immediately subjected to electrophoresis.

Electrophoresis

Vertical starch gel electrophoresis (Buchler Instruments, Inc.) was carried out in twelve percent gels (Connaught, starch-hydrolysed). Two buffer systems were routinely used: tris-citrate buffer and citrate-phosphate buffer.

Tris-citrate buffer: the gel buffer was 0.075 M tris and 0.02 M citrate (pH 7.3), while both electrode vessels were 0.3 M tris and 0.08 M citrate (pH 7.3). A voltage gradient of 4 V/cm was applied for 16 hours at 4°C.

Citrate-phosphate buffer: the gel buffer was 2.4 mM citrate and 8.5 mM Na_2HPO_4 (pH 6.0) while both electrode vessels were 8 mM citrate and 28 mM Na_2HPO_4 (pH 6.0). A voltage gradient of 3.2 V/cm was applied for 16 hours at 4°C.

Enzyme Staining

The gels were horizontally sliced prior to staining and incubated at 35°C in solutions containing reagents for specific enzyme staining. Zymograms were photographically documented on a MP-3 Polaroid camera.

The following enzymes were investigated:

Lactate Dehydrogenase (LDH; E.C. 1.1.1.27). Electrophoresis: tris-citrate or citrate-phosphate buffer. Enzyme staining: 50 ml L-lactate 1N (Boehringer), 80 mg NAD (Boehringer), 1.2 mg phenazine methosulfate (PMS)(Calbiochem) and 8 mg nitroblue tetrazolium (NBT)(Calbiochem).

Malate Dehydrogenase (MDH; E.C. 1.1.1.37). Electrophoresis: tris-citrate or citrate-phosphate buffer. Enzyme staining: 50 ml tris 0.2 M, pH 8.0, containing 60 mg L-malic acid (Mann), 100 mg NAD, 1.5 mg PMS and 7.5 mg NBT.

Malic enzyme (MOD; E.C. 1.1.1.40). (The abbreviation MOD is used according to Shows et al. 1970). Electrophoresis: citrate-phosphate buffer. Enzyme staining: 30 ml tris 0.2 M, pH 8.0, containing 250 mg L-malic acid, 7.5 mg NADP (Boehringer), 10 mg $MnCl_2 \cdot 4H_2O$, 1.5 mg PMS and 7.5 mg NBT.

Isocitrate Dehydrogenase (IDH; E.C. 1.1.1.42). Electrophoresis: tris-citrate buffer. Enzyme staining: 50 ml tris 0.1 M, pH 8.0, containing 50 mg DL-isocitric acid, trisodium salt (Mann), 30 mg NADP, 200 mg $MgCl_2 \cdot 6H_2O$, 2 mg PMS and 10 mg NBT.

6-Phosphogluconate Dehydrogenase (6-PGD; 1.1.1.44). Electrophoresis: tris-citrate or citrate-phosphate buffer. Enzyme staining: 50 ml tris 0.1 M, pH 8.0, containing 40 mg 6-phosphogluconate, trisodium-salt (Boehringer), 15 mg NADP, 0.5 mg PMS and 10 mg NBT.

Glucose-6-Phosphate Dehydrogenase (G6PD; E.C. 1.1.1.49) and Hexose-6-Phosphate Dehydrogenase (H6PD). Two separate enzymes were revealed in adult fish if gels were stained in solutions which contained glucose-6-phosphate as substrate. One of these enzymes reacted more actively with galactose-6-phosphate as substrate and is therefore referred to as hexose-6-phosphate dehydrogenase (H6PD). Electrophoresis: tris-citrate buffer. Enzyme staining: 50 ml tris 0.05 M, pH 8.0, containing 40 mg glucose-6-phosphate, disodium salt (Boehringer), 20 mg NADP, 0.6 mg PMS and 15 mg NBT. A second slice of the same gel was always incubated in a staining solution containing the same reagents as above, but 40 mg galactose-6-phosphate was used instead of glucose-6-phosphate.

Structural Variation of Enzymes

The number of animals of each strain analyzed for electrophoretic variation of these enzymes is indicated in Table 1. Enzyme tissue-specificity was investigated on a few specimens of each strain. Usually the tissue with highest activity of a particular enzyme was then used to score structural variation of this enzyme, as indicated in Table 2. The genetic bases of allozymic variation were verified by progeny studies. If homologous enzymes of different species differed in electrophoretic mobilities, interspecific hybrids of such species were analyzed in order to determine, whether variation results from epigenetic factors or whether it is genetically determined, i.e. results from variation at genetic loci which code for structurally different homologous enzymes.

Designation of Enzymes

Enzyme names are abbreviated in capital letters (cf. 6-PGD for 6-phosphogluconate dehydrogenase) as indicated under "enzyme staining" for each enzyme. Additional letters which are used, refer to the subunit composition of isoenzymes (cf. E_4-LDH is a homotetrameric LDH-isoenzyme which is composed of four LDH-E-polypeptides. A_1B_1-MDH is a heterodimeric MDH-isoenzyme composed of one MDH-A and one MDH-B-polypeptide). Designations for the polypeptide types (cf. A-, B- and E-polypeptides of LDH) are those which are accepted in the literature. Electrophoretically different homologous enzymes are designated with an index for the relative electrophoretic mobility as compared to the most anodally migrating enzyme (cf. 6-PGD^{85} migrates under our conditions of electrophoresis on zymograms 85% of the distance of the most anodally migrating 6-PGD^{100}).

Designation of Loci and Alleles

To designate a locus, enzyme names are abbreviated as before, but only the first letter is capital (cf. Ldh; 6-Pgd). Isoenzyme loci are numbered according to decreasing anodal mobility of their protein products (cf. Mdh-1, Mdh-2). Alleles are alphabetically designated according to decreasing anodal mobility of their protein products (cf. Ldh-1^{a}, Ldh-1^{b}).

In interspecific hybrids of the genus *Xiphophorus* genes for structurally different homologous enzymes behave like alleles (see Scholl, 1973b). For this reason we have made no distinction between genes for structurally different homologous enzymes in the *Xiphophorus* species and alleles at one locus (cf. Ldh-1 is monomorphic in *X. helleri*

Table 1. Number of animals of nine inbred strains of *Xiphophorus* species analyzed for electrophoretic variation of enzymes

	Platyfish				Swordtails					Total
Enzyme	1. *X. xiphidium*	2. *X. variatus*	3. *X. maculatus* (British Honduras)	4. *X. maculatus* (Mexico)	5. *X. helleri strigatus*	6. *X. helleri guentheri* (British Honduras)	7. *X. helleri guentheri* (Honduras)	8. *X. helleri helleri*	9. *X. montezumae cortezi*	
LDH	10	16	45	33	19	28	62	28	42	283
MDH	17	19	9	22	19	17	33	19	25	180
MOD	20	14	20	20	5	15	20	20	20	154
IDH	20	20	20	20	20	20	20	20	20	180
G6PD/H6PD	10	18	20	69	20	14	14	25	21	211
6-PGD	39	25	25	62	15	25	25	21	20	257

Table 2. Enzyme tissue specificity in the genus *Xiphophorus*

Enzyme	Locus	skeletal-muscle	liver	brain	testes	eye
monomolecular forms of enzymes:						
MOD	Mod	X	+	?	?	?
IDH	Idh	X	+	?	?	?
6PGD	6Pgd	±	+	±	+	X
multiple molecular forms of enzymes (isoenzymes):						
E_4-LDH	Ldh-1	O	O	±	±	X
A_4-LDH	Ldh-2	+	O	+	O	X
B_4-LDH	Ldh-3	±	+	±	+	X
A_2-MDH	Mdh-1	X	±	O	O	±
B_2-MDH	Mdh-2	X	+	±	±	±
G6PD	G6pd	±	X	±	±	+
H6PD	H6pd	O	X	O	±	O

Symbols: X = high activity, tissue usually taken to score locus
+ = high activity but usually not taken to score locus
± = minor activity
O = no activity
? = not tested

guentheri from British Honduras, but the locus is referred to as Ldh-1^b because an allele which codes for a more anodally migrating homologous enzyme is found in other species). (See Table 3).

4. Results

Enzyme Tissue-specificity (Table 2)

Tissue-specific differences with respect to enzyme activities are observed for each enzyme. Malic enzyme, isocitrate dehydrogenase and 6-phosphogluconate dehydrogenase appear as monomolecular forms in each tissue and have identical electrophoretic mobility in all tissues where they are observed. Multiple molecular forms (isoenzymes) which can be attributed to the codominant action of isoenzyme loci and which appear in tissue-specific patterns of distribution are observed for lactate dehydrogenase, malate dehydrogenase and glucose-6-phosphate dehydrogenase (G6PD and H6PD).

Isoenzymes

Lactate Dehydrogenase (Fig. 1). Lactate dehydrogenase exhibits a complex pattern of tissue-specifically distributed isoenzymes, as is usually found in teleosts and other vertebrates (Markert, 1968) and

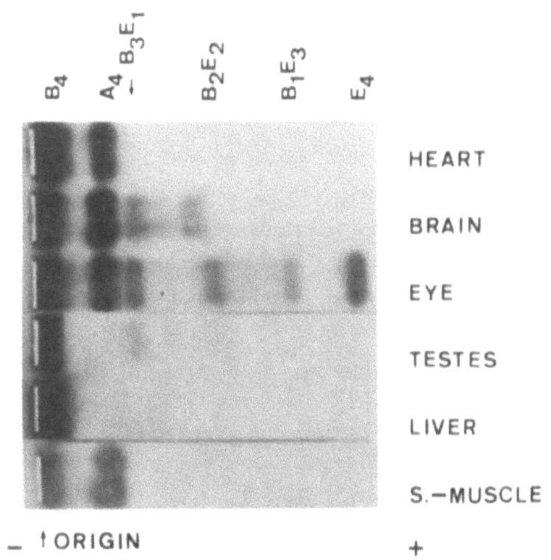

Fig. 1. LDH-zymogram of tissues of the platyfish *X. maculatus*. The subunit composition of the LDH-isozymes is indicated on top of the Figure

which results from the codominant action of separate isoenzyme loci and from the tetrameric structure of enzymatically active LDH-molecules. The tissue-specific LDH-isoenzyme patterns are identical in all *Xiphophorus* species, therefore only *X. maculatus* will be discussed.

The LDH-isoenzyme of platyfish and swordtails can be attributed to the codominant action of three Ldh-loci. All tissues express one isoenzyme which is close to the origin on zymograms. In addition a more anodally migrating isoenzyme is observed in most tissues. These two isoenzymes are designated B_4- and A_4-LDH, to indicate that they probably arise from two Ldh-loci coding for B- and A-polypeptides respectively, which are found in all vertebrates (Markert, 1968), and to indicate, that they are probably homotetrameric isoenzymes. Heterotetrameric isoenzymes which might be expected intermediate in mobility between the homopolymeric B_4- and A_4-LDH, are not observed. Subunit assembly of B- and A-polypeptides apparently is restricted in these fish like in other teleosts (Markert and Faulhaber, 1965; Rosenberg, 1971) to favor the formation of homotetrameric A_4- and B_4-LDH.

The more complex pattern of LDH isoenzymes in extracts from eyes can be attributed to a third Ldh-locus, coding for the LDH-E-polypeptide which is typically found in the retina of teleosts (Whitt, 1968, 1969, 1970a; Whitt and Horowitz, 1970). The most anodally migrating isoenzyme appears to represent the homotetrameric E_4-LDH. The three isoenzymes of minor anodal mobility are heterotetrameric LDH-isoenzymes which are most probably composed of B- and E-polypeptides, since these bands are also seen at very low activity in homogenates of testes, a tissue where the locus which codes for the A-polypeptide does not function. This interpretation of the genetic basis of LDH-isoenzymes in the genus *Xiphophorus* corresponds to the conclusions of Whitt and Booth (1970) which are based on immunological investigations. Since the isoenzyme loci are numbered according to decreasing anodal mobility of their protein products, Ldh-1 is the locus of the E-polypeptide, Ldh-2 the locus of the A-polypeptide and Ldh-3 the locus of the B-polypeptide.

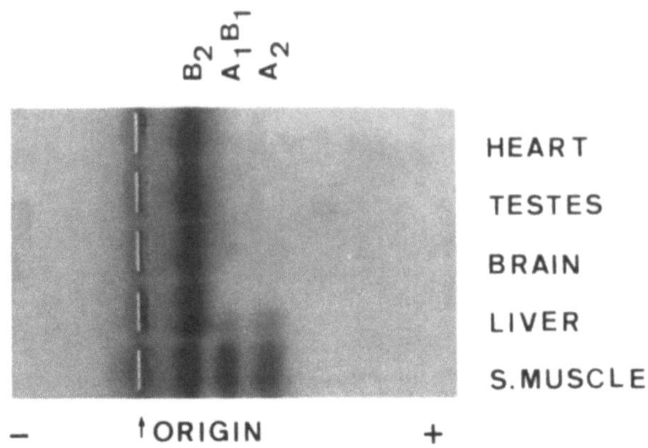

Fig. 2. MDH-zymogram of tissues of the platyfish *X. maculatus*. The subunit composition of the MDH-isozymes is indicated on top of the Figure

Malate Dehydrogenase (Fig. 2). Less complex patterns are observed for the isoenzymes of malate dehydrogenase as might be expected from the dimeric structure of enzymatically active MDH molecules (Bailey et al., 1969; Gerding and Wolfe, 1969). The tissue-specific patterns of MDH-isoenzymes are identical for all *Xiphophorus* species, therefore only *X. maculatus* will be discussed. Most tissue contain a single band of MDH activity only, but three widely spaced bands are seen in extracts from skeletal muscle. It is concluded that all three bands represent supernatant MDH since none of these bands stains if mitochondrial preparations are subjected to electrophoresis. Furthermore, none of these bands stains if NAD in the staining solution is replaced by NADP, thus ruling out the possibility that any of these bands might be MOD rather than MDH.

The MDH-isoenzymes in the genus *Xiphophorus* can be attributed to the codominant action of two Mdh-loci, Mdh-1 and Mdh-2, coding for MDH-polypeptides A and B respectively. The intermediate isoenzyme in skeletal muscle most probably represents a heterodimer which is composed of an A- and a B-subunit. Two codominant loci for supernatant MDH have previously been found in teleosts by other investigators (Whitt, 1970b; Wheat and Whitt, 1971; Wheat et al., 1972). Our interpretation of the genetic basis of MDH isoenzymes agrees with these investigators.

Glucose-6-Phosphate Dehydrogenase (G6PD) and Hexose-6-Phosphate Dehydrogenase (H6PD). Adult specimens (Fig. 3a and 3b). Two G6PD isoenzymes are resolved in liver extracts of adult specimens. The electrophoretic mobility of both isoenzymes is identical in all species. *X. maculatus* only will be discussed. The more anodally migrating enzyme is very active and is also seen in other tissues, however, it is most active in liver tissue. The second isoenzyme stains very faintly and is usually not seen in tissues other than liver. This second band stains always very active, if slabs of the same gel are incubated in staining solutions which contain galactose-6-phosphate instead of glucose-6-phosphate. The more anodally migrating isoenzyme reacts also with galactose-6-phosphate, though less active than with glucose-

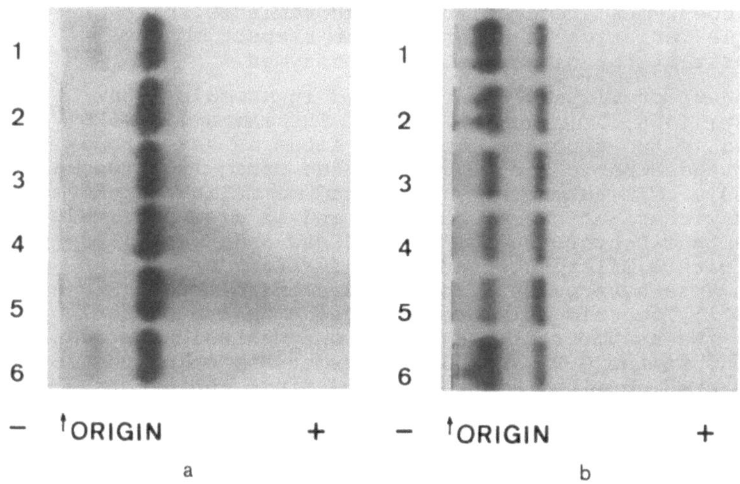

Fig. 3a and b. G6PD-isoenzymes in liver homogenates of the platyfish *X. maculatus*. a) and b) are duplicate slices of the same gel. Numbers refer to identical extracts in a) and b); a) was stained with glucose-6-phosphate as substrate; b) was stained with galactose-6-phosphate as substrate. The more anodally migrating isoenzyme which reacts more actively with glucose-6-phosphate, is referred to as G6PD in the text. The isoenzyme more close to the origin is designated H6PD, because it reacts more actively with galactose-6-phosphate

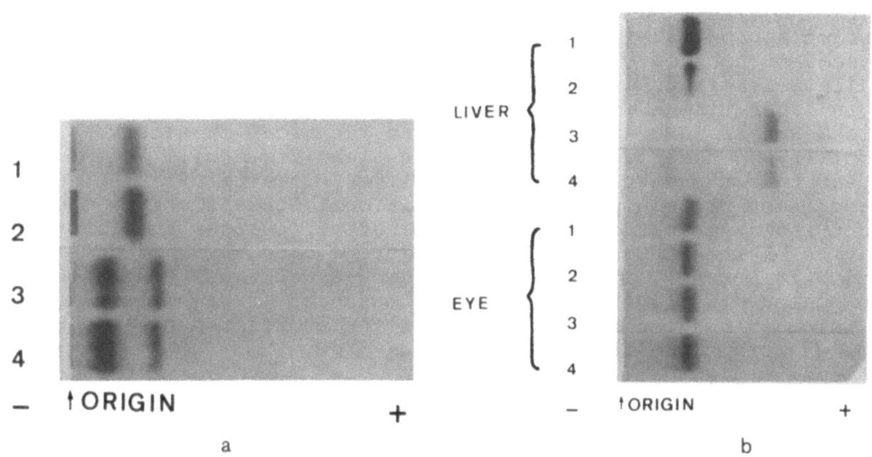

Fig. 4a and b. G6PD-isoenzymes of juvenile specimens of the platyfish *X. maculatus*. a) Extracts from liver. 1 and 2 are juvenile specimens (2.8 cm), 3 and 4 are sexually mature platyfish (>3.5 cm). Galactose-6-phosphate was used as substrate. b) Extracts from liver and eye. Numbers refer to extracts of the same animals. 1 and 2 are adult specimens (>3.5 cm), 3 and 4 are juvenile fish (2.3 cm). Glucose-6-phosphate was used as substrate

6-phosphate. Both isoenzymes are consistently scorable in spite of the fact that considerable variation is observed with respect to the amount of stain deposit at the sites of both isoenzymes.

Two G6PD-isoenzymes have previously been observed in mammals (Shaw and Barto, 1965; Shaw, 1966; Ohno et al., 1966). The mammalian G6PD isoenzyme are encoded in two separate loci, one locus is sex-chromosome linked (Kirkman and Hendrickson, 1963) and the other is autosomal (Shaw and Barto, 1965). The autosomally inherited mammalian isoenzyme reacts more actively with galactose-6-phosphate and is also referred to as "Hexose-6-Phosphate Dehydrogenase, H6PD" (Shaw and Koen, 1968). Two G6PD isoenzymes with similar substrate specificities as the mammalian G6PD-isoenzymes have recently been observed in trout (Stegeman and Goldberg, 1971), genetic polymorphism was demonstrated for H6PD of brook trout. We assume that the two isoenzymes which we describe, are homologous to the G6PD and H6PD of trout and other vertebrates and that they are inherited by two genes. Since the electrophoretic mobilities of both enzymes are the same in all species studied, we have not been able to definitely prove this assumption by genetic analysis in breeding experiments. We refer to the more anodally migrating isoenzyme as glucose-6-phosphate dehydrogenase (G6PD) and to the second band as hexose-6-phosphate dehydrogenase (H6PD).

Juvenile specimens of *X. maculatus* (Fig. 4a and 4b). A more complex pattern of glucose-6-phosphate dehydrogenase isoenzymes is seen in juvenile platyfish. At present some 50 juvenile *X. maculatus* only have been analyzed. A major portion of these animals revealed a different pattern of G6PD isoenzymes in liver extracts. Usually neither of the above mentioned G6PD nor H6PD is seen in liver extracts of sexually immature specimens. Instead a third isoenzyme is observed (Fig. 4a) which migrates intermediate between G6PD and H6PD on zymograms. In very young individuals a forth isoenzyme is observed which migrates more anodally than G6PD (Fig. 4b). Both isoenzymes of juvenile specimens react approximately equally well with glucose-6-phosphate and galactose-6-phosphate. There is no evidence that these isoenzymes represent allelic variants at the G6pd- or H6pd-locus, since they are only found in liver tissue of juvenile specimens, whereas other tissues of these fish express G6PD and H6PD which are identical in electrophoretic mobility to the isoenzymes of adult fish (Fig. 4b). Rather it does appear that the G6PD isoenzymes in liver extracts of juvenile specimens represent protein products of additional G6PD-isoenzyme loci. The different isoenzyme patterns scored may reflect changes in differential activities of the G6PD-isoenzyme loci during ontogeny. If this interpretation is correct, it may be of particular interest that such changes are observed in liver tissue at the onset of sexual maturation, since NADP which is reduced by the action of the pentose shunt, may be utilized for steroid biosynthesis.

Intraspecific and Interspecific Variation of Proteins

Invariant proteins. Intraspecific or interspecific variation in electrophoretic mobilities is only observed for a small fraction of the proteins investigated. A_4-LDH, B_4-LDH, IDH, G6PD and H6PD are invariant in their electrophoretic mobilities in all animals investigated.

Variant proteins. Variation in electrophoretic mobility is observed for E_4-LDH, A_2- and B_2-MDH and 6-PGD and can be attributed to allelic variation at the corresponding loci Ldh-1, Mdh-1, Mdh-2 and 6Pgd respectively, based on progeny tests and on examination of interspecific hybrids. In the genus *Xiphophorus* genes for homologous enzymes behave like alleles (see Scholl, 1973b).

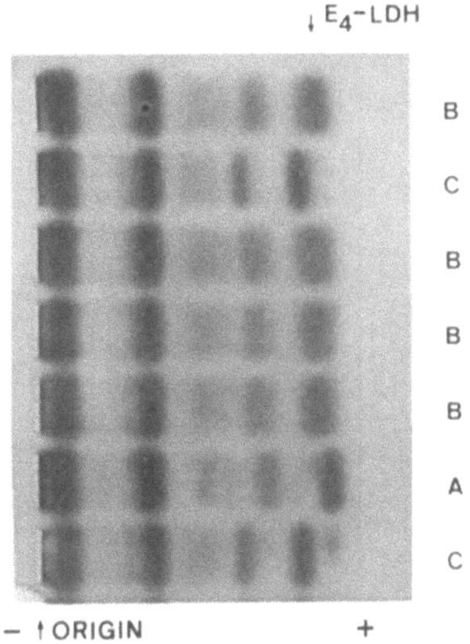

Fig. 5. LDH-E-polymorphism in the *X. helleri guentheri* eye. 7 specimens of *X. helleri guentheri* from Rio Lancetilla, Honduras, were used. The arrow indicates the E_4-LDH (compare with Fig. 1). Capital letters refer to the symbols used in the text to designate a particular phenotype. Electrophoretic mobility of the E_4-LDH of phenotype A is identical to the mobility of E_4-LDH in *X. maculatus* (Fig. 1)

E_4-LDH (Fig. 5). Three different phenotypes of E_4-LDH which are observed in the Honduras-strain of *X. helleri guentheri* are shown in Fig. 5 and are designated A, B and C respectively. Animals with exclusively phenotype B resulted from mating a phenotype A female to a phenotype C male, however in the F_2-generation phenotypes A, B and C were observed at an approximate ratio of 1:2:1. The data indicate that the three enzyme phenotypes result from allelic polymorphism at the Ldh-1 locus, two alleles Ldh-1^a and Ldh-1^b are involved. Animals with LDH-phenotype B (Fig. 5) are heterozygous at the Ldh-1 locus. The relative mobilities of the homopolymeric E_4-LDHs are 100 for phenotype A and 90 for phenotype C.

Dimorphism at the Ldh-1 locus is also found in *X. montezumae cortezi* (Table 3). Protein products of Ldh-1 alleles of *X. montezumae cortezi* and *X. helleri guentheri* are electrophoretically identical. Therefore it is concluded that this polymorphism at the Ldh-1 locus is phylogenetically old and existed prior to the divergence of swordtails into the species *X. helleri* and *X. montezumae*.

All other strains are monomorphic at the Ldh-1 locus and have either E_4-LDH^{100} or E_4-LDH^{90} (Table 3).

A_2- and B_2-MDH (Fig. 6). Each two electrophoretically different proteins are observed for A_2- and B_2-MDH and for the A_1B_1-heterodimeric MDH (Fig. 6). The homodimeric malate dehydrogenases have the relative electrophoretic mobilities of 100 and 80 for the A_2-MDH and 100 and 70 for the B_2-MDH. As is observed from an examination of *X. maculatus*/*X. montezumae cortezi* interspecific hybrids, variation in electrophoretic mobility of the supernatant MDHs results from variation at

Table 3. Allele frequencies and genic variability in the genus *Xiphophorus*

Enzyme	Relative Mobility	Locus Allele	Platyfish 1 *X.xiph.*	2 *X.var.*	3 *X.mac.* BH	4 *X.mac.* M	Swordtails 5 *X.h.s.*	6 *X.h.g.* BH	7 *X.h.g.* H	8 *X.h.h.*	9 *X.m.c.*
E_4-LDH	100	Ldh-1^a	1.0	1.0	1.0	1.0	1.0	-	0.23	1.0	0.31
	90	-1^b	-	-	-	-	-	1.0	0.77	-	0.69
A_4-LDH		Ldh-2				invariant					
B_4-LDH		Ldh-3				invariant					
A_2-MDH	100	Mdh-1^a	1.0	1.0	1.0	1.0	1.0	1.0	0.68	1.0	-
	80	-1^b	-	-	-	-	-	-	0.32	-	1.0
B_2-MDH	100	Mdh-2^a	1.0	1.0	1.0	1.0	1.0	1.0	1.0	1.0	-
	70	-2^b	-	-	-	-	-	-	-	-	1.0
MOD		Mod				invariant					
IDH		Idh				invariant					
G6PD		G6pd				invariant					
H6PD		H6pd				invariant					
6PGD	100	6Pgd-1^a	1.0	-	-	-	-	-	-	-	-
	93	-1^b	-	1.0	-	-	1.0	1.0	1.0	1.0	1.0
	85	-1^c	-	-	1.0	1.0	-	-	-	-	-

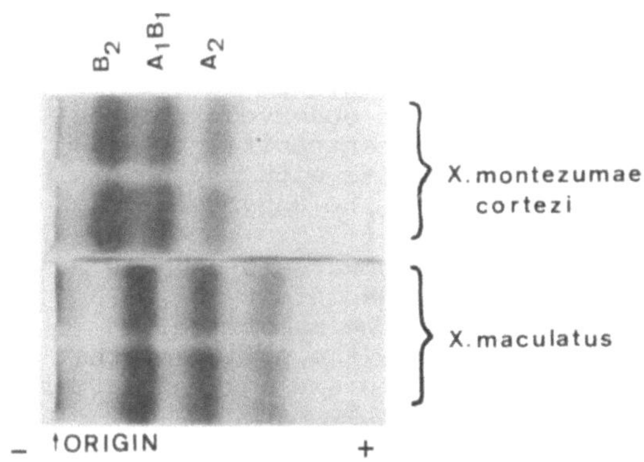

Fig. 6. Supernatant MDH-isoenzymes in extracts of skeletal-muscle of *X. montezumae cortezi* and *X. maculatus*. The subunit composition of the MDH-isoenzymes is indicated on top of the Figure

the Mdh-1 and Mdh-2 loci. Two alleles a and b are recognized at both loci (Table 3). Most strains are monomorphic for $Mdh\text{-}1^a$ and $Mdh\text{-}2^a$. *X. montezumae cortezi* is monomorphic for $Mdh\text{-}1^b$ and $Mdh\text{-}2^b$. The Honduras strain of *X. helleri guentheri* is dimorphic at the Mdh-1 locus, but monomorphic at the Mdh-2 locus for the a-allele (Table 3).

6-PGD (Fig. 7a and 7b). If gels are stained for 6-PGD a single band of activity is revealed and this band has identical electrophoretic

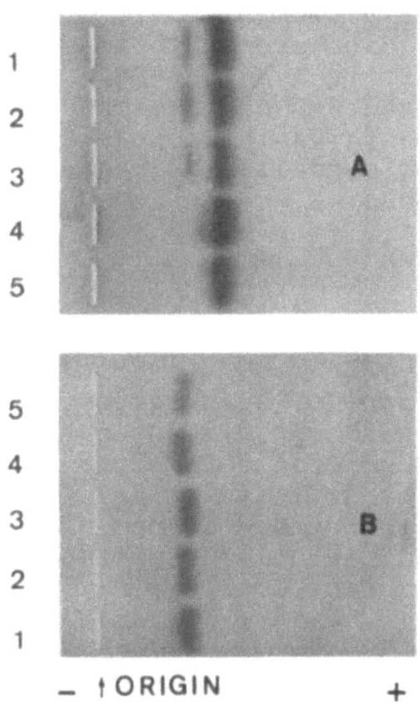

Fig. 7a and b. 6-PGD zymogram of liver tissue of the swordtail *X. helleri guentheri*. a) and b) are duplicate slices of the same gel. Numbers refer to identical extracts in a) and b); a) was stained with 6-phosphogluconate; b) was stained with glucose-6-phosphate. Comparison of both gels shows that the enzyme band of minor activity in a) is G6PD which also reacts if 6-phosphogluconate is used as substrate

mobility in all tissues. On some occasions a second band of minor anodal mobility is seen in liver tissue. This second band however does not appear to represent 6-PGD since it has identical electrophoretic mobility with G6PD and since it is only observed in specimens with very high levels of G6PD. Furthermore, as is observed from the examination of *X.maculatus/X.helleri guentheri* interspecific hybrids (see next chapter) this band does not share subunits with the generally observed 6-PGD band. Therefore we assume that the enzyme band of minor anodal mobility represents G6PD rather than 6-PGD.
Stain deposit at the site of G6PD if 6-phosphogluconate is used as substrate may result from overlapping substrate specificity of G6PD or from contaminations of glucose-6-phosphate in the substrate. According to specific information from Boehringer, the purity of the 6-phosphogluconate substrate used is high enough to exclude the latter alternative.

Three electrophoretically different 6-phosphogluconate dehydrogenases are observed in the genus *Xiphophorus*. They have the relative electrophoretic mobilities of 100, 93 and 85 respectively (Table 3). 6-PGD^{100} is only found in *X. xiphidium*, 6-PGD^{85} is found in both strains of *X. maculatus* whereas all other strains have 6-PGD^{93}.

Variation in electrophoretic mobility of 6-PGD is not due to epigenetic factors such as substrate or coenzyme binding which might influence the mobility of the enzyme, but rather can be attributed to variation at the 6-Pgd locus, as is observed from an examination of interspecific hybrids. Interspecific *X. xiphidium/X. maculatus* hybrids (see Scholl, 1973b) or *X. helleri guentheri/X. maculatus* F_1-hybrids exhibit three 6-PGD allozymes (hybrid phenotype). In addition to the 6-PGD of *X. xiphidium* and *X. maculatus* or *X. helleri guentheri* and *X. maculatus*, a third 6-PGD allozyme is seen in the interspecific hybrid. This third allozyme migrates on zymograms intermediate between the 6-PGD bands of both parents, as would be expected from the dimeric structure of 6-PGD (Carter et al., 1969; Ritter et al., 1969). Two 6-PGD phenotypes are observed at an approximate 1:1 ratio in back-cross hybrids. In *X. xiphidium/X. maculatus* back-cross hybrids either hybrid-6-PGD phenotype and 6-PGD^{100} or hybrid-6-PGD phenotype and 6-PGD^{85} are observed depending on whether *X. xiphidium* or *X. maculatus* is used as the recurrent parent. This would be expected from mendelian segregation of two codominant alleles of an autosomal locus.

5. Discussion

Genetic Interpretation of Isoenzyme Patterns

The genetic interpretation of isoenzyme patterns scored is in agreement with conclusions of other investigators and has been discussed for each enzyme. Additional isoenzyme loci are suspected for enzyme bands of minor activity or for additional enzyme bands in juvenile specimens for IDH, G6PD and possibly LDH. At present there is not sufficient evidence, whether these additional bands actually represent the enzymes they stain for, or whether they result from overlapping substrate specificities of other enzyme proteins.

Extent of Polymorphism in the *Xiphophorus* Species

A low degree of polymorphism may be expected for this material, since inbred strains of fish were used. In fact, seven out of nine strains are monomorphic at all loci, one strain (*X. montezumae cortezi*) is dimorphic at one locus (Ldh-1) and one strain (*X. helleri guentheri* from Honduras) is dimorphic at two loci (Ldh-1 and Mdh-1), as is indicated in Table 3.

This very low degree of polymorphism is probably attributable to inbreeding, since considerable polymorphism is usually observed at enzyme loci in natural populations. For comparative purposes some representative examples from the literature are indicated in Table 4, A. Avise and Selander (1972) observed an average number of 3.5 alleles per locus in the Mexican tetra (*Astyanax mexicanus*). These data are based on an examination of protein products from 17 loci in 257 individuals collected at six different localities. Some of the alleles however occurred at very low frequencies and were not observed in all populations. It does appear that the extent of polymorphism in *Astyanax mexicanus* is rather high. But considerable polymorphism has also been observed by McKinney et al.,(1972) in the side-blotched lizard (*Uta stansburiana*) (412 individuals from 17 local populations) and by Selander et al.,(1971) in the old-field mouse (*Peromyscus polionotus*) (745 individuals from 30 different localities).

However, in all these examples the average number of alleles per locus is much lower, if only a single population is analyzed (Table 4, B). In spite of the fact that some alleles are observed at low frequencies and not at all in every population, it does not appear that the average number of alleles per locus per population (Table 4, A) as opposed to the average number of alleles per locus per species (Table 4, B) is lower because of smaller sample sizes for each population. But rather this difference indicates the extent of biochemical differentiation at the species level, since it is also observed for the more common alleles that their frequencies may significantly change between several populations.

Since our *Xiphophorus* strains are each derived from a single local population, it is more appropriate, to compare the extent of polymorphism in these strains to the average number of alleles per locus per population and it still does appear that inbreeding of this material has been very effective.

Comparison of *Xiphophorus* Strains Derived from Conspecific Populations

Highly inbred material may not be very suitable to discuss aspects of biochemical evolution. But some information on the degree of biochemical diversity may still be gained from inbred material. If natural populations are usually very polymorphic, as is exemplified in Table 4, one might expect that inbred strains which are derived from a single population will not be identical, since most probably different alleles at some loci may have become monomorphic by inbreeding. The order of magnitude of biochemical differences will be greater, if strains are derived from several conspecific populations since natural populations already differ with respect to allele frequencies at some loci. A suitable example of the order of magnitude of such differences is provided for by an examination of 17 protein loci in cave populations of *Astyanax mexicanus*. These fish are comparable to inbred strains, since the small cave populations were probably isolated in the Pleistocene. While surface populations of the *Astyanax* are highly polymorphic (Table 4), two cave populations were almost exclusively monomorphic and they were identical at only 13 out of 17 loci investigated (Avise and Selander, 1972).

Each two strains of conspecific populations of *Xiphophorus helleri guentheri* and *Xiphophorus maculatus* were analyzed in this investigation. Both strains of *X. helleri guentheri* are only slightly different. The British Honduras strain is monomorphic at all loci whereas the strain from Honduras is dimorphic at two loci (Table 3). Each one allele of these dimorphic loci is identical to an allele in the British

Table 4. Average number of alleles per locus in several vertebrate species

Organism	Number of loci examined	A Average number of alleles per locus per species	B Average number of alleles per locus per population	Range	Source
Astyanax mexicanus	17	3.5	2.13	1.71-2.71	Avise and Selander, 1972
Uta stansburiana	18	3.4	1.38[a]	1.00-2.06	McKinney et al., 1972
Peromyscus polionotus	32	2.4	1.60[b]	-	Selander et al., 1971

[a] calculated from the data of McKinney et al., 1972
[b] calculated from the data of Selander et al., 1971

Honduras strain. The two *X. maculatus* strains are biochemically identical. This result is somewhat unexpected, but it parallels the very low degree of interspecific variation in the genus *Xiphophorus*.

Interspecific Variation of Proteins in the Genus *Xiphophorus*

As is demonstrated in Table 3, each two strains of the *Xiphophorus* species differ usually at one enzyme locus only, if they differ at all. This observation is very unexpected since it was estimated (Shaw, 1970) that closely related species which occur in the same genus differ at about 50 - 80% of their genes. This estimate has now been confirmed by electrophoretic studies on structural variation of proteins in congeneric animal species of both vertebrates (Johnson and Selander, 1971; Johnson et al., 1972; Smith et al., 1972; Webster et al., 1972) and invertebrates (Hubby and Throckmorton, 1965, 1968; Ayala et al., 1970, 1971; Ayala and Powell, 1972). We have calculated coefficients of genetic similarity for the *Xiphophorus* strains (Table 5), which are based on biochemical data. These coefficients indicate the fraction of alleles that appear to be identical in any two species and it is observed that they range between 0.8 and 1.0 for most of the strains. The order of magnitude of these coefficients of genetic similarity in the genus *Xiphophorus* is approximately the same as usually observed in conspecific animal populations and it is considerably higher than coefficients calculated for other congeneric vertebrate and invertebrate species (Table 6)[2].

There might be a simple explanation why we have not been able to demonstrate a higher degree of structural variation of proteins in the genus *Xiphophorus*. The sample of loci examined by us may comprise to a larger extent proteins whose net-charges are critical for enzyme function. Mutations which lead to an altered net-charge and thereby to electrophoretically recognizable structural differences, might not be tolerated at these loci. Esterases which have not been studied by us are usually very variable (Avise and Selander, 1972). Ayala and Powell (1972) observed in the *Drosophila willistoni* group that enzymes involved in glucose metabolism were only about half as polymorphic as other proteins. We have therefore compared the extent of structural variation of homologous proteins in the genus *Xiphophorus* with data on structural variation of these proteins in other conspecific vertebrate populations and congeneric vertebrate species (Table 7)[3].

---

[2]It is well appreciated that Rogers' coefficients of genetic similarity which have been calculated by Selander and collaborators and which are indicated in Table 6 are not the same measure of genetic relatedness which we have used and which is indicated in Table 5 for the *Xiphophorus* species (see Rogers, 1972). But Johnson and Selander (1971, page 399) "have determined that, for comparisons of pairs of populations, values of Rogers' coefficients of genetic similarity are generally rather similar to those obtained by calculating the proportion of loci at which all alleles are shared".

[3]No effort has been made to ascertain whether the isoenzyme loci actually code for homologous isoenzymes. In the protein surveys referred to, isoenzyme loci are designated in the order of decreasing anodal mobility. But the relative electrophoretic mobilities of homologous isoenzymes (cf. the A_4-LDHs or the B_4-LDHs) may be reversed if phylogenetically distant species are compared (Markert and Faulhaber, 1965). Thus, in some species Ldh-1 may be the locus of the A-polypeptide, but in other species the locus of the B-polypeptide. Furthermore, some isoenzymes may be unique to certain classes of vertebrates or even within one class

As is observed from comparison of Table 4 and Table 7, the average number of alleles found at the loci which have been studied by us is not very much different from the average number of alleles which are observed for a larger sample of loci. Therefore it does not appear that the enzymes which we have examined are particularly conservative in other vertebrates.

For these reasons we conclude that platyfish and swordtails have biochemically less evolved than would be expected based on the fact that they are systematically subdivided into several species. There is now a conflicting situation that platyfish and swordtails are morphologically well differentiated but biochemically they differ like conspecific animal populations rather than congeneric animal species.

It has previously been shown that morphological similarities do not imply little genetic differentiation. Ayala et al., (1970) observed that sibling species of the *Drosophila willistoni* group are morphologically very similar but genetically very different. One may speculate that the opposite will also be found, little genetic differentiation in spite of major morphological dissimilarities. We feel that our results indicate this situation for the genus *Xiphophorus*. Relatively little genetic differentiation for this genus may also be inferred from the wellknown fact that fertile interspecific hybrids are obtained from platyfish and swordtails by crossing in the aquarium or by artificial insemination.

Morphological dissimilarities may depend on a relatively small fraction of the genome. They may evolve in conspecific animal populations if special selection pressure operates at the genes which determine morphological characters, consequently morphological differentiation may falsely imply considerable evolutionary divergence in some cases. We feel that our data suggest that such a situation existed at some time in the evolution of the genus *Xiphophorus*.

6. Summary

1. Electrophoretic variation of enzyme proteins which are encoded in 10 enzyme loci is compared in platyfish and swordtails. Nine inbred strains which belong to seven species or subspecies are investigated.

2. A very low degree of structural variation of proteins is observed as compared to other congeneric animal species.

3. The data indicate little biochemical differentiation in the genus *Xiphophorus*.

4. Since the *Xiphophorus* species are morphologically well differentiated, it is suspected that special selection pressure may have operated at some time during the evolution of the genus *Xiphophorus* on a relatively small fraction of the genome which mainly determines morphological characters.

Footnote 3 (p. 293) continued
to certain major groups, like the LDH-E-polypeptide of teleosts (Horowitz and Whitt, 1972). Fish have usually two isoenzyme loci for supernatant MDH (Wheat et al., 1972) whereas one locus only is observed in higher vertebrates (Karig and Wilson, 1971). But all isoenzyme loci referred to code for supernatant forms of enzymes, loci for mitochondrial MDH and mitochondrial IDH are not considered.

Table 5. Coefficients of genetic similarity in platyfish and swordtails. (Calculated from the data in Table 3)

	Platyfish				Swordtails				
	1. *X.xiph.*	2. *X.var.*	3. *X.mac.*	4. *X.mac.*	5. *X.h.s.*	6. *X.h.g.*	7. *X.h.g.*	8. *X.h.h.*	9. *X.m.c.*
Platyfish:									
1. *X. xiphidium*	-	0.90	0.90	0.90	0.90	0.80	0.81	0.90	0.63
2. *X. variatus*		-	0.90	0.90	1.00	0.90	0.91	1.00	0.73
3. *X. maculatus* (British Honduras)			-	1.00	0.90	0.80	0.81	0.90	0.63
4. *X. maculatus* (Mexico)				-	0.90	0.80	0.81	0.90	0.63
Swordtails:									
5. *X. helleri strigatus*					-	0.90	0.91	1.00	0.73
6. *X. helleri guentheri* (British Honduras)						-	0.95	0.90	0.77
7. *X. helleri guentheri* (Honduras)							-	0.91	0.80
8. *X. helleri helleri*								-	0.73
9. *X. montezumae cortezi*									-

Table 6. Rogers' coefficients of genetic similarity in conspecific animal populations and congeneric animal species

Organism	Number of loci	Range of coefficients	Source
A. Conspecific animal populations:			
Dipodomys (11 species)	18	0.92 - 1.00	Johnson and Selander (1971)
Sigmodon[a] (2 species)	23	0.98 - 1.00	Johnson et al. (1972)
Peromyscus[a] *polionotus*	32	0.82 - 0.99	Selander et al. (1972)
Anolis[a] (4 species)	25	0.69 - 0.82	Webster et al. (1972)
Astyanax mexicanus	17	0.77 - 0.98	Avise and Selander (1972)
B. Congeneric animal species:			
Dipodomys (11 species)	18	0.31 - 0.89	Johnson and Selander (1971)
Sigmodon[a] (2 species)	23	0.76 - 0.77	Johnson et al. (1972)
Peromyscus[a] (2 species)	32	0.31 - 0.33	Smith et al. (1972)
Anolis[a] (4 species)	25	0.16 - 0.29	Webster et al. (1972)

[a]data cited from Avise and Selander (1972)

Table 7. Number of alleles at several enzyme loci in conspecific vertebrate populations

	Astyanax mexicanus[a]	*Uta stansburiana*[b]	*Peromyscus polionotus*[c]
Ldh-1	3	5	4
Ldh-2	2	6	1
Ldh-3	-	-	3
Mdh-1	2	2	2
Mdh-2	1	2	-
Idh-1	3	-	1
6-Pgd-1	2	-	3
Average	2.17	3.75	2.3

[a]data from Avise and Selander (1972) based on an examination of 393 fishes, collected at 9 different localities.

[b]data from McKinney et al.,(1972), 412 specimens, representing 17 local populations.

[c]data from Selander et al.,(1971), 745 individuals from 30 different localities.

References

Avise, J.C., Selander, R.K. (1972): Evolutionary genetics of cave-dwelling fishes of the genus *Astyanax*. Evolution 26, 1-19.

Ayala, F.J., Mourão, C.A., Pérez-Salas, S., Richmond, R., Dobzhansky, T. (1970): Enzyme variability in the *Drosophila willistoni* group. I. Genetic differentiation among sibling species. Proc. Nat. Acad. Sci. 67, 225-232.

Ayala, F.J., Powell, J.R. (1972): Enzyme variability in the *Drosophila willistoni* group. VI. Levels of polymorphism and the physiological function of enzymes. Biochem. Genet. 7, 331-345.

Ayala, F.J., Powell, J.R., Dobzhansky, T. (1971): Polymorphism in continental and island populations of *Drosophila willistoni*. Proc. Nat. Acad. Sci. 68, 2480-2483.

Bailey, G.S., Cocks, G.T., Wilson, A.C. (1969): Gene duplication in fishes: malate dehydrogenase of salmon and trout. Biochem. Biophys. Res. Communs. 34, 605-612.

Bellamy, A.W. (1922): Breeding experiments with the viviparous teleosts, *Xiphophorus helleri* and *Platypoecilus maculatus*. Anat. Rec. 23, 98-99.

Carter, N.D., Fildes, R.A., Fitch, L.I., Parr, C.W. (1968): Genetically determined electrophoretic variations of human phosphogluconate dehydrogenase. Acta genet. 18, 109-117.

Gerding, R.K., Wolfe, R.G. (1969): Malic dehydrogenase. VIII: Large scale purification and properties of supernatant pig heart enzyme. J. Biol. Chem. 244, 1164-1171.

Gerschler, M.W. (1914): Über alternative Vererbung bei Kreuzung von Cyprinodontiden-Gattungen. Z. ind. Abst. u. Vererb. Lehre 12, 73-96.

Horowitz, J.J., Whitt, G.S. (1972): Evolution of a nervous system specific lactate dehydrogenase isozyme in fish. J. Exp. Zool. 170, 13-32.

Hubby, J.L., Throckmorton, L.H. (1965): Protein differences in *Drosophila*. II. Comparative species genetics and evolutionary problems. Genetics 52, 203-215.

Hubby, J.L., Throckmorton, L.H. (1968): Protein differences in *Drosophila*. IV. A study of sibling species. Amer. Nat. 102, 193-205.

Johnson, W.E., Selander, R.K. (1971): Protein variation and systematics in the kangaroo rats (genus *Dipodomys*). Syst. Zool. 20, 377-405.

Johnson, W.E., Selander, R.K., Smith, M.H., Kim, Y.J. (1972): Biochemical genetics of sibling species of the cotton rat (*Sigmodon*). Studies in Genetics VII, Univ. Texas Publ. 7213, 297-305.

Kallman, K.D., Atz, J.W. (1966): Gene and chromosome homology in fishes of the genus *Xiphophorus*. Zoologica 51, 107-135.

Karig, L.M., Wilson, A.C. (1971): Genetic variation in supernatant malate dehydrogenase of birds and reptiles. Biochem. Genet. 5, 211-221.

Kirkman, H.N., Hendrickson, E.M. (1963): Sex-linked electrophoretic difference in glucose-6-phosphate dehydrogenase. Am. J. Human. Genet. 15, 241-249.

Kosswig, C. (1927): Über Bastarde der Teleostier *Platypoecilus* und *Xiphophorus*. Z. ind. Abst. u. Vererb. Lehre 44, 253.

Markert, C.L. (1968): The molecular basis for isozymes. Ann. N.Y. Acad. Sci. 151, 14-40.

Markert, C.L., Faulhaber, I. (1965): Lactate dehydrogenase isozyme patterns of fish. J. Exp. Zool. 159, 319-332.

McKinney, C.O., Selander, R.K., Johnson, W.E., Yang, S.Y. (1972): Genetic variation in the side-blotched lizard (*Uta stansburiana*). Studies in Genetics VII. Univ. Texas Publ. 7213, 307-318.

Ohno, S., Payne, H.W., Morrison, M., Beutler, E. (1966): Hexose-6-phosphate dehydrogenase found in human liver. Science 153, 1015-1016.

Prakash, S. (1969): Genic variation in a natural population of *Drosophila persimilis*. Proc. Nat. Acad. Sci. 62, 778-784.
Ritter, H., Baitsch, H., Wolf, U. (1969): Zur formalen Genetik von Isoenzymen, dargestellt am Beispiel der 6-PGD (E.C. 1.1.1.44). Humangenetik 7, 1-4.
Rogers, J.S. (1972): Measures of genetic similarity and genetic distance. Studies in genetics VII. Univ. Texas Publ. 7213, 145-153.
Rosenberg, M. (1971): Epigenetic control of lactate dehydrogenase subunit assembly. Nature New Biol. 230, 12-14.
Scholl, A. (1973a): Electrophoretic studies on structural variation of enzyme proteins in platyfish and swordtails (Poeciliidae, Teleostei). Submitted for publication.
Scholl, A. (1973b): Tissue-specific preferential expression of the *Xiphophorus xiphidium* allele for 6-phosphogluconate dehydrogenase in interspecific hybrids of platyfish (Poeciliidae, Teleostei). This issue.
Selander, R.K., Yang, S.Y., Lewontin, R.C., Johnson, W.E. (1970): Genetic variation in the horseshoe crab (*Limulus polyphemus*), a phylogenetic "relic". Evolution 24, 402-414.
Selander, R.K., Smith, M.H., Yang, S.Y., Johnson, W.E., Gentry, J.B. (1971): Biochemical polymorphism and systematics in the genus *Peromyscus*. I. Variation in the old-field mouse (*Peromyscus polionotus*). Studies in genetics VI, Univ. Texas Publ. 7103, 49-90.
Shaw, C.R. (1966): Glucose-6-phosphate dehydrogenase: homologous molecules in deer mouse and man. Science 153, 1013-1015.
Shaw, C.R. (1970): How many genes evolve? Biochem. Genet. 4, 275-283.
Shaw, C.R., Barto, E. (1965): Autosomally determined polymorphism of glucose-6-phosphate dehydrogenase in *Peromyscus*. Science 148, 1099-1100.
Shaw, C.R., Koen, A.L. (1968): Glucose-6-phosphate dehydrogenase and hexose-6-phosphate dehydrogenase of mammalian tissues. Ann. N.Y. Acad. Sci. 151, 149-156.
Shaw, C.R., Prasad, R. (1970): Starch gel electrophoresis of enzymes. - A compilation of recipes. Biochem. Genet. 4, 297-320.
Shows, T.B., Chapman, V.M., Ruddle, F.H. (1970): Mitochondrial malate dehydrogenase and malic enzyme: mendelian inherited electrophoretic variants in the mouse. Biochem. Genet. 4, 707-718.
Stegeman, J.J., Goldberg, E. (1971): Distribution and characterization of hexose-6-phosphate dehydrogenase in trout. Biochem. Genet. 5, 579-589.
Stone, W.S., Wheeler, M.R., Johnson, F.M., Kojima, K. (1968): Genetic variation in natural island populations of members of the *Drosophila nasuta* and *Drosophila ananassae* subgroups. Proc. Nat. Acad. Sci. 59, 102-109.
Webster, T.P., Selander, R.K., Yang, S.Y.: Genetic variability and similarity in the Anolis lizards of Bimini. Evolution, in press.
Wheat, T.E., Whitt, G.S. (1971): In vivo and in vitro molecular hybridization of malate dehydrogenase isozymes. Experientia 27, 647-648.
Wheat, T.E., Whitt, G.S., Childers, W.F. (1972): Linkage relationships between the homologous malate dehydrogenase loci in teleosts. Genetics 70, 337-340.
Whitt, G.S. (1968): Developmental genetics of lactate dehydrogenase isozymes unique to the eye and brain of teleosts. Genetics 60, 237.
Whitt, G.S. (1969): Homology of lactate dehydrogenase genes: E gene function in the teleost nervous system. Science 166, 1156-1158.
Whitt, G.S. (1970a): Developmental genetics of the lactate dehydrogenase isozymes of fish. J. Exp. Zool. 175, 1-35.
Whitt, G.S. (1970b): Genetic variation of supernatant and mitochondrial malate dehydrogenase in the teleost *Fundulus heteroclitus*. Experientia 26, 734-736.

Whitt, G.S., Booth, G.M. (1970): Localization of lactate dehydrogenase activity in the cells of the fish (*Xiphophorus helleri*) eye. J. Exptl. Zool. 174, 215-224.
Whitt, G.S., Horowitz, J.J. (1970): Evolution of a retina specific lactate dehydrogenase isozyme in teleosts. Experientia 26, 1302-1304.
Zander, C.D. (1967): Ökologische und morphologische Beiträge zur Systematik und geographischen Verbreitung der Gattung *Xiphophorus* (Pisces). Mitt. Hamburg. Zool. Mus. Inst. 64, 87-125.

Tissue-Specific Preferential Expression of the *Xiphophorus xiphidium* Allele for 6-Phosphogluconate Dehydrogenase in Interspecific Hybrids of Platyfish (Poeciliidae, Teleostei)

A. Scholl and F. Anders

1. Introduction

In fishes of the genus *Xiphophorus* sensu Gordon and Rosen (1951) (including platyfish and swordtails) interspecific hybrids can be easily obtained by crossing under experimental conditions in the aquarium (Gerschler, 1914; Bellamy, 1922; Kosswig, 1927; Häussler, 1928; Gordon, 1931; Breider, 1934) or by artificial insemination (Clark, 1950; Zander, 1961). Since most of these hybrids are fertile, they offer unique opportunities for biological research.

One of the most interesting features which is exhibited by such hybrids in the formation of melanomas (Gordon, 1927, 1931; Kosswig, 1927, 1929; Häussler, 1928). These tumors originate from small black spots. In the purebred species these spots consist of some hundred macromelanophores. They are mostly inherited by sex chromosome linked genes (Kallman and Atz, 1967). Considerable attention has been given to the phenomenon of tumor formation in these hybrids by many investigators, especially by the research groups of Breider, Kosswig and Gordon (for recent reviews see Anders, 1967, 1968, 1970; Anders, Klinke and Vielkind, 1972).

Tumor formation in *Xiphophorus* hybrids is assumed to be a problem of the regulation of the macromelanophore spot genes (Anders, 1967). It can formally be described as an increased expression of the spot genes, which is mediated through an overproduction of precursor cells of macromelanophores (Gordon, 1959; Vielkind, Vielkind and Anders, 1971; Anders, Klinke and Vielkind, 1972). The genetic basis of normal and abnormal pigment cell growth is relatively well known (Anders, Anders and Klinke, 1973) but there is no information as to what happens at the primary sites of gene action, since the primary products of macromelanophore genes are unknown.

However, one may assume that the levels of primary products of macromelanophore genes are abnormally high or low in the hybrid, to result in a failure of normal macromelanophore differentiation. If this assumption is correct, primary products of macromelanophore genes may not be the only ones that are expressed at unusual rates in the hybrid. From the molecular point of view a better understanding of tumor formation in these hybrids might therefore be provided for by an investigation on the expression of genes which are responsible for the production of specific proteins such as enzymes.

By the use of electrophoretic techniques we have been able to demonstrate structural variation of homologous enzymes in the genus *Xiphophorus* (Scholl, 1973; Scholl and Anders, 1973). These enzymes now may serve as biochemical markers for an investigation of the expression of homologous genes in the hybrid at primary sites of gene action.

In this contribution we should like to present some results on the expression of 6-phosphogluconate dehydrogenase (6-PGD) genes in

Xiphophorus xiphidium/Xiphophorus maculatus hybrids. It should be mentioned that these hybrids do not produce premelanomas or melanomas, since they lack special macromelanophore genes, which are capable of tumor formation. These non-tumorous hybrid genotypes were used for the present research, because tumor formation could falsify the results by secondary influences, due to the destructive capacity of their cells and due to degradation of tumor cells themselves, which results in general illness of the animals.

2. Material

Inbred strains of the platyfish species *Xiphophorus xiphidium* (Gordon, 1932) and *Xiphophorus maculatus* Guenther, 1866) were used (Fig. 1 and 2). The strains are derived from collections of Gordon from Rio Soto

a b

Fig. 1a and b. *Xiphophorus xiphidium*. a) female, b) male. Special attention is given in the text to the autosomal dorsoventral melanophore stripes which are only expressed in the male

la Marina, Mexico (*X. xiphidium*) and Rio Jamapa, Mexico (*X. maculatus*) and were inbred in the aquarium for more than 40 generations. In other publications these species are referred to as *Platypoecilus xiphidium* and *Platypoecilus maculatus* (Mexico); *X. xiphidium* (Gordon, 1932) and *X. variatus xiphidium* (Rosen, 1960) are identical. For systematics of the genus *Xiphophorus* see Zander (1967).

Artificial insemination was necessary to obtain offspring from *X. xiphidium* females and *X. maculatus* males. The reciprocal cross was obtained without artificial insemination. Since these hybrids are

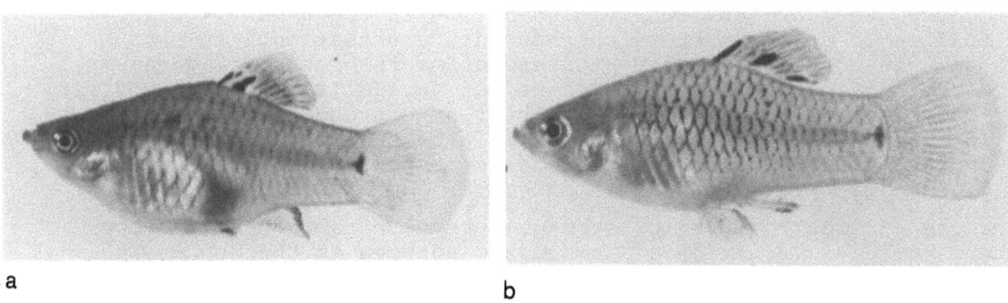

a b

Fig. 2a and b. *Xiphophorus maculatus*. a) female, b) male. Special attention is given in the text to the X-chromosome linked melanophore spots in the dorsal fin of both male and female and to the Y-chromosome linked melanophore network on the body side of the male

fertile, back-crossings with both *xiphidium* and *maculatus* parents were performed.

3. Methods

Homogenization

The tissues of the fish were removed immediately upon the sacrifice of the animals and were hand homogenized in special microhomogenizers. Two to three volumes of buffer (0.1 M tris-HCL, pH 8.0, at 4°C) were used for skeletal muscle, eye, brain and testes, four to nine volumes of buffer were used for liver. The homogenates were centrifuged in a special centrifuge for small volumes (Ole Dich, Denmark) at 20.000. g for 15 minutes at 4°C. The final supernatants were immediately subjected to electrophoresis.

Electrophoresis

Vertical starch gel electrophoreses (Buchler Instruments, Inc.) were carried out in 12% gels (Connaught, starch-hydrolysed). Two buffer systems were used and gave comparable results: Tris-citrate buffer: the gel buffer was 0.075 M tris and 0.02 M citrate (pH 7.3) while both electrode vessels were 0.3 M tris and 0.08 M citrate (pH 7.3). A voltage gradient of 3.6 - 4.0 V/cm was applied for 16-18 hours at 4°C, on some occasions for up to 24 hours at 4°C. Citrate-phosphate buffer: the gel buffer was 2.4 mM citrate and 8.5 mM Na_2HPO_4 (pH 6.0) while both electrode vessels were 8 mM citrate and 28 mM Na_2HPO_4 (pH 6.0). A voltage gradient of 3.2 V/cm was applied for 16 to 20 hours at 4°C.

Staining

The gels were sliced horizontally prior to staining. The 6-PGD-activity was detected in 50 ml 0.05 M tris-HCL containing 20 mg 6-phosphogluconate (Boehringer), 15 mg NADP (Boehringer), 10 mg p-nitro-blue tetrazolium (Calbiochem) and 0.5 mg phenazine methosulfate (Calbiochem). The zymograms were photographically documented on a MP-3 Polaroid camera.

4. Results

Morphological Comparisons of Parental Species and Their Hybrids

X. xiphidium (Fig. 1), *X. maculatus* (Fig. 2) and their interspecific F_1-hybrids (Fig. 3) can readily be distinguished by major morphological differences. There is neither maternal nor paternal effect in the F_1-offspring but both males and females exhibit a preferential manifestation of morphological characters which are contributed by *X. xiphidium*. Therefore, the F_1-hybrids exhibit more similarities to *X. xiphidium* than to *X. maculatus*. In addition, some of the characters of *X. xiphidium* such as the autosomal dorsoventral melanophore stripes which in the purebred species occur only in males (Fig. 1b), are enhanced in both male and female hybrids (Fig. 3a and 3b). To the contrary, the X-chromosome linked melanophore gene *Sd* (= spotted dorsal) of *X. maculatus*, which in the purebred species is responsible for spots in the dorsal fin (Fig. 2a and 2b), is not expressed in the hybrid (Fig. 3a), and the Y-chromosome linked melanophore gene *Sr* (= stripe sided), which in the purebred *X. maculatus* is responsible for the network on the bodyside (Fig. 2b) exhibits a diminished expression in the hybrid

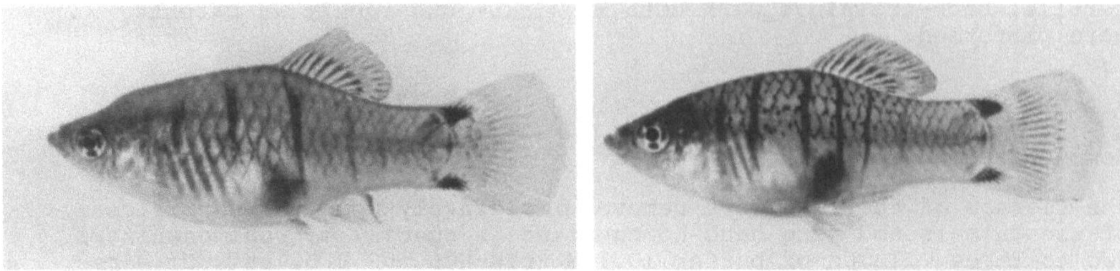

a b

Fig. 3a and b. *X. xiphidium/X. maculatus* F_1-hybrids. a) female, b) male. *X. xiphidium* characters are preferentially expressed in these hybrids. The autosomal dorsoventral melanophore stripes which are contributed by the *X. xiphidium* genome, are enhanced and are observed in both male and female. The female carries the X-chromosome linked gene *Sd* (= spotted dorsal) of *X. maculatus*, but the melanophore spots in the dorsal fin are not expressed. The male carries the Y-chromosome linked gene *Sr* (= stripe sided) of *X. maculatus*, but the melanophore network on the body side is diminished in expression

(Fig. 3b). In addition to this preferential expression of many of the *X. xiphidium* features, all F_1-hybrids exhibit hybrid vigor

Since F_1-hybrids are fertile, backcrossings with both *X. xiphidium* and *X. maculatus* parents were performed. The resulting B_1-hybrids exhibit variability in morphological characters, particularly with respect to body size.

6-PGD-Phenotypes in Inbred Strains of *X. xiphidium* and *X. maculatus* (Fig. 4)

The levels of 6-PGD-activity differ tissue-specifically. Very high levels are found in liver tissue, but all tissues which were analyzed

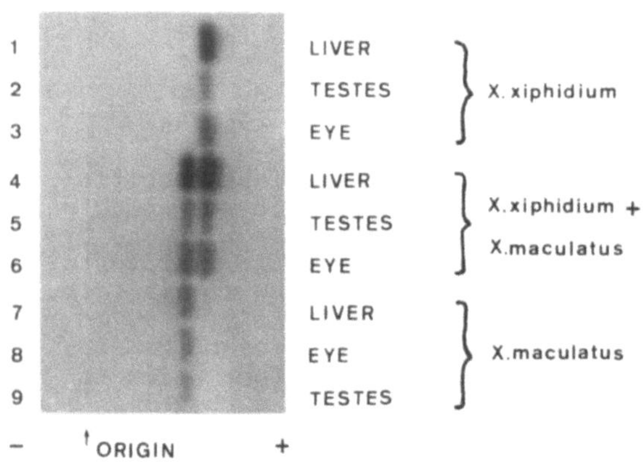

Fig. 4. 6-PGD zymogram of tissues of *X. xiphidium* and *X. maculatus*. 1 - 3: *X. xiphidium*, 1 = liver, 2 = testes, 3 = eye. 4 - 6: mixtures of homogenates from *X. xiphidium* and *X. maculatus*, 4 = liver, 5 = testes, 6 = eye. 7 - 9: *X. maculatus*, 7 = liver, 8 = eye, 9 = testes. 45 µl of supernatant fractions of homogenates were inserted

exhibit levels of 6-PGD-activity sufficiently high enough to demonstrate the enzyme on zymograms. In all tissues there is a single band of 6-PGD-activity and within each species this band has identical electrophoretic mobility in all tissues.

The following observations may be relevant for an interpretation of our subsequent findings: addition of NADP to either gels or tissue homogenates prior to electrophoresis does not alter the electrophoretic mobility of the enzyme or the 6-PGD-activity. Addition of Mn ions to the staining solution does not influence the enzyme activity, unless the tissue homogenates are prepared in buffer solutions containing EDTA. Since EDTA significantly reduces 6-PGD-activity, it was omitted from all buffer solutions.

No stain deposit at the sites of 6-PGD-activity is observed, if 6-phosphogluconate is omitted from the staining solution (blank) or replaced by either glucose-6-phosphate, galactose-6-phosphate or fructose-6-phosphate. Glucose-6-phosphate dehydrogenase isoenzymes of adult *X. xiphidium* and *X. maculatus* have a lower electrophoretic mobility than 6-PGD (Scholl, 1973). Therefore, the stain deposit which is observed on this zymogram (Fig. 4), is only due to 6-PGD-activity.

The electrophoretic mobility of 6-PGD is different in *X. xiphidium* and *X. maculatus*. This difference is not due to epigenetic factors such as substrate or coenzyme binding, which might influence the net-charge of the protein, since mixtures of tissue homogenates of both parental species are resolved into two 6-PGD-bands, which have the same electrophoretic mobility as the *xiphidium*- and *maculatus*-enzyme from the non-mixed homogenates respectively. Subsequent breeding experiments have clearly indicated, that the difference in electrophoretic behaviour of the enzyme originates from variation at the 6-PGD locus, due to homologous genes of *X. xiphidium* and *X. maculatus* which behave like alleles. Both species are monomorphic for allelic variants at the 6-PGD locus, as shown in previous experiments (Scholl, 1973).

6-PGD-Phenotypes in *X. xiphidium/X. maculatus* F_1-Hybrids (Fig. 5)

Male and female F_1-hybrids of both reciprocal crosses have identical 6-PGD phenotypes. As demonstrated in Fig. 5, the 6-PGD zymogram of the *X. xiphidium/X. maculatus* hybrid differs from the pattern of the mixed extracts of both parental species. A third enzyme is seen in the hybrids, which migrates intermediate between the *xiphidium*- and *maculatus*-6-PGD. Thus, the pattern of 6-PGD enzymes in the interspecific F_1-hybrids is not simply additive of the two enzyme bands of both parental species.

This result is to be expected, since the enzymatically active 6-PGD-protein is apparently composed of two polypeptide subunits (Carter et al., 1968; Ritter et al., 1969). The hybrid therefore does not have the information for two different enzymes, but rather the information for two different polypeptide subunits, a *xiphidium*- and a *maculatus*-type subunit. The subunits self-assemble to form enzymatically active 6-PGD proteins. Since 6-PGD genes of *X. xiphidium* and *X. maculatus* behave like alleles three 6-PGD allozymes are expected in the F_1-hybrid, which differ with respect to their subunit composition and their electrophoretic behaviour: homodimeric enzymes from two *xiphidium* type subunits, heterodimeric enzymes from one *xiphidium*- and one *maculatus*-type subunit and finally homodimeric enzymes of two *maculatus* subunits. The heterodimeric allozyme will be intermediate in electrophoretic mobility between both homopolymers. If both types of subunits are present at equal quantities, as would be expected for the

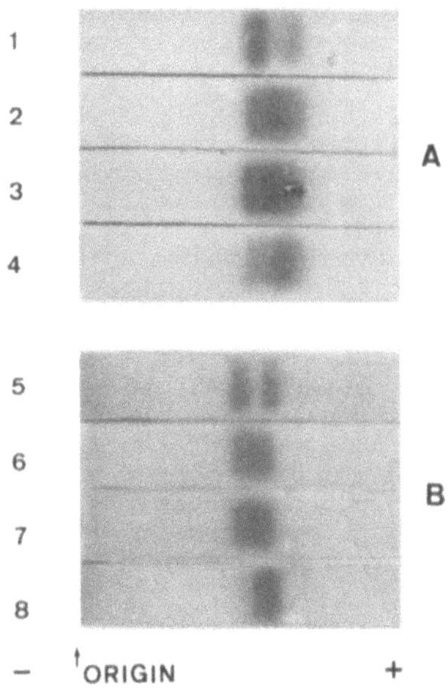

Fig. 5. 6-PGD zymogram of tissues of *X. xiphidium*/*X. maculatus* F_1-hybrids.
A and B are two different electrophoreses.
A: F_1-hybrid from *X. maculatus* female and *X. xiphidium* male.
1 = reference: mixture of liver supernatants from *X. maculatus* and *X. xiphidium*.
2 - 4: F_1-hybrid, 2 = eye, 3 = brain, 4 = liver.
B: F_1-hybrid from *X. xiphidium* female and *X. maculatus* male.
5 = reference: mixture of liver supernatants from *X. maculatus* and *X. xiphidium*.
6 - 8: F_1-hybrid, 6 = eye, 7 = brain, 8 = liver

F_1-hybrid, the three allozymes will occur at a ratio of 1:2:1, unless steric hindrance favours the formation of homodimers.

This ratio is seen in most tissues of the F_1-hybrids (Fig. 5, slots 2, 3, 6, 7). But a different pattern is observed for liver (Fig. 5, slots 4 and 8). This tissue exhibits significantly more stain deposit at the sites of the heterodimer and the *xiphidium*-type 6-PGD, irrespective of which species is the maternal parent. Thus, the expected ratio of allozymes is not confirmed for this tissue, rather it appears that the *xiphidium* allele is preferentially expressed in liver.

6-PGD-Phenotypes in Back-cross Hybrids

6-PGD phenotypes observed in back-cross hybrids are summarized in Table 1.

If *X. maculatus* is used as the recurrent parent (Table 1, A), the 6-PGD pattern of the offspring is either as in F_1-hybrids or as in *X. maculatus* (Fig. 6). This indicates two classes of genotypes, animals which are heterozygous and homozygous respectively at the 6-PGD locus. Both phenotypes are observed at an approximate 1:1 ratio (see Table 1, 1st and 2nd line and subtotal A). In those back-cross hybrids which

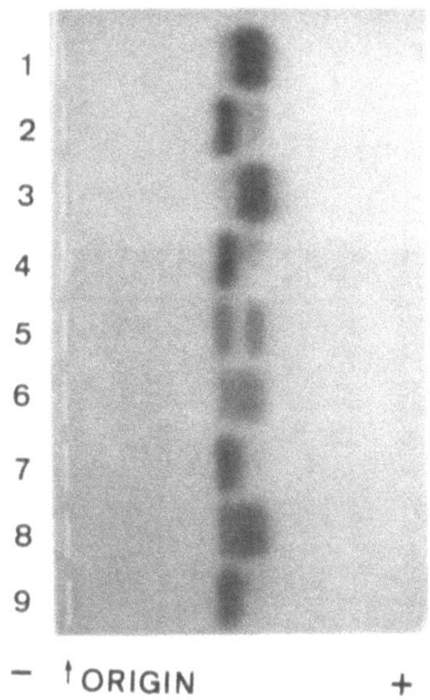

Fig. 6. 6-PGD zymogram of B_1 back-cross hybrids.
Recurrent parent: *X. maculatus*.
1 - 4 = liver supernatants of 4 B_1-hybrids.
5 = reference: mixture of liver supernatants from *X. maculatus* and *X. xiphidium*.
6 - 9 = eye supernatants of 4 B_1-hybrids (same animals as 1 - 4)

are heterozygous at the 6-PGD locus (hybrid 6-PGD phenotype), preferential expression of the *X. xiphidium* allele is always observed in liver extracts. In other tissues however, both alleles are expressed at equal rates. Representative zymograms are shown in Fig. 6. Slots 1 and 3 contain liver homogenates of two back-cross hybrids with hybrid-6-PGD phenotype. Slots 6 and 8 contain eye homogenates of the same animals. At the sites of the *maculatus* 6-PGD almost no stain deposit is observed in liver extracts. But the three 6-PGD allozymes are expressed at an approximate 1:2:1 ratio in extracts of eyes of these animals. In fact, if liver homogenates only had been analyzed, the 6-PGD phenotypes of back-cross hybrids could have been falsely interpreted to be either like in *X. maculatus* or in *X. xiphidium*.

If *X. xiphidium* is used as the recurrent parent (Table 1, B), the 6-PGD pattern of the offspring is either like in F_1-hybrids or like in *X. xiphidium* (Fig. 7). Again, this indicates the two expected classes of genotypes. There are more animals with hybrid phenotype than with *X. xiphidium* phenotype (see Table 1, 3rd and 4th line and subtotal B). However, the difference is not statistically significant. Preferential expression of the *X. xiphidium* allele is always observed in liver extracts of heterozygous animals. A representative example is shown in Fig. 7. Slots 1 and 2 contain tissue homogenates of a back-cross hybrid which is heterozygous at the 6-PGD locus, 1 is liver and 2 is eye. Almost no stain deposit is observed in liver at the site of the *X. maculatus* 6-PGD, however stain is heavily deposited at the sites of the heterodimeric 6-PGD and the *X. xiphidium* 6-PGD. Contrary to this pattern in liver, the three allozymes are observed at an approximate 1:2:1 ratio in the homogenate from eye.

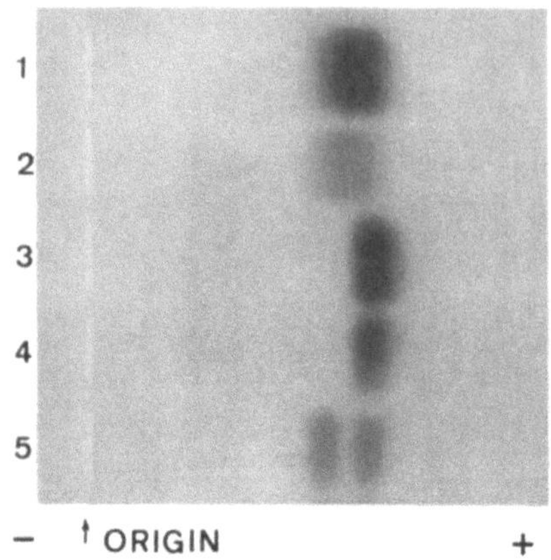

Fig. 7. 6-PGD zymogram of B_1 back-cross hybrids.
Recurrent parent: *X. xiphidium*.
1 - 2 = supernatants of the same B_1-hybrid, 1 = liver, 2 = eye.
3 - 4 = supernatants of a second B_1-hybrid, 3 = liver, 4 = eye.
5 = reference: mixture of liver supernatants from *X. maculatus* and *X. xiphidium*

Table 1. 6-PGD phenotypes in back-bross hybrids (all F_1-hybrids used for back-cross matings were from *X. maculatus* females x *X. xiphidium* males)

Mating Female	Male	6-PGD phenotype Hybrid	*Maculatus*	*Xiphidium*	Total
A: recurrent parent *X. maculatus*					
1. *X. maculatus*	x F_1	31	32	0	63
2. F_1	x *X. maculatus*	45	49	0	94
Subtotal A		76	81	0	157
B: recurrent parent *X. xiphidium*					
3. *X. xiphidium*	x F_1	10	0	8	18
4. F_1	x *X. xiphidium*	29	0	16	45
Subtotal B		39	0	24	63
Total		115	105		220

Heterozygosity at the 6-PGD Locus and Hybrid Vigor

At initial stages in the investigations on B_1-hybrids only 77 animals (females and males at equal proportions), which were the largest ones in the offspring of each back-cross mating, had been taken for analysis of the 6-PGD phenotypes. Only 25 of these hybrids had either the *xiphidium*- (homozygous for the *xiphidium* allele) or the *maculatus*-type of 6-PGD (homozygous for the *maculatus* allele), while 52 of these fish represented the hybrid enzyme-phenotype. Thus, a significant deviation (P = 0.2%) from the expected 1:1 ratio of 6-PGD phenotypes was observed in these large individuals. Later on, by analyzing the remaining smaller animals of the same matings, the approximate 1:1

ratio of homo- and heterozygosity at the 6-PGD locus was established (see total in Table 1). But with respect to the B_1-hybrids, where *X. xiphidium* is the recurrent parent, there still remains a deficit of animals homozygous at the 6-PGD locus (see Table 1, subtotal B). This agrees with the fact that dwarfs are observed, especially in this group of B_1-hybrids, and these dwarfs are not as vigorous as the B_1-hybrids of normal body size. Therefore we suppose that the deficit of animals with the *xiphidium* 6-PGD phenotype in this group is due to the failure of some dwarfs.

We do not know any substantial reasons for the occurrence of dwarfs with low viability, but these data indicate that heterozygosity at the 6-PGD locus or at least heterozygosity of a locus linked to the same chromosome, where the 6-PGD gene is located, may be responsible for some features of the hybrid vigor seen in all F_1-hybrids and in many of the B_1-hybrids. This aspect is presently investigated in more detail.

5. Discussion

Inheritance of the 6-PGD Phenotypes

The three 6-PGD enzyme bands observed in the *X. xiphidium/X. maculatus* F_1-hybrids are consistent with the fact, that both parents differ at the 6-PGD locus, for previous genetic analysis has shown, that 6-PGD behaves like a dimeric enzyme (Shaw, 1965; Parr, 1966; Ohno, 1967; Thuleine et al., 1967; Bender and Ohno,1968; Carter et al.,1968).Furthermore, as is observed from the analysis of the F_1- and B_1-hybrids, the 6-PGD phenotypes are inherited in a manner consistent with two homologous genes which behave like two different alleles at the 6-PGD locus. It has been pointed out that hybrid vigor may be associated with heterozygosity at the 6-PGD locus, as is observed from the ratio of 6-PGD phenotypes in B_1-hybrids. The 6-PGD locus is not sex-chromosome linked, because F_1-males and F_1-females of both reciprocal crosses (*X. xiphidium*-females x *X. maculatus*-males and *X. maculatus*-females x *X. xiphidium*-males) have identical enzyme phenotypes and because 6-PGD phenotypes of B_1-hybrids are distributed among males and females in approximately the proportions expected for mendelian segregation and dimerization of the enzyme.

Tissue Specific Preferential Expression of the *X. xiphidium* Allele of 6-PGD

The preferential expression of the *xiphidium* 6-PGD allele in interspecific hybrids is of considerable interest. Preferential expression of this allele is concluded from the observation of higher amounts of stain deposit on zymograms at the sites of *xiphidium*-6-PGD polypeptides. However, this does not necessarily indicate different levels of *xiphidium*- and *maculatus*-6-PGD polypeptides. Alternatively it might reflect a difference in catalytic properties. If this interpretation does apply to our observations, a preferential expression of the *xiphidium* allele would be expected to occur in all tissues. The observed preferential expression, however, is tissue specific for liver while other tissues express both alleles at equal rates. Therefore, it seems more likely that *xiphidium* and *maculatus*-6-PGD polypeptides have basically identical catalytic properties and that the observed preferential expression of the *xiphidium* allele in some tissues most probably indicates more *xiphidium*-6-PGD polypeptides in these tissues.

It would be speculative to conclude that the *xiphidium* allele is more active in transcription. Cellular levels of proteins depend on both their rates of synthesis and rates of degradation (Schimke, 1964; Fritz et al., 1969). The conformation of a protein appears to play an important role in determining its rate of degradation in any particular cellular environment (Schimke, 1969). Since we demonstrated that *xiphidium*- and *maculatus*-6-PGD polypeptides differ in their electrophoretic mobilities, it follows that these enzymes are not identical in their conformations. The conformational differences might lead to different rates of degradation in some cells. Therefore, there is no way to decide from these experiments, whether the *xiphidium* allele is more active in transcription in some cells of interspecific hybrids or whether the *X. xiphidium*-6-PGD polypeptides have a lower rate of degradation than *X. maculatus*-6-PGD polypeptides. Even though we are unable at present to distinguish between these alternative possibilities, we demonstrate at the molecular level, that a trait of one of both parental species is preferentially expressed in the hybrid.

Expression of Homologous Genes in Hybrids between other Species

The expression of homologous genes for enzyme proteins has previously been studied in other laboratories (Wright and Moyer, 1966; Hitzeroth et al., 1968; Klose and Wolf, 1970; Whitt et al., 1972). Most of the work was devoted to an analysis of the expression of maternal and paternal genes in stages of embryonic development and mostly preferential expression of maternal alleles in interspecific hybrids during embryogenesis has been reported (Wright and Moyer, 1968; Hitzeroth et al., 1968). These observations could be interpreted as a result of nucleocytoplasmic interactions, which lead to a delay in the activation of the paternal genome. Our results, however, are different from these observations, since: 1. in our material differential expression of allelic enzymes is observed in adult fish. 2. Irrespective of which species is the maternal parent, it is always the *xiphidium*-6-PGD allele that is preferentially expressed. Similar observations have recently been made by Brown and Blackler (1972) on gene amplification in interspecific *Xenopus* hybrids. In both *Xenopus laevis* and *Xenopus mulleri*, genes for ribosomal RNA are amplified during ovogenesis. In interspecific hybrids, however, only *X. laevis* genes are amplified regardless of whether *X. laevis* is the paternal or maternal parent.

Preferential Gene Expression and Melanoma Formation

In the present research we demonstrate for the first time at the molecular level, what has previously been observed at higher levels of organization in this material. These previous observations include a preferential manifestation of several heritable morphological characters, which are contributed to the hybrid only by the *X. xiphidium* parent.

Furthermore, it has been pointed out that the autosomally inherited dorsoventral stripes of *X. xiphidium* are expressed only in males and are hard to be observed. In the hybrids however, these stripes are seen in both males and females, and they are intensified and multiplied (compare Fig. 1, 2 and 3). Thus, the melanophores of these stripes and their precursor cells whose determinating genes are contributed by *X. xiphidium*, are increased in number. Using other melanophore genes the increase of the number of melanophores ultimately leads to melanoma formation, which also can be interpreted as an increased expression of the melanophore determining genes (Anders, 1967). This kind of preferential gene expression has been shown to depend on elimination or disturbance of regulating elements of the melanophore determining gene. The elimination of regulatory elements can be achieved by combining melanophore determining genes of one species with the genome of another species, which has not evolved such melanophore genes and therefore is thought to be without the melanophore-gene-specific regulating elements (Anders, 1967). The disturbance of regulating elements has been achieved by X-irradiation (Anders, A. et al., 1971; Pursglove et al., 1971; Anders, F. et al., 1973). Most probably therefore this preferential expression of the melanophore determining genes is due to a deficiency of a regulatory gene product, but a certain level of this regulatory product is necessary for normal action of the melanophore gene, which results in a normal differentiation of melanophores.

At present there is no information on the nature of primary gene products of melanophore genes and regulating elements. Our observations on an increased manifestation of allelic variants of enzymes in interspecific *Xiphophorus* hybrids, therefore, may serve as a model for a better understanding of events at the molecular level, which may lead to tumors in these fishes.

6. Summary

1. 6-phosphogluconate dehydrogenase (6-PGD) of an inbred strain of *Xiphophorus xiphidium* is different in its electrophoretic mobility from the mobility of this enzyme in inbred strains of *Xiphophorus maculatus*. Each strain is monomorphic at the 6-PGD locus. A biochemical marker has thus been found which may be used in studies on the expression of maternal and paternal genes in interspecific hybrids.

2. The homologous genes for 6-PGD of *X. xiphidium* and *X. maculatus* behave like alleles in *X. xiphidium/X. maculatus* hybrids. The enzyme phenotypes observed on back-cross hybrids are approximately in the proportions expected by mendelian segragation. Back-cross hybrids which are heterozygous at the 6-PGD locus, are more vigorous. Hybrid vigor may be associated with heterozygosity at the 6-PGD locus or a locus linked to the same chromosome where the 6-PGD gene is located.

3. The *X. xiphidium* allele for 6-PGD is preferentially expressed in liver tissue of *X. xiphidium/X. maculatus* hybrids, irrespective of whether *X. xiphidium* is the maternal or paternal parent. Both alleles are expressed at equal rates in other tissues. Preferential expression of the *X. xiphidium* allele in liver tissue is also observed in back-cross hybrids which are heterozygous at the 6-PGD locus, and occurs irrespective of which species is the recurrent parent.

4. Preferential expression has previously been observed at higher levels of organization in these fishes and may lead to tumors in some *Xiphophorus* hybrids. Since primary products of the genes which are responsible for tumor growth are not known, these observations on the tissue-specific preferential expression of the *X. xiphidium* allele for 6-PGD may serve as a model for a better understanding of events at the molecular level, which may lead to tumors in these fishes.

References

Anders, A., Anders, F., Pursglove, D.L. (1971): X-ray induced mutations of the genetically-determined melanoma system of xiphophorin fish. Experientia 27, 931-932.

Anders, F. (1967): Tumor formation in platyfish - swordtail hybrids as a problem of gene regulation. Experientia 23, 1-10.

Anders, F. (1968): Genetische Faktoren bei der Entstehung von Neoplasmen. Zbl. f. Vet. Med. B 15, 29-46.
Anders, F. (1970): Das Krebsproblem. In: Schwerpunkte von Prophylaxe und Therapie. Herausg. von G.W. Parade. 9-28. München-Gräfelfing.
Anders, F., Anders, A., Klinke, K.: Regulation of gene expression in the Gordon-Kosswig-melanoma system. This issue.
Anders, F., Klinke, K., Vielkind, U. (1972): Genregulation und Zelldifferenzierung im Melanomsystem der Zahnkarpfen. Biologie in unserer Zeit 2, 35-66.
Bellamy, A.W. (1922): Breeding experiments with viviparous teleosts, *Xiphophorus helleri* and *Platypoecilus maculatus*. Anat. Rec. 23, 98-99.
Bender, K., Ohno, S. (1968): Duplication of the autosomally inherited 6-phosphogluconate dehydrogenase gene locus in tetraploid species of cyprinid fish. Biochem. Genet. 2, 101-107.
Breider, H. (1934): Über das Auftreten von Rudimentär-hermaphroditismus nach Kreuzung getrenntgeschlechtlicher Arten. Verh. dt. zool. Ges. Greifswald 190-195.
Brown, D.D., Blackler, A.W. (1972): Gene amplification proceeds by a chromosome copy mechanism. J. Mol. Biol. 63, 75-83.
Carter, N.D., Fildes, R.A., Fitch, L.I., Parr, C.W. (1968): Genetically determined electrophoretic variations of human phosphogluconate dehydrogenase. Acta genet. 18, 109-117.
Clark, E. (1950): A method for artificial insemination of viviparous fishes. Science 112, 722-723.
Fritz, P.J., Vesell, E.S., White, L.E., Pruitt, K.M. (1969): The roles of synthesis and degradation in determining tissue concentrations of Lactate Dehydrogenase-5. Proc. Nat. Acad. Sci. 62, 558-565.
Gerschler, M.W. (1914): Über alternative Vererbung bei Kreuzung von Cyprinodontiden-Gattungen. Z. Ind. Abst. Vererb. Lehre 12, 73-96.
Gordon, M. (1927): The genetics of a viviparous Top-Minnow *Platypoecilus*; the inheritance of two kinds of melanophores. Genetics 12, 253-283.
Gordon, M. (1931): The heredity basis for melanosis in hybrids of mexican killifishes. Proc. Nat. Acad. Sci. 17, 276-280.
Gordon, M. (1959): The melanoma cell as an incompletely differentiated pigment cell. In: Pigment Cell Biology (M. Gordon, editor), 215-236, New York.
Gordon, M., Rosen, D.E. (1951): Genetics of species differences in the morphology of the male genitalia of Xiphophorin fishes. Bull. Am. Mus. Nat. Hist. 95, 413-464.
Gordon, M., Smith, G.M. (1938): Progressive growth stages of heritable melanotic neoplastic disease in fishes from the day of birth. Am. J. Cancer 34, 255-272.
Häussler, G. (1928): Über Melanombildung bei Bastarden von *Xiphophorus helleri* und *Platypoecilus maculatus* var. Rubra. Klin. Wschr. 7, 1561-1562.
Hitzeroth, H., Klose, J., Ohno, S., Wolf, U. (1968): Asynchronous activation of parental alleles at the tissue-specific gene loci observed on hybrid trout during early development. Biochem. Genet. 1, 287-300.
Kallman, K.D., Atz, J.W. (1967): Gene and chromosome homology in fishes of the genus *Xiphophorus*. Zoologica (N.Y.) 51, 107-135.
Klose, J., Wolf, U. (1970): Transitional hemizygosity of the maternally derived allele at the 6-PGD locus during early development of the cyprinid fish *Rutilus rutilus*. Biochem. Genet. 4, 87-92.
Kosswig, C. (1927): Über die Bastarde der Teleostier *Platypoecilus* und *Xiphophorus*. Z. ind. Abst. Vererb. Lehre 44, 253.
Kosswig, C. (1929): Das Gen in fremder Erbmasse. Nach Kreuzungsversuchen mit Zahnkarpfen. Züchter 1, 152-157.

Ohno, S. (1967): Sex chromosomes and sex linked genes. Edited by A. Labhard, T. Mann, and L.T. Samuels. Vol. 1, Monographs on Endocirnology. Heidelberg-Berlin-New York: Springer.
Parr, C.W. (1966): Erythrocyte phosphogluconate dehydrogenase polymorphism. Nature 210, 487-489.
Pursglove, D.L., Anders, A., Döll, G., Anders, F. (1971): Effects of X-irradiation on the genetically-determined melanoma-system of Xiphophorin fish. Experientia 27, 695-697.
Ritter, H., Baitsch, H., Wolf, U. (1969): Zur formalen Genetik von Isoenzymen, dargestellt am Beispiel der 6-PGD (E.C.: 1.1.1.44). Humangenetik 7, 1-4.
Scholl, A. (1973): Biochemical evolution in the genus *Xiphophorus* (Pisces, Poeciliidae). This issue.
Scholl, A., Anders, F. (1973): Electrophoretic studies on structural variation of enzyme proteins in platyfish and swordtails (Poeciliidae, Teleostei). Submitted for publication.
Schimke, R.T. (1964): The importance of both synthesis and degradation in the control of arginase levels in rat liver. J. Biol. Chem. 239, 3808-3817.
Schimke, R.T. (1969): On the roles of synthesis and degradation in regulation of enzyme levels in mammalian tissues. In: Current topics in cellular regulation (B.L. Horecker and E.R. Stadtman, editors) Vol. I, 77-124. New York: Academic Press.
Shaw, R.C. (1965): Electrophoretic variation in enzymes. Science 149, 936-943.
Thuline, H.C., Morrow, A.C., Norby, D.E., Motulsky, A.G. (1967): Autosomal phosphogluconic dehydrogenase polymorphism in the cat (*Felis catus L.*). Science 157, 431-432.
Vielkind, J., Vielkind, U., Anders, F. (1971): Melanotic and amelanotic melanomas in Xiphophorin fishes. Cancer Res. 31, 868-875.
Whitt, G.S., Cho, P.L., Childers, W.F. (1972): Preferential inhibition of allelic isozyme synthesis in an interspecific sunfish hybrid. J. exp. Zool. 179, 271-282.
Wright, D.A., Moyer, F.H. (1966): Parental influences on lactate dehydrogenase in the early development of hybrid frogs in the genus *Rana*. J. exp. Zool. 163, 215-221.
Zander, C.D. (1961): Künstliche Befruchtung bei lebendgebärenden Zahnkarpfen. Zool. Anz. 166, 81-87.
Zander, C.D. (1967): Ökologische und morphologische Beiträge zur Systematik und geographischen Verbreitung der Gattung *Xiphophorus* (Pisces). Mitt. Hamburg. Zool. Mus. Inst. 64, 87-125.

Investigations on the Serum Polymorphism of Trout and Carp[1]

H.-H. Reichenbach-Klinke

The extensive increase in trout production and therefore the growing deficiencies of these fish by infectious diseases have induced greater activities in the rearing of disease-resistent races. Stamping out of disease alone by chemicals is not possible, even ignoring the residual problem. About 4 years ago these reflections forced the author to attack the following question with the help of ichthyologi-cal colleagues: whether among the trouts of today in the Federal German Republic tribes or races exist which are producing antibodies

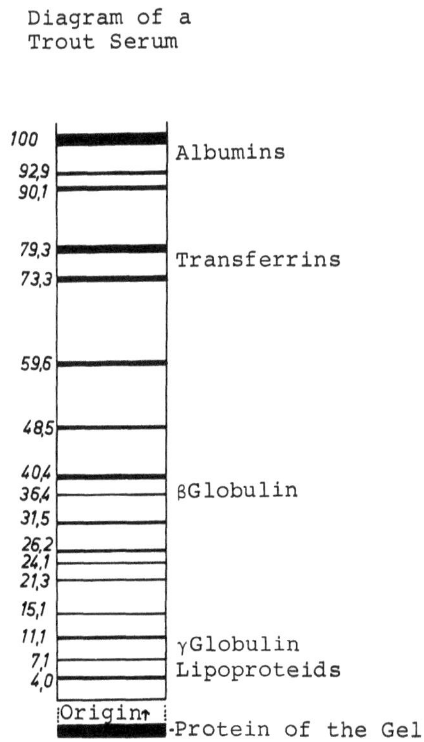

Fig. 1

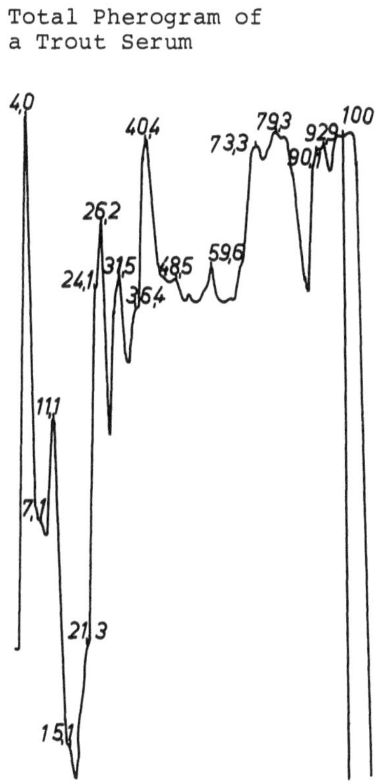

Fig. 2

[1]The investigations were realized by help of the Deutsche Forschungs-gemeinschaft, to whom we are indebted.

against the most important infectious agents in a most excessive manner. Creyssel et al. just as Langholz et al. were appointed to investigate these circumstances some time ago. Following these efforts the serum pattern of the rainbow trout (*Salmo gairdneri* Richardson) and the common carp (*Cyprinus carpio* L.) was analyzed by disc-electrophoresis. Significant variations were found which are to be discussed here.

I am very much obliged to my collaborators Miss G. Kolb and B. Ollenschläger, Dr. Vet. Med.

As material we were taking
126 rainbow trout } in the adult stage.
229 carps

The fish originated from pond cultures in Upper Bavaria, among others, from the pond culture station Wielenbach of the Bavarian Biological Experimental Institution. My thanks are also given to this organization.

The results of the serum-protein-investigations were published in a preliminary note 1971, and completed by the following figures:

Fig. 1 shows the diagram of the serum of a healthy rainbow trout. Above it demonstrates the albumins, postalbumins, transferrins, globulins and lipoproteids. The diagram is constructed with the help of a mechanical transcriber which fixed the peaks on Fig. 2. When we parallel the picture of the serum of a trout diseased by the

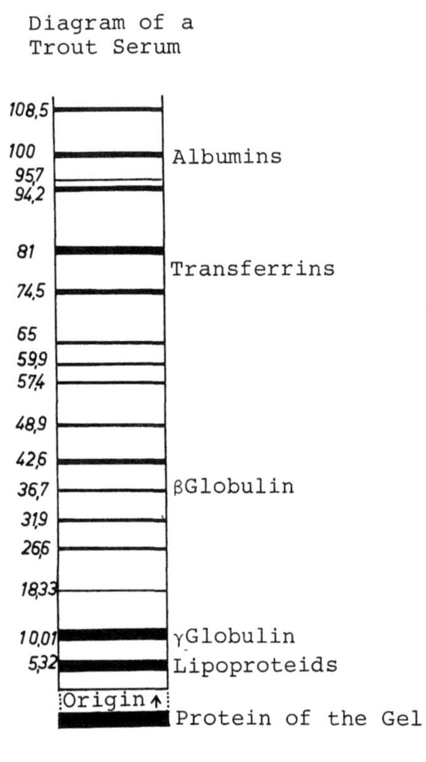

Fig. 3

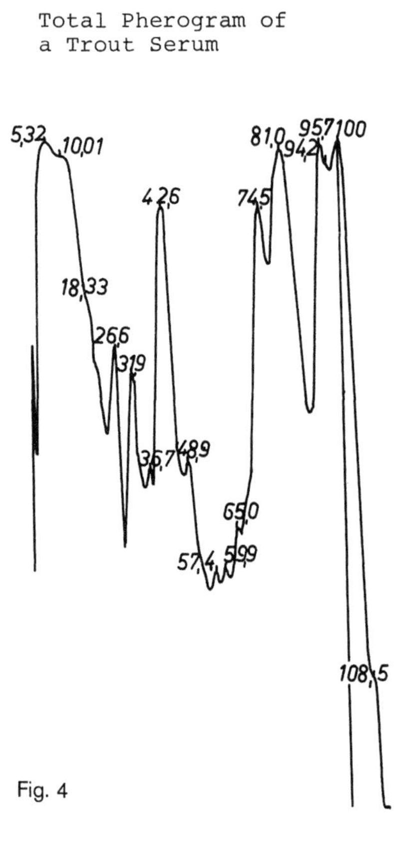

Fig. 4

hemorrhagic viral septicemy, we note that the quantity of distinct albumin patterns decrease in relation to the γ-globulins. It can be remarked that the transferrin-patterns which are identified by Fe^{59} are localized in a different manner so that we conclude a polymorphism.

More than 300 specimen were tested. Among their blood figures different groups could be isolated in view of localization and the range of the transferrin patterns. Distinct types could be gathered which were more or less frequent. In Fig. 5 one has tried to characterize

Transferrin-Polymorphism
of rainbow trouts
n = 126
observed combinations

Allele	Frequency	1	2	3	4	5
A	0,1429	▭	▭			
B	0,1555	▭				▭
C	0,2794	▭	▭	▭	▭	▭
D	0,1143	▭		▭		
E	0,2508	▭	▭	▭	▭	
F	0,0571					▭

Fig. 5

these types. So we can acknowledge that type 1 contains 5 patterns of transferrin, whereas the types 2 to 5 contain only 2 or 3, in different loci. In view of this, 36 combinations are possible; however, only 5 are realized in this strain. The frequency is varied. One quarter of all specimen possess the line allele C, another quarter that of the allele E, whereas f.i. the line F is forthcoming only in 5% of all cases.

The results admittedly are preliminary. Greater material must be tested, in addition to the stability of the transferrin-patterns against temperature, oxygen, etc. The results prove only that within the species "rainbow trout" different strains are contained which are fixed genetically.

Similar structures we find in the blood of carps *Cyprinus carpio* L. Also in this species we found polymorphism in the transferrin pattern, which enabled to postulate 6 types within 36 possibilities (Fig. 6).

The preliminary results demonstrate that we have the possibility to isolate several favoured polymorphisms among the current existing German strains. It would be another question surely if these characteristics would be recessive or dominant in the inheritance and if they could be realized.

There would be a significant differencc if the resistance is directed against a certain agent, a certain infectious group, or the resistance against the capacity or eliminating behaviour of frequent pollutant toxicants (detergents, oils etc.) should be favoured.

Transferrin-Polymorphism of carps

n = 229

observed combinations

Allele	Frequency	1	2	3	4	5	6
A	0,2301					▬	
B	0,2920	▬			▬	▬	
C	0,2478		▬	▬	▬	▬	▬
D	0,1593		▬				▬
E	0,0553			▬		▬	▬
F	0,0155						▬

Fig. 6

The goal is to isolate distinct combinations of the transferrin pattern in order to select groups with a constant polymorphism. Only with these would we be able to work in a genetically correct method and sense.

Summary

Investigations were carried forth on the polymorphism in the transferrin pattern in the blood of trouts and carps. By disc-electrophoresis with 300 specimen within the German trout and carp strains it could be determined that more than 6 types from 36 possibilities are existing of which 5-6 are realized. It is desirable to concentrate on the combinations with high resistance against certain diseases or pollutant toxicants.

References

Creyssel, R., Silberzahn, P., Richard, G., Manuel, Y. (1966): Transferrin variants in carp serum. Nature 212, 1362.

Langholz, H.J., Dinklage, H. (1970): Über die Serum-Protein-Auftrennung von Forellen. Arb. Deutscher Fischereiverb. 14, 67-69.

Reichenbach-Klinke, H.-H. (1971): Genetische Untersuchungen zur Förderung krankheitsresistenter Zuchtfische. Fish. Res. Inst. Vodnany Symposium 1971, 73-87.

Inheritance of the S-Form of NADP-Dependent Isocitrate Dehydrogenase Polymorphism in Rainbow Trout

H.-H. Ropers, W. Engel, and U. Wolf

1. Introduction

There are various pieces of evidence complementing each other that salmonid fish have evolved from a diploid ancestor by tetraploidization during the more recent past. The pertaining data, put together by Ohno (1970) and Engel et al., (1971), are the genome size, the number of chromosomes, and the number of gene loci for various protein marker systems which appear to be duplicated as compared to clupeoid fish.

Although the morphology and chromosome number of the salmonid karyotype has been changed by Robertsonian fusions, the occurrence of multivalents during meiosis indicates the persistance of large homologies of the duplicated genome (Ohno et al., 1965). This finding has been substantiated by the discovery that in trout several enzyme systems are inherited tetrasomically (Wolf et al., 1970; Engel et al., 1970; Stegeman and Goldberg, 1972). For other enzyme systems investigated, duplicated genes could be demonstrated as the result of diverging evolution, but their gene products still show close biochemical relatedness (Bailey and Wilson, 1968; Bailey et al., 1969; Morrison and Wright, 1966). In addition, an enormous variety of electrophoretically different alleles at these loci has been demonstrated. This finding can be expected in functionally tetraploid species.

Morrison (1970) reported a peculiar segregation pattern of two LDH-loci in trout which could not be explained by linkage alone; preferentially in male trouts the two loci segregated independently, whereas in females, linkage predominated. In addition, in certain matings he observed a statistically significant surplus of hybrid types compared to the expected number of parental combinations among the offspring. Although Morrison offered some alternative genetic models for these findings, a convincing solution for this problem has thus far not been presented.

Stegeman and Goldberg (1972) investigated the polymorphism of hexose-6-phosphate-dehydrogenase in brook trout; they conclude that in one population this enzyme is inherited tetrasomically while in another it follows a disomic mode of inheritance. This latter finding is explained by either Robertsonian fusion or by loss of the duplicated gene.

The extensive polymorphisms of the S-form of NADP-dependent isocitrate dehydrogenase in rainbow trout led to the assumption that this species is tetraploid with respect to the S-form of NADP-dependent IDH (Wolf et al., 1970). In order to substantiate this hypothesis, breeding experiments were performed; the results of which are reported here.

It will be shown that the mode of inheritance of IDH genes confirms the existence of duplicated genes. However, they may segregate in meiosis in a varying and unforeseeable combination assumingly on the basis of the respective Robertsonian fusions involved.

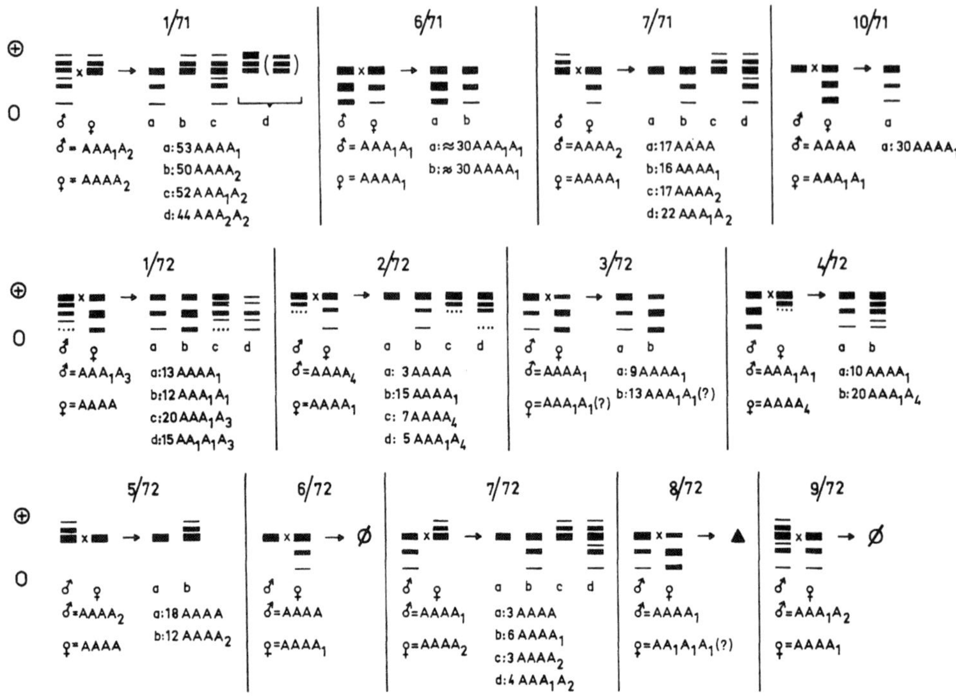

Fig. 1

2. Material and Methods

The soluble form of NADP-dependent IDH is mainly expressed in liver. Since an earlier study (Wolf et al., 1970) had indicated a high frequency of various electrophoretically different types of this enzyme in rainbow trout, we felt encouraged to perform the matings with randomly chosen individuals whose liver IDH-types were determined after spawning by sacrificing the animals. Twenty-four different matings were performed in 1971 and 1972. Rainbow trouts were supplied in part by A. Wehrle in Gengenbach/Baden, and in part by Mr. Hofer in Oberndorf/Neckar; breeding was performed by A. Wehrle at his trout-hatchery, to whom we are deeply indebted. The animals were typed immediately after crossing; methods applied were identical with those previously reported (Wolf et al., 1970). Typing gave clear results at about the fingerling size of the offspring. Due to the large number of crosses the size of the samples was limited to between 30 and 200 individuals per mating, depending on the parental phenotypes. In very few of the crosses a still smaller number of specimens was examined because of the low survival rate.

Nomenclature: The interpretation of the IDH phenotypes is based on the assumption of a dimeric enzyme structure (Henderson, 1968).

In rainbow trout liver, up to 6 electrophoretical bands were observed, some of which showed relative staining intensities strikingly similar to a 9:6:1 ratio. On the other hand, single band patterns also occurred. These findings and the occurrence of multivalents in meio-

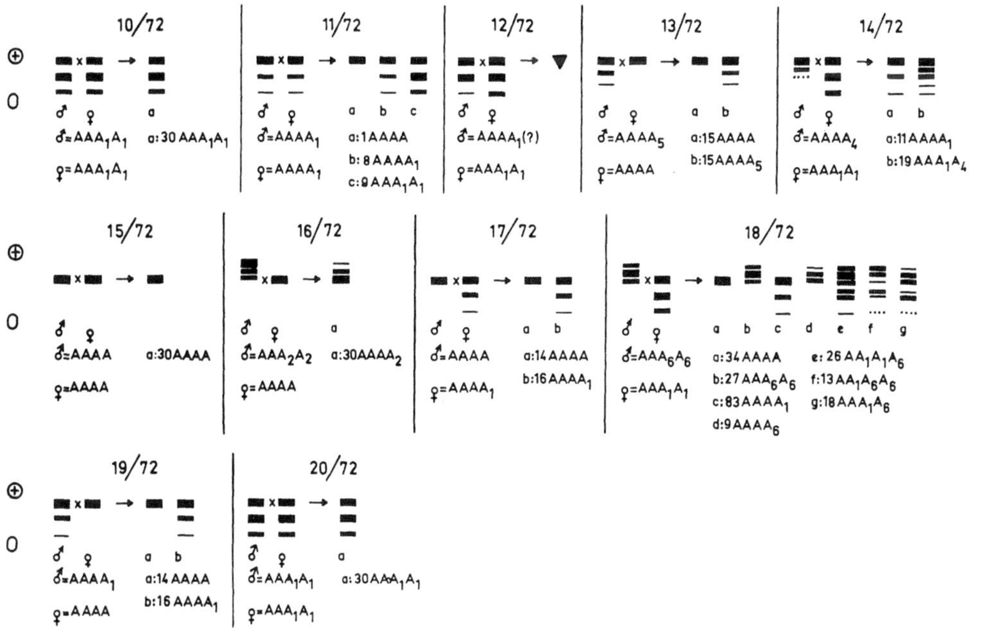

Fig. 2

Fig. 1. and 2. Parental and offspring IDH patterns of 24 matings. ø = no offspring. ▲ = perhaps 3 kinds of offspring ($AAAA_1$, AAA_1A_1 and $AA_1A_1A_1$?). ▼ = two kinds of offspring ($AAAA_1$ and AAA_1A_1?)

sis led Wolf et al., (1970) to the conclusion that the underlying genetics followed a tetrasomic model. According to this assumption, the phenotypes occurring in the breeding experiments reported here have been designated by a nomenclature fitting this genetic model (see Fig. 1 and 2).

3. Results and Interpretation

A first glance at Fig. 1 and 2 reveals that the banding patterns observed are not in keeping with a genetic model of a single disomic gene locus: several phenotypes show more than 3 bands. In addition, patterns with identical positions of bands, but differing relative staining intensities are depicted.

The question which arises is whether the complex patterns are, in fact, due to tetrasomic inheritance or whether they can be explained by some other genetic model including the presence of two (or more) separate disomic loci.

Between these borderlines of regular segregation, preferential pairing of two of the four homologous chromosomes at meiosis has to be taken into consideration. On the other hand, linkage between two IDH-loci might occur, which could indicate a corresponding chromosomal rearrangement. In addition, the mode of inheritance of IDH could be influenced by cytogenetic peculiarities found in trout, such as

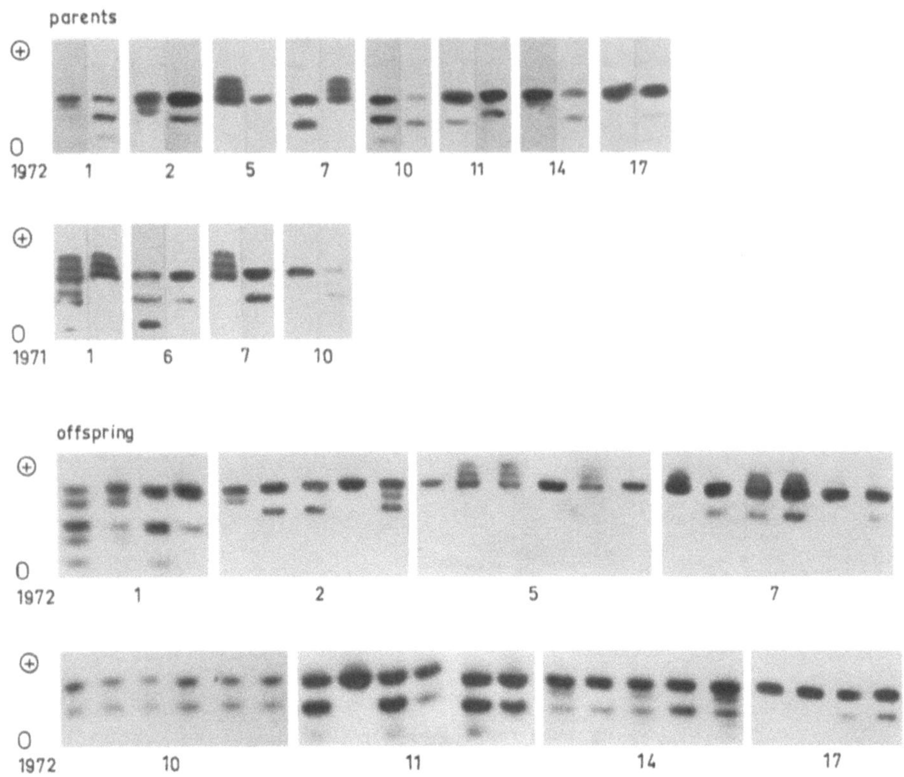

Fig. 3. Examples for the IDH patterns shown diagrammatically in Fig. 1 and 2. The first position of each mating, indicated by the figures below, represents the pattern of the male parent. For designation of the offspring see Fig. 1 and 2

Robertsonian fusion of homologous or heterologous chromosomes and a corresponding meiotic segregation of the gene loci involved.

Among the 24 matings examined, not all are informative with respect to the mode of inheritance. In two crosses, (6/72; 9/72) no animals reached fingerling size; two others (8/72 and 12/72) have not been discussed in the text because we were not sure of having found the correct interpretation of the parental and offspring patterns.

The remaining 20 parental combinations may be classified into four categories:

1. AAAA x $AAAA_m$
2. $AAAA_l$ x $AAAA_m$
3. $AAAA_m$ x AAA_nA_n and similar crosses
4. others.

<u>Category 1: AAAA x $AAAA_m$</u>. Five crosses fall into this category, namely 5, 13, 15, 17 and 19/72. In each mating, the heterozygous parent shows a three-banded pattern, the relative staining intensities being similar to a 9:6:1 ratio, which is indicative for a 3:1 ratio of subunits. The pattern of the other parent consists of one single band in the main position (designated AAAA). Whereas, the homozygous

parent should produce uniform gametes, the other parent is expected to bring about two forms of gametes, AA and AA_n, irrespective of the genetic model considered. So, two kinds of offspring resembling the parents' phenotypes are to be expected. Fig. 1 shows that the results fit this prediction. A notable fact is an apparent disequilibrium in the production of gametes favouring the AA_n types for which no obvious explanation can be offered.

Because no additional information can be gained from this type of crosses, the results are not further discussed (see Fig. 1 and 2).

Category 2: $AAAA_1$ x $AAAA_n$. Four matings belong to this category, the numbers 7/71, 2/72, 7/72 and 11/72. As pointed out above, each parent of these crosses is expected to produce two forms of gametes, resulting in 4 different kinds of offspring. (The above is true only if the parents each have one different allele; otherwise three offspring types are expected).

In fact, for crosses 7/71, 2/72 and 7/72 four types of offspring are observed. Mating 11/72 (identical parental types) shows 3 different kinds of progeny.

Thus, category 2 also allows no discrimination between the different possible modes of inheritance. In mating 11/72 ($AAAA_1$ x $AAAA_1$) the deviation from the expected 1:2:1 ratio of phenotypes among the offspring is obvious; AA_1-gametes are favoured.

Category 3: $AAAA_m$ x AAA_nA_n and similar matings. A parental type consisting of two pairs of different alleles allows determination of the mode of inheritance. The three possibilities, strictly tetrasomic inheritance, preferential pairing, and the possibility that two of the four "alleles" each belong to separate unlinked loci, may be differentiated.

According to tetrasomic inheritance, a parental type AAA_nA_n gives rise to three forms of gametes AA, AA_n and A_nA_n in a 1:4:1 ratio. Disomic inheritance, A and A_n representing alleles at separate loci, produces just one kind of gametes, AA_n. Preferential pairing between chromosomes carrying identical alleles increases the number of AA_n-gametes exceeding the 1:4:1 ratio.

For example, preferential pairing leading to the formation of a chromosome pair ratio of

$$4AA : 1AA_n : 4A_nA_n$$

instead of the random ratio

$$1AA : 4AA_n : 1A_nA_n,$$

which is expected to occur at meiosis in a tetraploid organism. This causes a 1:34:1 ratio, on the gametic level.

Six different matings fall into this category, these are number 6/71, 10/72, 3/72, 4/72, 14/72 and 16/72. As they all contribute additional information, we shall deal with them more detailed.

Mating 6/71 ($AAAA_1$ x AAA_1A_1) has produced 2 forms of offspring resembling the parent types. Since we had not been aware of the differences of both kinds, an accurate ratio of both types of offspring cannot be given; it, approximately, was 1:1.

Mating 10/71 (AAAA x AAA_1A_1) - one parent showing one band, the other three bands - produces only one kind of progeny with a banding pattern differing from that of both parents, indicating the $AAAA_1$ type. All animals examined were completely identical.

In cross 3/72 the parents were designated $AAAA_1$ and AAA_1A_1, respectively. Two kinds of offspring occurred among the individuals investigated; 16 belonged to the $AAAA_1$, 14 to the AAA_1A_1 type.

The parent types of mating 4/72 and 14/72 are identical and so are the progeny. In both crosses one parent is of the AAA_1A_1 type while the other shows a three-band pattern indicative of an $AAAA_4$-type. Again two classes of offspring are observed which can be described by the designation $AAAA_1$ and AAA_1A_4.

Mating 16/72 consists of one obviously homozygous parent (AAAA); the other shows a three-band pattern with slightly increased staining intensity of the two additional bands towards the anode. This type tentatively has been designated AAA_2A_2. The offspring is clearly uniform and is in keeping with the description $AAAA_2$.

In two further matings, namely 10/72 and 20/72, both representing the combination AAA_1A_1 x AAA_1A_1, only individuals of the AAA_1A_1-type are found.

From these findings it follows that regular tetrasomic inheritance can be excluded. The results strongly support a genetic model of two separate genetic loci, since in all parents of the AAA_nA_n type only AA_n gametes were produced.

However, as pointed out above, very strict preferential pairing would also fit these results.

Category 4: other matings. In two of the remaining crosses, one parent carries 3 alleles (1/71 and 1/72); the third mating has been put into this category because it brings about a rather confusing combination of different offspring.

The banding patterns of mating 1/72 indicate that the male parent carries 3 alleles A, A_1 and A_3 suggesting an AAA_1A_3 genotype. The female parent looks like the AAA_1A_1-type which has been proven to produce uniform gametes. If so, this mating can yield conclusive information about the arrangement of IDH-loci. If A_1 and A_3 are alleles at one locus, no more than two kinds of offspring are possible. If tetrasomic segregation in the male parent is taken into consideration, the number of possible gametes is four. The same would be the case, if the alleles A_1 and A_3 belonged to separate gene loci. As Fig. 1 indicates, 4 kinds of offspring in an approximate 1:1:1:1 ratio were observed which were compatible with the four types $AAAA_1$, AAA_1A_1, AAA_1A_3 and $AA_1A_1A_3$. Therefore, we can conclude that the subunits A_1 and A_3 are not produced at one and the same disomic gene locus. The other two models - A_1 and A_3 being alleles at separate loci, and tetrasomic inheritance - remain possible explanations. Though a tetrasomic mode of inheritance could not be inferred from the other matings, this possibility has not necessarily to be excluded for this reason. In the case that A_1 and A_3 belong to separate loci, close linkage can be eliminated.

Mating 1/71 may be tentatively characterized as AAA_1A_2 x $AAAA_2$. Assuming that A_1 and A_2 belong to separate unlinked loci, the male parent is expected to produce 4 kinds of gametes. If A_1 and A_2 are alleles at one locus, only two forms of gametes are produced. In the latter case the offspring should consist of four classes, $AAAA_1$, $AAAA_2$, AAA_1A_2 and AAA_2A_2 occurring in the ratio 1:1:1:1. Cum grano salis, this was observed, but, to confuse the situation, in the class designated AAA_2A_2 two types of banding patterns are discernible occurring in an approximate 2:1 ratio. The first shows about equal intensities of all three bands, and the other indicates increasing enzyme activity towards the anode. The simplest explanation for this peculiarity could be confusion of the different types in the early stages of our investigation, the types of apparently equal banding intensities representing overstained $AAAA_2$-types. An indication for this could be the outcome of mating 16/72 demonstrating that the gradient-type male parent produces uniform gametes of the AA_2-type. It should also be mentioned that, if A_1 and A_2 would represent alleles at separate loci, close linkage would also account for the occurrence of just 4 types of offspring.

The analysis of mating 18/72 reveals the most confusing results of all the crosses investigated. The paternal pattern seems distinctive for the genotype AAA_6A_6 showing a 1:2:1 ratio of relative staining intensities. Thus, it may represent a homozygote at two loci, because the idea of two rare identical alleles occurring at separate loci must be considered unlikely. But, as an analysis of the offspring shows, in this animal segregation of the alleles into three gametic types AA, AA_6 and A_6A_6 has to be postulated. Therefore, tetrasomic segregation would be a satisfying explanation. The maternal phenotype being $AAAA_1$ or AAA_1A_1 cannot readily be classified the one or the other way. Regarding the progeny, two A_1-alleles seem to be necessary in the parents; three forms of maternally derived gametes (AA, AA_1 and A_1A_1) must be postulated to explain the varieties of offspring observed. In contrast, all other parents of the AAA_1A_1 type included in this study produced uniform AA_1 gametes, whereas the $AAAA_1$ types produced AA and AA_1 gametes. Calculation of the relative numbers of gametes to be expected from the offspring observed, indicate that there is a marked deviation from the expected ratios. Deviations from a 1:4:1 or 1:2:1 ratio (which would be expected under the assumption of tetrasomic inheritance or adequately arranged alleles at two disomic loci) concerning only the "heterozygous" gametes can be explained by preferential pairing at meiosis or, in case of separate loci, by linkage.

In this mating, however, even the "homozygous" gametes do not occur in a 1:1 ratio. For this phenomenon we consider other mechanisms responsible than those hitherto discussed, e.g. gametic or postzygotic selection mechanisms.

Summarizing our findings we can state the following:

1. The majority of informative matings indicates the existence of at least two separate loci for the S-form of NADP-dependent IDH. For some matings tetrasomic inheritance remains one of the most likely explanations.

2. In a number of cases, the gene products are electrophoretically not discernible, suggesting close similarity or identity of both loci. This strongly suggests the occurrence of a recent duplication of the IDH-locus.

3. The determination of linkage between both loci by calculating the frequency ratios of the different types of offspring leads to contradictory results. Apparently, close linkage or strict preferential pairing - a possible explanation for several mating results - can be eliminated in others.

4. Some controversial segregation patterns are obviously the result of factors other than hitherto discussed.

4. Concluding Remarks

In a previous study, the S-NADP-IDH was examined in a random sample of 135 specimens of the rainbow trout (Wolf et al., 1970). The great variety of different phenotypes seemed best to be interpreted by the assumption of a tetrasomic gene locus for this enzyme. Only, some phenotypes expected on the basis of the different alleles present were never observed; and furthermore, the frequencies of the various alleles were far from the expectation assuming a Hardy-Weinberg equilibrium of $(p+q)^4$, though the sample size was rather small. For these reasons, preferential pairing of two definite chromosomes each of the original four homologies seemed to be a probable explanation. Thus, the behaviour of an enzyme IDH would confirm the hypothesis that the trout is a tetraploid species in the process of diploidization (Ohno, 1970).

In order to get a closer insight into the underlying mechanism of allelic segregation, it seemed promising to analyze the offspring of definite matings. The results of the present study, however, do not give unequivocal information, and this may be the clue to the interpretation of the underlying mechanisms.

The outcome of the majority of matings is compatible with the assumption of two separate disomic loci coding for electrophoretically identical gene products. Thus, in these cases, the chromosomes involved should be diploidized by some cytogenetic mechanisms, e.g. Robertsonian fusion of homologous chromosomes. In a few matings, tetrasomic inheritance could fully explain the phenotypes of the offspring.

Here, the original 4 chromosomes may either not be involved in a translocation, or fused in part to heterologous chromosomes. In the latter case random segregation must be guaranteed. Thus, the process of diploidization has not yet been settled in these chromosomes.

In the majority of crosses, the frequency of gametic types as calculated from the offspring is not compatible with either regular disomic or tetrasomic segregation. The observation of Ohno et al. (1965) that meiotic figures from the same fish contain varying numbers of multivalents seems pertinent in this connection.

If the chromosomes carrying the IDH genes, are involved in those translocations to a varying extent, it would not be possible to predict the frequencies of the different gametic types to be formed, as it is, indeed, the case here.

The frequencies of some offspring phenotypes, however, cannot be explained by these mechanisms, even if the formation of unbalanced gametes due to irregular segregation of multivalents is taken into consideration. For this finding, some postmeiotic selection mechanism has to be postulated. There may be either gametic selection to a certain degree, or some offspring genotypes may be more viable than others.

5. Summary

In the order *Isospondyli*, salmonid fish appear tetraploid with respect to the DNA content per cell and the number of chromosomes as compared to other members of this order. The analysis of various protein markers revealed that in salmonids also the number of genes coding for these markers is duplicate. For an enzyme, S-NADP-isocitrate dehydrogenase, the polymorphism observed indicated a tetrasomic mode of inheritance. In order to test whether this mechanism was responsible for the phenotype distribution, breeding experiments were performed. In the offspring of the majority of matings, the phenotypes can be interpreted assuming a disomic mode of inheritance; in others, however, tetrasomic segregation is more likely to have occurred. The frequency distribution of the phenotypes reflects the occurrence of varying multivalents in meiosis due to Robertsonian fusion and persisting homologies of the duplicated genome. In addition, selection mechanisms also seem to be involved. Thus, the results confirm that salmonid fish are originally tetraploid species in the process of diploidization.

6. Acknowledgment

The skillful technical assistance of Miss Heidrun Strohhäker is gratefully acknowledged.

References

Some pertaining references are added which are not mentioned in the text.

Bailey, G.S., Wilson, A.C. (1968): Homologies between isozymes of fishes and those of higher vertebrates. J. Biol. Chem. 243, 5843-5853.

Bailey, G.S., Cocks, G.T., Wilson, A.C. (1969): Gene duplication in fishes: malate dehydrogenase of salmon and trout. Biochem. Biophys. Res. Commun. 34, 605-612.

Bailey, G.S., Wilson, A.C., Halver, J.E., Johnson, C.L. (1970): Multiple forms of supernatant malate dehydrogenase in salmonid fishes. J. Biol. Chem. 245, 5927-5940.

Engel, W., Op'T Hof, J., Wolf, U. (1970): Genduplikation durch polyploide Evolution: die Isoenzyme der Sorbitdehydrogenase bei herings- und lachsartigen Fischen (Isospondyli). Humangenetik 9, 157-163.

Engel, W., Faust, J., Wolf, U. (1971): Isoenzyme polymorphism of the sorbitol dehydrogenase and the NADP-dependent isocitrate dehydrogenase in fish family Cyprinidae. Anim. Blood Grps. Biochem. Genet. 2, 127-133.

Henderson, N.S. (1965): Isozymes of isocitrate dehydrogenase: subunit structure and intracellular location. J. exptl. Zool. 158, 263-274.

Henderson, N.S. (1968): Intracellular location and genetic control of isozymes of NADP-dependent isocitrate dehydrogenase and malate dehydrogenase. Ann. N.Y. Acad. Sci. 151, 429-440.

Morrison, W.J., Wright, J.E. (1966): Genetic analysis of three lactate dehydrogenase isozyme systems in trout: evidence for linkage of genes coding subunits A and B^1. J. exptl. Zool. 163, 259-270.

Morrison, W.J. (1970): Nonrandom segregation of two lactate dehydrogenase subunit loci in trout. Transactions Amer. Fish. Soc. 99, 193-206.

Ohno, S., Stenius, C., Faisst, E., Zenzes, M.T. (1965): Postzygotic chromosomal rearrangement in rainbow trout (*Salmo irideus* Gibbons). Cytogenetics 4, 117-129.

Ohno, S., Wolf, U., Atkin, N.B. (1968): Evolution from fish to mammals by gene duplication. Hereditas 59, 169-187.

Ohno, S. (1970): Evolution by gene duplication. Berlin-Heidelberg-New York: Springer.

Stegeman, J.J., Goldberg, E. (1971): Distribution and characterization of hexose 6-phosphate dehydrogenase in trout. Biochem. Genet. 6, 579-589.

Stegeman, J.J., Goldberg, E. (1972): Inheritance of hexose 6-phosphate dehydrogenase polymorphism in brook trout. Biochem. Genet. 7, 279-288.

Wheat, T.E., Whitt, G.S., Childers, W.F. (1972): Linkage relationships between the homologous malate dehydrogenase loci in teleosts. Genetics 70, 337-340.

Wolf, U., Engel, W., Faust, J. (1970): Zum Mechanismus der Diploidisierung in der Wirbeltierevolution: Koexistenz von tetrasomen und disomen Genloci der Isocitratdehydrogenase bei der Regenbogenforelle, *Salmo irideus*. Humangenetik 9, 150-156.

Wright, J.E., Atherton, L.M. (1970): Polymorphism for LDH and transferrin loci in brook trout populations. Transactions Amer. Fish. Soc. 99, 179-192.

Biochemical Variants in Pacific Salmon and Rainbow Trout: Their Inheritance and Application in Population Studies

F. M. Utter, H. O. Hodgins, F. W. Allendorf, A. G. Johnson, and J. L. Mighell

1. Abstract

Data are presented supporting hypotheses of Mendelian inheritance for biochemical genetic variation in three species of Pacific salmon (*Oncorhynchus* spp.) and in rainbow trout (*Salmo gairdneri*). Variants studied included: chinook salmon (*O. tshawytscha*)--tetrazolium oxidase; sockeye salmon (*O. nerka*)--phosphoglucomutase; coho salmon (*O. kisutch*) --transferrin; rainbow trout--alpha glycerophosphate dehydrogenase, lactate dehydrogenase, malate dehydrogenase, tetrazolium oxidase and transferrin. Variation in the frequencies of these polymorphisms among populations indicates a usefulness of these variants for the identification and characterization of populations.

2. Introduction

The use of biochemical genetic variants for characterizing populations of fishes has accelerated in recent years (see De Ligny, 1969, 1972). In most of these studies the allelic nature of particular variations had to be inferred because it was difficult or impossible to carry out direct breeding studies. In our investigations we have inferred a genetic basis for the variation that we have seen by lines of evidence including: 1. starch gel electrophoretic patterns for particular proteins that are consistent with genetic variants of the same proteins in other species, 2. repeatability of expression from duplicate samplings of a given individual, 3. stability of expression over long developmental periods, and 4. conformance of frequencies of phenotypes to a Hardy-Weinberg statistical distribution. While the above criteria cumulatively provide strong evidence for allelism, the strongest data are from breeding experiments.

This paper presents family data for biochemical genetic variants in three species of Pacific salmon (*Onchorhynchus* spp.) and in rainbow trout (*Salmo gairdneri*) and gives population data for these variants and similar variants in related species. Hypotheses of Mendelian inheritance are supported and differences within species for many of these variants are demonstrated and discussed.

3. Experimental Procedures

Parents of progeny used in this study were obtained as follows: coho salmon (*O. kisutch*)--Washington State Department of Fisheries; chinook (*O. tshawytscha*) and sockeye (*O. nerka*)--adult fish returning to the NMFS Northwest Fisheries Center Laboratory in Seattle; rainbow trout--adult fish reared at the Seattle Center; and anadromous rainbow trout (steelhead) from the Chambers Creek Hatchery of the Washington State Department of Game. Crosses were made after the parental phenotypes were determined from electrophoresis. The methods used for handling eggs and sperm were those reported by Poon and Johnson (1970). All

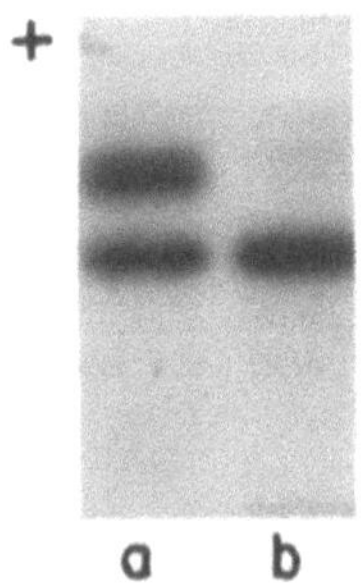

Fig. 1

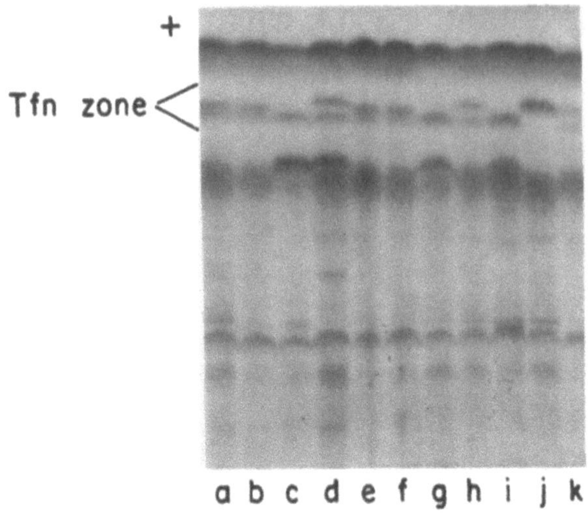

Fig. 2

Fig. 1. PGM variation in sockeye salmon. a - heterozygote, b - common homozygote

Fig. 2. Transferrin phenotypes in coho salmon. a, b, e, f - BC; c, g, i - CC; d, h, k - AC; j - AA. The same phenotypes also occur in rainbow trout

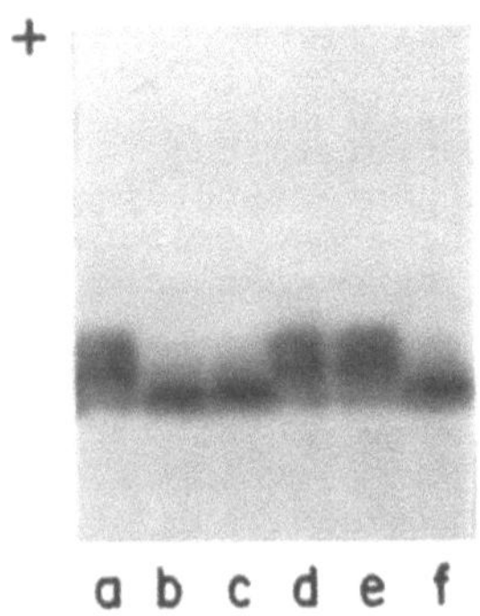

Fig. 3

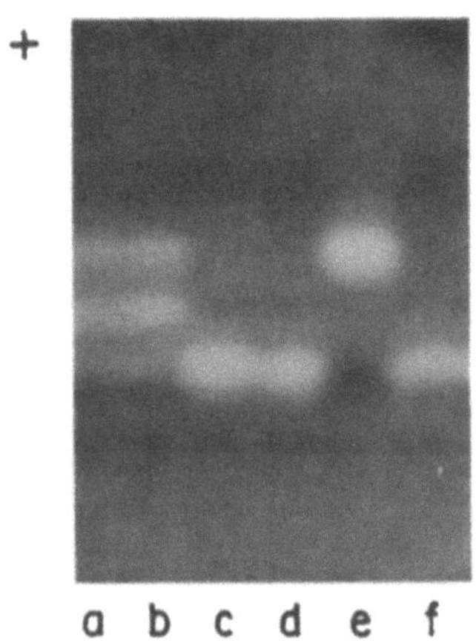

Fig. 4

Fig. 3. AGPD variants in rainbow trout: a, d, e - heterozygotes; b, c, f - common homozygotes

Fig. 4. TO phenotypes in rainbow trout: a, b - AB; c, d, f - BB; e - AA

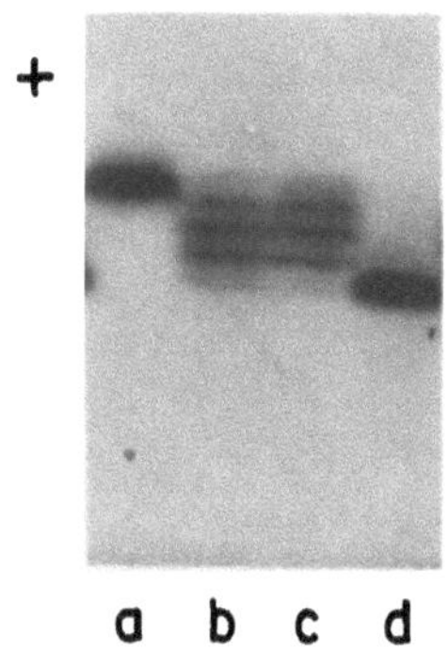

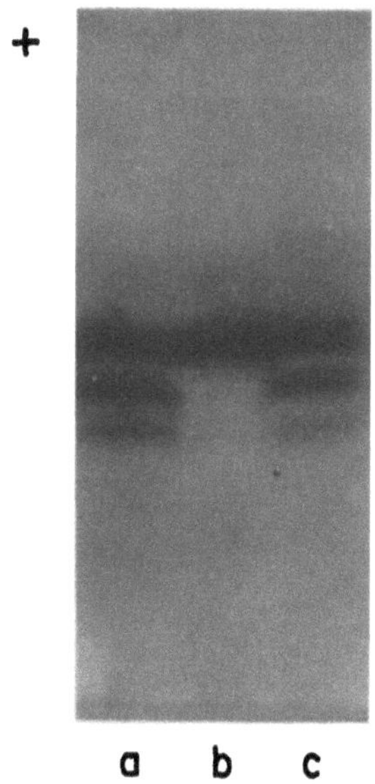

Fig. 5 Fig. 6

Fig. 5. LDH phenotypes of rainbow trout livers: a - $B^{2\prime}B^{2\prime}$; b, c - $B^{2\prime}B^{2\prime\prime}$; d - $B^{2\prime\prime}B^{2\prime\prime}$

Fig. 6. MDH variants of rainbow trout: a, c - BB'; b - B'B'. Note asymmetry of heterozygous bands

progeny were hatched and reared under similar conditions at this Center. Chinook and sockeye salmon progeny were tested between 2 and 4 months after hatching. Coho salmon and rainbow trout were tested between 6 and 9 months after hatching; somewhat larger fish were required from these species because a blood sample (difficult to obtain from small fish) was needed to test for transferrin.

All biochemical systems but transferrin (Tfn) were found in skeletal muscle extracted as described by Utter and Hodgins (1970). Blood plasma for Tfn typing was obtained by withdrawing blood from the pericardial cavity of freshly killed fish with a capillary pipette. Approximately one drop of whole blood was expressed into two drops of Alsever's solution (a citrate-dextrose-saline anticoagulant) in 10 x 75-mm culture tubes. Each sample was centrifuged at 1 000 x g for 3 minutes before testing.

Details of electrophoresis were described by Utter and Hodgins (1969). Buffer systems for the respective biochemical variants were those described by Utter and Hodgins (1972). Specific staining methods for enzymes followed those described by Shaw and Prasad (1970). Tfn was detected by a non-specific protein staining method using a 0.1% nigrosin-buffalo black solution dissolved in a 5:4:1 water-methanol-acetic acid mixture. Destaining was carried out with the water-methanol-acetic acid solution.

4. Biochemical Variants

Descriptions of each of the biochemical systems studies in this report have been given elsewhere (Hodgins, Ames and Utter, 1969; Utter, Ames

Table 1. PGM phenotypes of parents and progeny in sockeye salmon matings[a]

	Progeny phenotypes			Phenotypes of parents	
Lot	AA	AB	BB	Male	Female
1	0(0)	90(99.5)	109(99.5)	AB	BB
2	0(0)	90(98.5)	107(98.5)	BB	AB
3	0(0)	78(84)	90(84)	BB	AB
Control	0	0	100	BB	BB

[a]Parenthetical figures in Tables 1 through 8 represent expected numbers assuming Mendelian inheritance

Table 2. TO phenotypes of parents and progeny in chinook salmon matings

	Progeny phenotypes[a]			Phenotypes of parents	
Lot	EE	EF	FF	Male	Female
2-10	12(10.5)	9(10.5)	0(0)	EF	EE
2-11	12(10)	21(20)	7(10)	EF	EF
2-13	10(10)	20(20)	10(10)	EF	EF
4-14	8(11)	14(11)	0(0)	EE	EF
5-10	0(0)	29(30)	1(0)	EE	FF
5-12	0(0)	40(40)	0(0)	FF	EE
7-13	0(0)	14(14.5)	15(14.5)	FF	EF
Control	100	0	0	EE	EE

[a]The designation of TO alleles conforms to that of Utter, Allendorf, and Hodgins (1973) and differs from that originally described by Utter (1971)

Table 3. Transferrin phenotypes of parents and progeny in coho salmon matings

Lot	Progeny phenotypes						Phenotypes of parents	
	AA	AC	CC	AB	BB	BC	Male	Female
21	0	60(63.5)	67(63.5)	0	0	0	CC	AC
22	0	61(64.5)	68(64.5)	0	0	0	CC	AC
23	0	28(28)	29(28)	30(28)	0	25(28)	BC	AC
32	0	0	45(35.5)	0	0	26(35.5)	CC	BC
34	0	36(45.5)	48(45.5)	45(45.5)	0	53(45.5)	AC	BC
51	0	0	57(64)	0	0	71(64)	BC	CC
53	0	64(52.5)	41(52.5)	0	0	0	AC	CC
Control	0	0	20	0	0	0	CC	CC

Table 4. TO phenotypes of parents and progeny in rainbow trout matings

Lot	Progeny phenotypes						Phenotypes of parents	
	AA	AB	BB	AC	BC	CC	Male	Female
1301	0(0)	19(20)	21(20)	0(0)	0(0)	0(0)	BB	AB
A10	0(0)	65(60)	55(60)	0(0)	0(0)	0(0)	BB	AB
A17	0(0)	60(50)	40(50)	0(0)	0(0)	0(0)	BB	AB
SH2	0(0)	20(21)	22(21)	0(0)	0(0)	0(0)	AB	BB
A23	0(0)	105(104)	103(104)	0(0)	0(0)	0(0)	AB	BB
A24	0(0)	67(65)	63(65)	0(0)	0(0)	0(0)	AB	BB
A29	0(0)	38(44)	50(44)	0(0)	0(0)	0(0)	AB	BB
4347	0(0)	0(0)	18(15)	0(0)	26(30)	16(15)	BC	BC
Control	0	0	200	0	0	0	BB	BB

Table 5. Tfn phenotypes of parents and progeny in rainbow trout matings

	Progeny phenotypes			Phenotypes of parents	
Lot	AA	AC	CC	Male	Female
1301	7(8.3)	22(16.4)	4(8.3)	AC	AC
5641	32(30)	28(30)	0(0)	AC	AA
Control	50	0	0	AA	AA

Table 6. AGPD phenotypes of parents and progeny in rainbow trout matings

	Progeny phenotypes			Phenotypes of parents	
Lot	AA	AB	BB	Male	Female
SH2	0(0)	22(21)	20(21)	AB	BB
A24	0(0)	62(64.5)	67(64.5)	AB	BB
A28	0(0)	10(10)	10(10)	BB	AB
Control	0	0	140	BB	BB

Table 7. LDH phenotypes of parent and progeny in rainbow trout matings

	Progeny phenotypes			Phenotypes of parents	
Lot	$B^{2'}B^{2'}$	$B^{2'}B^{2''}$	$B^{2''}B^{2''}$	Male	Female
SH8	0(0)	45(45)	0(0)	$B^{2''}B^{2''}$	$B^{2'}B^{2'}$
SH14	0(0)	11(11)	0(0)	$B^{2''}B^{2''}$	$B^{2'}B^{2'}$
A5	0(0)	28(28)	0(0)	$B^{2''}B^{2''}$	$B^{2'}B^{2'}$
A10	52(55)	58(55)	0(0)	$B^{2'}B^{2'}$	$B^{2'}B^{2''}$
A17	58(50)	42(50)	0(0)	$B^{2'}B^{2''}$	$B^{2'}B^{2'}$
A18	11(12)	13(12)	0(0)	$B^{2'}B^{2''}$	$B^{2'}B^{2'}$
A20	25(27)	29(27)	0(0)	$B^{2'}B^{2''}$	$B^{2'}B^{2'}$
A23	129(129)	129(129)	0(0)	$B^{2'}B^{2''}$	$B^{2'}B^{2'}$
A28	15(16.5)	18(16.5)	0(0)	$B^{2'}B^{2''}$	$B^{2'}B^{2'}$
A29	50(44)	38(44)	0(0)	$B^{2'}B^{2''}$	$B^{2'}B^{2'}$
A4	0(0)	51(50)	49(50)	$B^{2''}B^{2''}$	$B^{2'}B^{2''}$
Control	200	0	0	$B^{2'}B^{2'}$	$B^{2'}B^{2'}$

Table 8. MDH phenotypes of parents and progeny in rainbow trout matings

Lot	Progeny phenotypes B'B'	BB'	BB	Phenotypes of parents Male	Female
A4	43(50)	57(50)	0(0)	B'B'	BB'
A10	52(55)	58(55)	0(0)	BB'	B'B'
A17	54(50)	46(50)	0(0)	BB'	B'B'
A20	16(19.5)	23(19.5)	0(0)	BB'	B'B'
A23	103(109)	115(109)	0(0)	BB'	B'B'
A24	70(63.5)	57(63.5)	0(0)	BB'	B'B'
A16	14(20)	45(40)	21(20)	BB'	BB'
Control	68	0	0	B'B'	B'B'

and Hodgins, 1970; Utter and Hodgins, 1970; Hodgins and Utter, 1971; Utter, 1971; Utter and Hodgins, 1972). These include lactate dehydrogenase (LDH) and phosphoglucomutase (PGM, Fig. 1) in sockeye salmon; tetrazolium oxidase (TO) in chinook salmon; Tfn (Fig. 2) in coho salmon and in rainbow trout; alpha glycerophosphate dehydrogenase (AGPD, Fig. 3), TO (Fig. 4), LDH (Fig. 5), and malate dehydrogenase (Fig. 6).

The phenotypes of most of the systems studied here are codominantly expressed on starch gel electrophoresis. Thus the presumed genotype can be interpreted directly from a given phenotype. The data for the fully codominant systems are given in Tables 1 through 8. In these systems, the observed phenotypes of the parents and progeny are presented and the expected numbers of progeny--assuming simple Mendelian inheritance--are shown in parentheses. In systems involving two alleles, all three possible phenotypes (the two homozygous phenotypes and the heterozygote) are listed, although some of these phenotypes may not be expected to occur from these matings. Similarly, all six possible genotypes are listed in systems involving three alleles. Controls are data from one or more matings where both parents have the same homozygous genotype.

Data from most crosses conform to expected Mendelian proportions based on parental phenotypes. The only qualitative exception to Mendelian inheritance in all of the matings is seen in Table 2 for TO phenotypes in chinook salmon progeny, lot 5-10. In the EE x FF cross, only heterozygous progeny are expected; however, one FF individual was found in this lot. It seems most likely that this individual is from another lot and had been placed in the wrong holding tank, because the remaining individuals were all of the expected phenotype. A quantitative deviation from expected ratios was observed in the cumulative totals of the sockeye salmon PGM crosses (Table 1) where an excess of homozygous progeny--though not significant in any single mating--became significant when the data were pooled (X^2 - 4.04, d.f. - 1, .05>P>.01), suggesting a selective factor favoring the BB phenotype.

A possible exception to the rule of codominance occurs in MDH variants of rainbow trout (Table 8). Bailey et al. (1970) observed that two loci of rainbow trout appear to code for MDH subunits (B') giving rise to active enzymes having the same electrophoretic mobility. One of these loci also coded for an allelic subunit (B). Because the B' subunit was synthesized in all individuals, it was impossible to qualitatively differentiate between BB' heterozygotes and BB homozygotes. However, they were able to quantitatively differentiate these two genotypes on the basis of different intensities of staining of bands having the same mobilities. Our MDH phenotypes are similar to those described by Bailey et al. (1970) and support their hypothesis of duplicate loci.

The finding that data from each of our crosses are consistent with an assumption of Mendelian inheritance for the genes controlling the observed biochemical variants was expected and virtually eliminated the possibility that these variants could be artifacts reflecting different environmental stresses or exposures (although they may indirectly reflect environmental differences through processes of natural selection). The data also indicate that these variants do not vary between the juvenile life history stage and the time of spawning. Finally, the data give strong support for hypotheses of Mendelian inheritance for similar (probably homologous) variants of other salmonid species where family data have not yet been obtained.

5. Population Studies

Much of our effort has been directed towards studies of the variants described above (and their homologs in related species) in natural populations for use as genetic markers (see Utter et al., 1972). We summarize some of these findings below.

Sockeye Salmon

We have found two-allele polymorphisms for LDH (B^2 locus) and for PGM in sockeye salmon (Hodgins, Ames and Utter, 1969; Hodgins and Utter, 1971; Utter and Hodgins, 1970). A distinct cline has been observed for both systems (Table 9). A virtual absence of the variant LDH allele (B^2) has been observed in fish taken from the Skeena River (B.C.) southward, while both alleles ($B^{2'}$ and B^2) have been regularly observed in samples taken from southeastern Alaska westward through the Kamchatka Peninsula, Siberia. The geographic variation in the distribution of the PGM alleles is more quantitative, but clear differences between regions are observed.

Table 9. Distribution of LDH and PGM variants in sockeye salmon

Variant	Area	Frequency of less common allele	Number of fish
LDH (B^2 allele)	Kamchatka Pen.	.084	101
	Bristol Bay	.120	750
	Copper River	.117	115
	Southeastern Alaska	.022	90
	Skeena R., southward	.002	591
PGM (A allele)	Bristol Bay	.292	406
	Southeastern Alaska	.205	90
	Puget Sound	.080	87

Chinook Salmon

Polymorphisms have been reported for three systems in chinook salmon: TO (Utter, 1971), MDH (Bailey et al., 1970), and sorbitol dehydrogenase (Utter et al., 1972). We report here only on the TO variants because of the low frequencies of the variants in the other systems. A third TO allele (B) has been observed in chinook salmon (Utter et al., 1972); data for this allele have been pooled with E allele data because of its infrequent occurrence. Considerable heterogeneity is seen in the frequency of the E and F alleles in different areas (Table 10). Although no clear geographic patterns are evident, the lower TO allele frequencies of spring run fish suggests a possible relationship between TO alleles and ecological factors.

Coho Salmon

A three-allele transferrin system was reported in coho salmon (Utter et al. 1970) that had a markedly different distribution in populations sampled from the Columbia River and Puget Sound. These differences have persisted in samples tested from additional areas and for different year classes (Table 11). The gene frequencies of the Puget Sound and Washington Coast samples are generally similar but are very different from those of the Columbia River. The B allele, having a frequency between 10% and 35% in other areas, is completely absent in Columbia River samples. The A allele is the sole or predominant gene in Columbia River fish, whereas the C allele is found most frequently in other areas.

It is difficult to explain these major differences between Columbia River fish and those taken from other areas on the basis of random factors alone. The Willipa, Nemah, and Chehalis Rivers are as close or closer to the Columbia River tributaries as they are to other coastal or Puget Sound streams, yet their gene frequencies are typical of the latter group. It may be that the A allele favorably or the B allele infavorably interacts with selective forces associated with the Columbia River system while such interactions are not operating in the other (much smaller) river systems. Regardless of cause, these differences of distribution of transferrin alleles appear to have much potential for identifying Columbia River coho salmon in areas of the Pacific Ocean where they mix with fish originating from other areas.

Table 10. Frequencies of the TO^F allele in populations of chinook salmon

Area	F allele frequency	Number of fish
Columbia River		
Rapid River (spring run)	.088	98
Wind River (spring run)	.156	80
Kalama River (fall run)	.512	207
Puget Sound		
Skagit River (fall run)	.416	77
Lake Washington (fall run)	.300	60
Green River (fall run)	.274	42
Skykomish River (fall run)	.198	58
Alaska		
Taku River (spring run)	.103	97

Table 11. Frequencies of transferrin alleles of coho salmon taken from three areas of Washington State

Area	Gene Frequency A	B	C	Number of Fish
Puget Sound				
Quilcene River	.200	.125	.675	60
Issaquah Creek	.235	.170	.595	100
Green River	.282	.155	.563	87
Minter Creek	.357	.140	.503	84
Skykomish River	.149	.104	.747	67
Washington Coast				
Dungeness River	.287	.266	.447	47
Soleduck River	.190	.220	.590	50
Chehalis River	.373	.216	.411	51
Willipa River	.280	.350	.370	50
Nemah River	.300	.180	.520	50
Columbia River				
Elokomin River	.786	.000	.214	98
Toutle River	.980	.000	.020	100
Speelyai Creek	1.000	.000	.000	99
Cowlitz River	1.000	.000	.000	50

Rainbow Trout

Genetic variants at six loci were described for four populations of rainbow trout by Utter and Hodgins (1972); these observations are extended here (Table 12). A revised notation for transferrin alleles is introduced here because it was found that the steelhead (i.e. anadromous rainbow trout) had an allele (B) not found in other groups tested. It was previously assumed that the common transferrin allele of other rainbow trout populations studied (A) was fixed in steelhead.

Table 12. Frequencies of most common alleles[a] of various polymorphisms found in rainbow trout

Area	LDH	TO	PGM	AGPD	MDH	Transferrin A	B	C	Number of fish
Chambers Cr., Wash., steelhead	.83	.72	1.00	.99	.80	0	1.00	0	168
Clearwater R., Idaho, steelhead	.29	1.00	.99	1.00	.97	0	1.00	0	72
Entiat, Wash.	1.00	.36	.96	.79	1.00	1.00	0	0	45
Quilcene, Wash.	1.00	.70	.90	.81	.99	1.00	0	0	45
Arlington, Wash.	1.00	.85	.99	1.00	.51	.93	0	.07	85
Soap Lake, Wash.	.68	.85	1.00	1.00	.92	.89	0	.11	37

[a]Excepting the transferrin locus where frequencies of all alleles are given.

Biochemical genetic variants clearly have much potential for characterizing populations of rainbow trout. Each trout population may be distinguished from any other listed in Table 12 on the basis of a characteristic biochemical genetic profile. The two steelhead populations are the only ones having the Tfn B allele and are distinguishable from each other by very different LDH allelic frequencies. The Entiat and Quilcene fish are the only groups having a sizeable degree of AGPD polymorphism and differ from each other in distribution of TO alleles. The Arlington fish have a much greater amount of MDH variation than any other group and the Soap Lake fish are the only non-steelhead group having LDH polymorphism.

6. Summary

In summary, we have presented data demonstrating the Mendelian inheritance of biochemical genetic variants at six loci in four species of Pacific salmonid fishes and have presented data indicating that this variation appears to be useful for separating and characterizing salmonid populations.

References

Bailey, G.S., Wilson, A.C., Halver, J.E., Johnson, C.L. (1970): Multiple forms of supernatant malate dehydrogenase in salmonid fishes. J. Biol. Chem. 245, 5927-5940.

Hodgins, H.O., Ames, W.E., Utter, F.M. (1969): Variants of lactate dehydrogenase isozymes in sera of sockeye salmon (*Oncorhynchus nerka*). J. Fish. Res. Board Can. 26, 15-19.

Hodgins, H.O., Utter, F.M. (1971): Lactate dehydrogenase polymorphism of sockeye salmon (*Oncorhynchus nerka*). Cons. Perma. Int. Explor. Mer., Rapp. Proc.-Verb. Réun. 161, 100-101.

Ligny, W. de (1969): Serological and biochemical studies on fish populations. Oceanogr. Mar. Biol. Annu. Rev. 7, 411-513.

Ligny, W. de (1972): Blood groups and biochemical polymorphisms in fish. In: G. Kovacs and M. Papp (Editors), XIIth European Conference on animal blood groups and biochemical polymorphism, pp. 55-65. The Hague: Dr. W. Junk N.V., Publ.

Poon, D.C., Johnson, A.K. (1970): The effect of delayed fertilization of transported salmon eggs. Prog. Fish-Cult. 32, 81-84.

Shaw, C.R., Prasad, R. (1970): Starch gel electrophoresis of enzymes: a compilation of recipes. Biochem. Genet. 4, 297-320.

Utter, F.M. (1971): Tetrazolium oxidase phenotypes of rainbow trout (*Salmo gairdneri*) and Pacific salmon (*Oncorhynchus* spp.). Comp. Biochem. Physiol. 39B, 891-895.

Utter, F.M., Allendorf, F.W., Hodgins, H.O.: Genetic variability and relationships in Pacific salmon and related trout species based on protein variations. Syst. Zool. 22, 257-270.

Utter, F.M., Ames, W.E., Hodgins, H.O. (1970): Transferrin polymorphism in coho salmon (*Oncorhynchus kisutch*). J. Fish. Res. Board Can. 27, 2371-2373.

Utter, F.M., Hodgins, H.O. (1969): Lactate dehydrogenase isozymes of Pacific hake (*Merluccius productus*). J. Exp. Zool. 172, 59-67.

Utter, F.M., Hodgins, H.O. (1970): Phosphoglucomutase polymorphism in sockeye salmon. Comp. Biochem. Physiol. 36, 195-199.

Utter, F.M., Hodgins, H.O. (1972): Biochemical genetic variation at six loci in four stocks of rainbow trout. Trans. Am. Fish. Soc. 101, 494-502.

Utter, F.M., Hodgins, H.O., Johnson, A.G. (1972): Biochemical studies of genetic differences among species and stocks of fish. Int. N. Pac. Fish. Comm., Annu. Rep. 1970, 98-101.

Subject Index

This multi-author volume, dedicated to Prof. C. Kosswig on his seventieth birthday, opens with his own historical review of the contribution research on fish has made to the study of genetics. The material is arranged, conveniently if perhaps rather artificially, according to the main features of ichthyogenetics:
experimental transformation of embryonic DNA, mechanisms determining sex, tumor genesis, mutagenesis, chromosome analysis, cell differentiation, behavioral genetics, the analysis of biochemical features, and evolutionary mechanisms.

MIX
Papier aus verantwortungsvollen Quellen
Paper from responsible sources
FSC® C105338

FSC
www.fsc.org

If you have any concerns about our products,
you can contact us on
ProductSafety@springernature.com

In case Publisher is established outside the EU,
the EU authorized representative is:
Springer Nature Customer Service Center GmbH
Europaplatz 3, 69115 Heidelberg, Germany

Printed by Libri Plureos GmbH
in Hamburg, Germany